NETWORK OPTIMIZATION PROBLEMS

SERIES ON APPLIED MATHEMATICS

Editor-in-Chief: Frank Hwang
Associate Editors-in-Chief: Zhong-ci Shi and Kunio Tanabe

Vol. 1 International Conference on Scientific Computation
ed. T. Chan and Z.-C. Shi

ies on
olied Mathematics
ume 2

NETWORK OPTIMIZATION PROBLEMS

ALGORITHMS, APPLICATIONS AND COMPLEXITY

Editors

Ding-Zhu Du

Department of Computer Science
University of Minnesota
and
Institute of Applied Mathematics
Academia Sinica, Beijing

Panos M. Pardalos

Department of Industrial and Systems Engineering
University of Florida

World Scientific
Singapore • New Jersey • London • Hong Kong

Published by

World Scientific Publishing Co. Pte. Ltd.
5 Toh Tuck Link, Singapore 596224
USA office: 27 Warren Street, Suite 401-402, Hackensack, NJ 07601
UK office: 57 Shelton Street, Covent Garden, London WC2H 9HE

Library of Congress Cataloging-in-Publication Data
Network optimization problems : algorithms, applications, and
complexity / editors, Ding-Zhu Du, Panos M. Pardalos.
p. cm. -- (Series on applied mathematics; v. 2)
Includes bibliographical references.
ISBN 9810212771
1. System analysis. 2. System design. 3. Mathematical
optimization. I. Du, Dingzhu. II. Pardalos, P. M. (Panos M.),
1954- . III. Series.
T57.6.N47 1993
003--dc20 93-16337
CIP

British Library Cataloguing-in-Publication Data
A catalogue record for this book is available from the British Library.

ISBN-10 981-02-1277-1 (hardcover)
ISBN-13 978-981-02-1277-3 (hardcover)
ISBN-13 978-981-279-819-0 (ebook for institutions)
ISBN-13 978-981-4504-58-4 (ebook for individuals)

For any available supplementary material, please visit
https://www.worldscientific.com/worldscibooks/10.1142/1920#t=suppl

Preface

The field of networks is a lively one, both in terms of theoretical developments and in terms of the diversity of its applications. Many problems of design and analysis of large systems can be formulated and solved using techniques of network theory. Such problems include communication systems, electrical networks, computer networks, transportation, scheduling of industrial processes, facility location, and modeling of combinatorial optimization problems.

Network theory originated many years ago, before our information age. In the eighteenth century, Euler solved the famous Königsberg Bridge problem and later Kirchoff initiated the theory of electrical networks. But it was not until late last century, when Bell invented the telephone, that many areas of network theory were stimulated.

After the appearance of the first graph theory book (by D. König) in 1936, there was tremendous development regarding the theory and applications of networks. Hitchcock proposed the first complete algorithm for the transportation problem in 1941, Dantzig proposed the simplex algorithm for linear programming in 1947, and algorithms for the minimum spanning tree (Kruskal, 1956) and shortest path problems were proposed (Prim, 1957). During the same period, the first commercial computers became available. As it happened with many other areas of research, the fields of computer science and networks influenced each other in many respects. In 1962 the book by Ford and Fulkerson on "Flows in Networks" appeared. With the development of new data structure techniques and the theory of computational complexity we entered a new era of algorithmic developments in networks.

During the second half of our century we saw major technological developments in all areas of human endeavor and particularly in information processing. Computer networks play a vital role in providing fast, reliable, cost-effective means of communication and information sharing. In addition, network techniques and computer technology enable us to solve large-scale network models that appear in applications such as transportation and telecommunications.

It is clear that the theory and applications of networks is so great that this book could not give a full account and systematic treatment of the subject in its entirety. It is our intention to introduce a number of special topics in order to show the spectrum of recent research activities and the richness of ideas in the development of algorithms and the applications of networks. While we were able to provide only a glimpse of this expansive field, we felt that this glimpse would allow the reader to sense the breadth and the depth of the field.

We would like to take the opportunity to thank the authors of the papers, the anonymous referees, and the publisher for helping us to produce this excellent collection of papers.

Ding-Zhu Du and Panos M. Pardalos

University of Minnesota and University of Florida
October 1992

Contents

Network Optimization Problems, pp. 1-22
Eds. D.-Z. Du and P.M. Pardalos

Greedily Solvable Transportation Networks and Edge-Guided Vertex Elimination

Ilan Adler
IEOR Department, University of California, Berkeley, CA 94720 USA.

Ron Shamir
Department of Computer Science, Sackler Faculty of Exact Sciences, Tel Aviv University, Tel-Aviv 69978, ISRAEL.[1]

Abstract

The greedy algorithm for the transportation problem repeatedly picks an edge, maximizes flow on it and updates the supplies and demands. If, with the same order of edges, the greedy algorithm gives an optimal (resp., feasible) solution for every feasible supply and demand functions, we call that order an optimality (resp., feasibility) sequence. We show that with a feasibility sequence, one can guarantee a feasible solution which is also a spanning tree, or give a certificate on infeasibility. Furthermore, with an optimality sequence, one can guarantee an optimal basic solution for every feasible problem. We also show how to obtain a spanning tree with a given signature or a dual feasible basis with a given signature, using feasibility and optimality sequences. Our results build on some interesting properties of vertex elimination algorithms which are guided by edge orders.

1 Introduction

The transportation problem can be stated as follows: A commodity which is available at certain sources is demanded at some destinations. Shipping costs from each source

[1]Supported by AFOSR grants 89-0512 and 90-0008, and by NSF grant STC88-09648.

to each destination are known. Given the amount available at each source and the amount demanded at each destination, the problem is to determine how much to send directly from each source to each destination so that all supplies and demands are met and the total shipping cost incurred is minimum. When not every source can ship to every destination, we say that the problem is *restricted.* The transportation problem is one of the fundamental problems in combinatorial optimization, and has been studied intensively over the last fifty years. The excellent surveys [2] and [11] describe several algorithms for transportation and related network flow problems.

One natural approach for solving the transportation problem is by a greedy algorithm, guided by a predetermined order of the edges. This algorithm repeatedly picks the next edge in that order, sends the maximum possible amount of flow along it and updates the supplies and demands accordingly. In general, this approach does not guarantee an optimal solution, but is often used to generate an initial feasible solution for more elaborate methods in unrestricted problems. The "north-west corner rule" and the "minimum C_{ij} rule" are two well-known examples of this approach (see, e.g., [13].)

The greedy algorithm has the obvious advantage that it is very fast, hence the interest in identifying when it guarantees an optimal solution. Research has focused on cases when there exists a *single* order of the edges which - when used by the greedy algorithm - is guaranteed to produce an optimal solution *for all* feasible problems on that network. Such order of the edges is called an *optimality sequence.* The knowledge of an optimality sequence is useful in particular if one needs to solve several problems with the same costs, but with varying supplies and demands: Given an optimality sequence, each problem is solvable in linear time. Similarly, if the greedy algorithm guided by some edge-order produces a feasible solution for every feasible problem, that order is called a *feasibility sequence.*

Hoffman [15] gave a characterization of bipartite networks which admit an optimality sequence, in terms of the Monge property. (The property will be defined in the next section.) Dietrich and Shamir [8, 18] (see also [19]) generalized the characterization to restricted transportation problems. Many families of problems which fit those characterizations or closely related ones have been studied in operations research, computational geometry and molecular biology applications. (See [1] for a list of references.) Alon et. al. [3] provided an efficient algorithm which finds an optimality sequence or determines that no such sequence exists for unrestricted problems. Dietrich and Shamir [8, 18, 19]) generalized the algorithm to restricted transportation problems. Adler, Hoffman and Shamir [1] characterized the bipartite graphs which admit a feasibility sequence in graph-theoretic terms and gave very efficient algorithms for recognizing them. They also extended the characterizations and algorithms for feasibility and optimality sequences to general, non-bipartite networks.

In this paper we introduce a requirement that the solution obtained from the greedy algorithm will also be *basic.* This requirement has several motivations: In many linear programming situations, the optimal solution of a problem is also required to be basic. This is needed when dual information is required, or for post-optimality analysis. Such complete information on the problem is not always available from an optimal solution, unless it is also a dual feasible basis. For example, a solution to the dual problem may not be immediately available in that situation. In case the problem is infeasible, it may also be important to obtain a proof of infeasibility by pointing out a "bottleneck" which caused the infeasibility in the original problem. For example, methods using linear programming solvers as subroutines (or "oracles"), require a separating hyperplane in case the problem is infeasible (cf. [10]). How much then do we have to sacrifice in terms of complexity, and in terms of the size of the class of greedily solvable problems, in order to guarantee that we get a basis? As it turns out, by modest modifications of the algorithms, we can guarantee obtaining basic information without increasing the complexity and without decreasing the size of the class.

It is well known in linear programming theory that proper perturbation of the right-hand side and/or the objective function generates a problem in which the only optimal solution is also an optimal basis. However, it is interesting to show how to achieve an optimal basis directly, by equipping an optimization algorithm with an appropriate mechanism (typically some sort of tie breaker). For example, Bland [7] showed how to prevent cycling in the simplex method, by endowing the simplex algorithm with some combinatorial properties, instead of using the geometric 'trick' of perturbation. Similarly, for the greedy algorithm in the transportation problem, we construct here edge selection rules which guarantee *independently* that the final selected set is a basis, primal feasible and dual feasible. One immediate consequence of this setup is that imposing the intersection of these selection rules guarantees that the final set of edges is an optimal basis.

The "modular" approach outlined above yields additional results: Since these selection rules are independent, one can combine a subset of these rules with other rules so that the final selected set of edges will satisfy additional desired properties. For example, an additional rule which enforces a given signature leads the greedy algorithm (using an optimality sequence) to produce a dual feasible spanning tree with a a prescribed signature. This type of analysis is used to obtain several additional results. Our main results are the following:

- In section 3 we define a very simple generic vertex elimination algorithm, which scans the edges in predetermined order and eliminates one vertex in each step. We show that with any edge order and on any graph, that algorithm generates a spanning forest. These results do not require the transportation and greedy algorithm setting, and may be of interest by themselves. Next we show that for bipartite graphs, if the edge order forms a feasibility sequence, then the generic algorithm forms a spanning tree and therefore a basis. Moreover, if the order is an optimality sequence, then the resulting basis is dual feasible.

- In section 4 we slightly modify the greedy algorithm so that it will have the properties of the vertex elimination algorithm defined in section 3. We show that with an optimality sequence this new algorithm guarantees an optimal basis for every feasible problem.

- In section 5 we concentrate on the consequences on negative termination in the greedy algorithm: Whenever the algorithm uses a feasibility sequence and determines that a problem is infeasible, we can identify a "bottleneck" set of sources (or destinations) whose total supply exceeds the total demand of all their neighbors in the original problem.

- As a by-product, we show in section 6 that the tools developed here yield some interesting results on signatures of transportation problems. Recall that the signature of a tree on a bipartite graph is the vector of the degrees of the sources in that tree. We give an algorithm which when guided by a feasibility sequence, produces a spanning tree with a given signature or determines that the signature is invalid. The same algorithm, when guided by an optimality sequence, produces a dual feasible basis with that signature, if it is valid. Finally, we give new characterizations for feasibility and optimality sequences using signatures.

2 Preliminaries

Let $G = (I, J; E)$ be a bipartite graph whose two vertex sets (or *sides*) are I and J, where $E \subseteq I \times J$, $|I| = m$, $|J| = n$ and $|E| = p$. The vertices in I and J are numbered $1, \ldots, m$ and $m+1, \ldots, m+n$, respectively. We assume throughout that graphs are undirected and connected. For convenience, ij and ji are used interchangeably to denote the edge between i and j. Denote $V = I \cup J$. The set $N(v) = \{j \in V | vj \in E\}$ is called the *neighborhood* of v, and $d_v = |N(v)|$ is called the *degree* of vertex v. v is called an *end-vertex* if $d_v = 1$.

For each edge $ij \in E$, a *cost* $C_{ij} \geq 0$ is assigned. Again, we use both C_{ij} and C_{ji} for the cost of the edge between i and j. Edge costs may also be represented by an $m \times n$ matrix, in which case we define $C_{ij} = \infty$ for each $ij \notin E$. The graph together with its edge-costs are called a *network* and are denoted $N = (G, C)$. Finally, for each vertex $v \in V$, a non-negative *excess* e_v is given. The bipartite network together with an excess vector specify the input to a transportation problem. The problem is called *restricted* if the underlying bipartite graph is incomplete.

The origin of the model is in planning of shipping, where the sets I and J correspond to sources and destinations, respectively. A commodity available at the sources must be shipped to satisfy demands at the destinations. C_{ij} is the cost of shipping each unit of the commodity on edge ij. e_v is the supply available at source v if $v \in I$, or the demand at destination v if $v \in J$. The *transportation problem* is to determine the amount x_{ij} to be shipped each edge ij, so that all supplied and demands are met at minimum overall cost, i.e.,

$$\begin{aligned} \min \ & \sum_{ij\in E} C_{ij}x_{ij} \\ \text{s.t.} \ \sum_{\{j|ij\in E\}} x_{ij} & = e_i \qquad i \in V \\ x_{ij} & \geq 0 \qquad ij \in E \end{aligned} \tag{P(N,e)}$$

We say that the excess vector e is *feasible* if $P(N,e)$ has a feasible solution, i.e., a solution in which all supplies and demands are satisfied. An obvious necessary condition for feasibility is that $\sum_{i\in I} e_i = \sum_{j\in J} e_j$. Given a spanning tree T in G, the (unique) solution of $\sum_{ij\in T} x_{ij} = e_i \;\; i \in V, x_{ij} = 0 \;\; ij \in E - T$ is called the *the primal basic solution associated with* T. (A spanning tree is sometimes called a basis, since the set of columns corresponding to its edges forms a maximal independent set in the coefficient matrix of the equations in $P(N,e)$.) If the primal basic solution is non-negative then it is called a feasible solution, and T is called *primal feasible.*

The dual of the transportation problem is defined as :

$$\begin{aligned} max \ \sum_{i\in I} e_i u_i \ & + \sum_{j\in J} e_j v_j \\ \text{s.t. } u_i + v_j \ & \leq C_{ij} \qquad ij \in E \end{aligned} \tag{D(N,e)}$$

Given a spanning tree T in G, the (unique) solution of the system $u_i + v_j = C_{ij} \;\; ij \in T$ is called the *the dual basic solution associated with* T. If the dual basic solution satisfies the rest of the inequalities in D(N,e) then it is called dual feasible solution, and T is called *dual feasible.* A tree which is both primal and dual feasible is called *optimal.*

Let $S = (S_1, S_2, \ldots S_p)$ be a permutation of the edges. For a problem $P(N,e)$, the *greedy algorithm* maximizes each variable in turn, according to the order given in S. A formal description is the following. (We assume that initially $x_{ij} = 0$ for all $ij \in E$).

```
algorithm GREEDY(S);
begin
    For i = 1,...,p do :
    begin pick the next edge S_i = rs.
        x_rs ← min{e_r, e_s}
        e_r ← e_r − x_rs
        e_s ← e_s − x_rs
    end
    if e ≠ 0 then output "infeasible" else output x
end
```

A permutation S of the edges is a *feasibility sequence* for the graph G if for every feasible excess vector e, GREEDY(S) provides a feasible solution to $P(N,e)$. S is

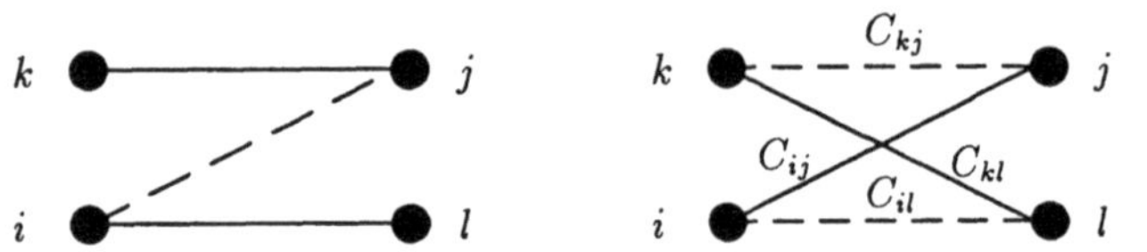

The Z Property $\quad$ The Hoffman Property

$kl \notin E$ $\quad$ $C_{ij} + C_{kl} < C_{il} + C_{kj}$

Figure 1: demonstration of the Z and Hoffman properties.
A broken edge cannot appear first in S among the edges in its figure.

called *an optimality sequence* for the network N if for every feasible excess vector e, algorithm GREEDY(S) provides an optimal solution to $P(N, e)$. For an order S of the edges, write $ij \overset{S}{\prec} kl$ if ij precedes kl in S. To simplify notation, we shall use $ij \prec kl$ when there is no ambiguity about the order. We say that S has *the Z property* for the graph G if the following holds:

The Z Property:
For every $i, k \in I$ and $j, l \in J$, if $ij \overset{S}{\prec} il$ and $ij \overset{S}{\prec} kj$ then $kl \in E$.

Note that the condition implicitly requires that ij, il and kj are edges, by the fact that they are included in S. An equivalent statement to that above is that if $kl \notin E$ then the order $li \overset{S}{\succ} ij \overset{S}{\prec} jk$ is impossible (see figure 1).

We say that S has *the Hoffman property* for the network N if the following property holds (see figure 1):

The Hoffman Property:
For every $i, k \in I$ and $j, l \in J$, such that ij, il, kj and kl are all edges in E, if $ij \overset{S}{\prec} il$ and $ij \overset{S}{\prec} kj$ then $C_{ij} + C_{kl} \leq C_{il} + C_{kj}$.

A sequence is called a *Monge sequence* for the network N if it satisfies both the Z property and the Hoffman property. Note that in a complete bipartite graph every permutation trivially satisfies the Z property, hence the Hoffman property and the Monge property are equivalent on unrestricted transportation networks.

The following theorem summarizes some of the known characterizations and complexity results for optimality and feasibility sequences in bipartite networks. (Recall that a bipartite graph is called chordal bipartite if it does not contain an induced cycle of length six or more [12].) For the complexity results we assume $m \leq n$:

Theorem 2.1
(a) [15, 19] S is an optimality sequence for a network if and only if it is a Monge sequence.

(b) [19, 1] S is a feasibility sequence for G if and only if it has the Z property.
(c) [1] A bipartite graph admits a feasibility sequence if and only if it is chordal bipartite.
(d) [3, 19] In a bipartite network, one can construct an optimality sequence or determine that no such sequence exists in $O(pm \log n)$ steps.
(e) [1] In a bipartite graph, one can construct a feasibility sequence or determine that no such sequence exists in $O(p \log n)$ steps.

3 Vertex Elimination and Spanning Trees

First, we consider vertex elimination in a general setting. A basic step which will be used in several algorithms in the sequel is eliminating a vertex together with all the edges incident on it from the graph $G(V, E)$. A formal description is the following:

```
procedure ELIMINATE(v);
begin
      V ← V − {v}
      E ← E − { vj ∈ E | j ∈ N(v) }
end
```

In all the algorithm we shall discuss, we call a vertex *active* at a certain stage if it is in the current set of vertices V. An edge is called *active* if it is in the current set E, i. e., if both of its endpoints are active.

The generic algorithm described below builds a set of edges B which under certain conditions will be shown to form a spanning tree. Initially, B is empty. Part of the input of the algorithm is a permutation of the edges $S = (S_1, S_2, \ldots, S_p)$, which is used to guide the algorithm: The algorithm examines the edges one at a time, according to their order in S. Whenever the examined edge is active, it is added to the set B and one of the vertices incident on it is eliminated. If exactly one vertex incident on the examined edge is an end-vertex, that vertex is eliminated. Otherwise, the eliminated vertex is chosen arbitrarily. This algorithm is described below:

```
algorithm GENERATE-B(S);
begin
      B ← ∅;
      For i = 1,...,p do :
      begin  pick the next edge S_i = rs.
            if rs is active then
            begin  B ← B ∪ {rs}
                  if exactly one endpoint of rs is an end-vertex then call it v.
                  else pick either v ← r or v ← s endif
                  ELIMINATE(v)
            end
```

end
end

This algorithm satisfies two elementary properties, which we call the *vertex elimination properties*:

- VE.1: For each edge added to B, one of its endpoints is eliminated and no other edge incident on that endpoint may be added later into B.
- VE.2: If an edge added to B is incident on at least one end-vertex, then one of its endpoints which is an end-vertex is eliminated.

The freedom to choose the permutation S and the eliminated vertex is quite large. We shall show in the sequel several ways of exploiting this freedom, in designing algorithms which carry out additional tasks while maintaining these vertex elimination properties.

Vertex and edge elimination schemes have been studied in the past, beginning with the work of Rose [17], in conjunction with efficient Gaussian elimination in sparse matrices (see [12, ch. 3 and 12] and the references thereof). In that context, vertex elimination procedures use a permutation of the *vertices*, and in each step one vertex is eliminated. In edge elimination, an ordering of *a subset of the edges* in sought, and *both endpoints* are always eliminated with the current edge. The vertex elimination properties defined here require that the algorithm eliminate *one* endpoint of a selected edge in each step, but be guided by an ordering of *all* the edges. The reason for this tighter definition will become clear in the following sections.

We first note some useful properties of the algorithm which are true for any ordering S of the edges.

Lemma 3.1 *Let B be a set of edges generated by algorithm GENERATE-B. If $(i_1i_2, i_2i_3, \ldots, i_{k-1}i_k)$ is a path of edges in B, then either $i_1i_2 \prec i_ri_{r+1}$ for $r = 2, \ldots k-1$, or $i_{k-1}i_k \prec i_ri_{r+1}$ for $r = 1, \ldots k-2$.*

Proof. Assume that i_ti_{t+1} is the first among the edges in the path to appear in S, and $1 < t < k-1$. Then on the same step when that edge was introduced into B, by property VE.1, either i_t or i_{t+1} became inactive. In the former case, $i_{t-1}i_t$ could not be added later to B. In the latter, $i_{t+1}i_{t+2}$ could not be added later to B. In both cases we obtain a contradiction. ∎

Corollary 3.2 *If B was generated by algorithm GENERATE-B, then for every path of edges in B, the order of appearance of these edges in S is unimodal. That is, if the path is $(i_1i_1, i_2i_3, \ldots, i_{k-1}i_k)$, then for some r, $1 \le r \le k-1$,*

$$i_1i_2 \prec i_2i_3 \prec \ldots \prec i_{r-1}i_r \prec i_r, i_{r+1} \succ i_{r+1}i_{r+2} \succ \cdots \succ i_{k-1}i_k.$$

Proof. Apply Lemma 3.1 to subpaths of that path. ∎

Corollary 3.3 *The set B generated by the algorithm is acyclic.* ■

Note that for the above only the first vertex elimination property (VE.1) was required. Note also that in every stage of the algorithm, every connected component of B contains exactly one active vertex.

So far we did not require the bipartiteness of the graph. We shall use the results above for graphs of the transportation problems, which are bipartite. Generating a spanning forest or a spanning tree in any graph is an easy, well-studied problem. It is the additional requirements from the spanning tree (e.g., dual feasibility, optimality or correspondence to a given signature) which make the problem of interest to us.

In general, the final set B generated by the above algorithm may contain several connected components. We now show that if S is a feasibility sequence, the final set B will be a spanning tree. The proof relies on the following lemma:

Lemma 3.4 *Let S be a feasibility sequence for G. Algorithm GENERATE-B(S) maintains the following invariants: (i) the set E is connected, (ii) the set $B \cup E$ is connected, and (iii) the set $B \cup E$ spans the original graph.*

Proof. The fact that in each step $B \cup E$ spans the original graph is immediate from the algorithm. The proof of (i) and (ii) is inductive. We give here the proof of (i) only, since the proof of (ii) in similar. Originally G is connected. Assume that E is connected before the insertion of edge rs into B. Denote the set of active edges after that step by E'. We need to prove that E' is connected. If $d_r = 1$ or $d_s = 1$, by property VE.2, $E' = E - \{rs\}$, so the statement is clearly true. Otherwise the degrees of both endpoints are at least two.

Let $q \neq s$ and $t \neq r$ be such that $rq \in E$ and $ts \in E$. Since $rs \prec rq$, $rs \prec ts$, by the Z property of sequence S, $tq \in E$. More generally, by the Z property, the set $N(r) \cup N(s)$ induces a complete bipartite graph in E. Without loss of generality suppose r is eliminated. Then $N(s) - \{r\} \neq \emptyset$, and the set $N(r) \cup N(s) - \{r\}$ induces a complete bipartite graph in E'. If E' is not connected, then since E was connected, and since all the edges which were inactivated in that step were incident on r, each component of E' must contain a vertex in $N(r)$. But any two neighbors of r are also neighbors of t, so they remain connected in E', a contradiction. ■

Proposition 3.5 *If S is a feasibility sequence, then the final set B produced by algorithm GENERATE-B(S) is a spanning tree.*

Proof. When the algorithm terminates, $E = \emptyset$. Hence by Lemma 3.4, B is connected and spans the original graph. By Corollary 3.3 B is cycle-free. Hence it is a spanning tree. ■

Let us now show that in a sense, the converse of Proposition 3.5 also holds: Call a permutation S a *spanning tree sequence* for G if for every induced connected subgraph $G'(V', E')$ of G, algorithm GENERATE-B(S) operating on G' generates a spanning tree. (The algorithm can be viewed as operating on G while considering all vertices which are not in V' as inactive from the beginning. An equivalent interpretation is that the induced subsequence of S which is obtained by restricting S to edges in E' is used on G'.)

Proposition 3.6 *S is a spanning tree sequence for G if and only if it is a feasibility sequence for G.*

Proof. If S is a feasibility sequence then for every induced subgraph G', the corresponding induced subsequence of S is also a feasibility sequence on G', by the Z property. Hence by Proposition 3.5, S is a spanning tree sequence. If S is not a feasibility sequence, then there exist some i, j, k, l such that $kl \notin E$ and $kj \succ ji \prec il$ (cf. figure 1). On the subgraph induced by those four vertices, S does not form a spanning tree. ∎

We now show that when S is a Monge sequence, the spanning tree generated by the algorithm is also dual feasible:

Theorem 3.7 *If S is a Monge sequence for N, GENERATE-B(S) produces a dual feasible spanning tree.*

Proof. Since every Monge sequence is a feasibility sequence, by Proposition 3.5, B is a spanning tree. To show it is dual feasible, it suffices to prove that no cycle which is formed by adding a non-basic edge to B is a negative cost cycle (see, e.g., [2]). That is, if $i_k i_{k+1} \in B$ $k = 1, \ldots 2l-1$, and $i_{2l} i_1 \in E - B$,

$$C_{i_1 i_2} + C_{i_3 i_4} + \cdots + C_{i_{2l-1} i_{2l}} \leq C_{i_2 i_3} + C_{i_4 i_5} + \cdots + C_{i_{2l} i_1}. \tag{3.1}$$

The proof is by induction on l: For $l = 2$, let the cycle be $(st,\ tu,\ ur,\ rs)$ where rs is the non-tree edge. Applying lemma 3.1 to the path $(st,\ tu,\ ur)$ we get that tu cannot precede both st and ur in S. Clearly, rs cannot precede both st and ur, since otherwise rs would have been in B. Hence by the Hoffman property, $C_{st} + C_{ur} \leq C_{rs} + C_{tu}$, and equation (3.1) is satisfied.

Assume now that the equation (3.1) is true for all cycles of length up to $2l - 2$, and let $(\, i_1 i_2,\ i_2 i_3, \ldots,\ i_{2l-1} i_{2l},\ i_{2l} i_1 \,)$ be a cycle of $2l$ edges, where only $i_{2l} i_1 \notin B$ (see figure 2). Using Lemma 3.1 again, we can assume without loss of generality that $i_1 i_2$ appeared first in S among the cycle edges. By property Z of the Monge sequence this implies that $i_{2l} i_3 \in E$. Moreover, by the Hoffman property,

$$C_{i_1 i_2} + C_{i_{2l} i_3} \leq C_{i_2 i_3} + C_{i_{2l} i_1}. \tag{3.2}$$

Using the induction hypothesis for the cycle $(\, i_3 i_4,\ i_4 i_5,\ \ldots,\ i_{2l-1} i_{2l},\ i_{2l} i_3 \,)$ whose length is $2l - 2$, we get that

$$C_{i_3 i_4} + C_{i_5 i_6} + \cdots + C_{i_{2l-1} i_{2l}} \leq C_{i_4 i_5} + C_{i_6 i_7} + \cdots + C_{i_{2l} i_3}. \tag{3.3}$$

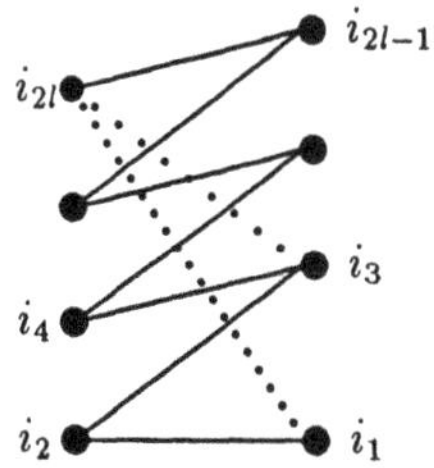

Figure 2: diagram for the proof of Theorem 3.7.
Solid lines: edges in B. Dotted lines: edges in $E - B$.

Summing equations (3.2) and (3.3) together, we obtain the desired proof of (3.1) for l. ∎

We can again show that the above condition leads to a new characterization of Monge sequences, in a fashion similar to Proposition 3.6: Call a sequence a *dual feasibility sequence* for N if GENERATE-B(S) produces a dual feasible tree on every induced connected subnetwork.

Proposition 3.8 *S is a dual feasibility sequence for N if and only if it is an optimality sequence for G.*

Proof. Analogous to the proof of Proposition 3.6. ∎

4 Generating an Optimal Spanning Tree

The algorithm above used only the graph structure and the order S to select a subset of the edges which forms a spanning tree. We now describe an algorithm which also receives as input the excesses at the vertices, and attempts to generate a feasible or optimal solution to the transportation problem. The algorithm greedily maximizes the flow on each active edge in turn, according to the order S, and eliminates one of its two endpoints. If one endpoint has strictly smaller excess then that vertex is eliminated. If both have equal excess and one is an end-vertex, then that vertex is eliminated. Otherwise, (if excesses at the two endpoints are equal, and either none is an end-vertex or both are,) we can choose the eliminated vertex arbitrarily. Infeasibility is detected whenever a vertex with positive excess becomes isolated. The algorithm is described below:

algorithm GREEDY*(S);
begin
$B \leftarrow \emptyset$; $k \leftarrow 0$;

```
        For i = 1,...,p do :
        begin  pick the next edge S_i = rs.
              if rs is inactive then return
              else begin  k ← k + 1 /* step k begins */
                    if one endpoint has less excess, say e_r < e_s then
              begin  ELIMINATE(r); if d_s = 0 then output "infeasibility" and stop end
                    else (e_r = e_s)
              begin  if d_r = 1 then ELIMINATE(r) else ELIMINATE(s) end
                    B ← B ∪ {rs}
                    x_rs ← min{e_r, e_s}
                    e_r ← e_r − x_rs
                    e_s ← e_s − x_rs
              end
        end
    end
```

Successful termination occurs when step $k = m + n - 1$ is completed without detecting infeasibility. In that case a single vertex remains active and has excess zero.

The above algorithm is a specialization of algorithm GREEDY which was described in section 2, modified so that it will also have the vertex elimination properties. In other words, it has both the properties of the greedy algorithm and of algorithm GENERATE-B. An immediate consequence of this, using Theorem 2.1(b) and Proposition 3.5, is the following:

Proposition 4.1 *If the sequence S satisfies the Z property, then for every feasible problem, algorithm GREEDY*(S) produces a primal feasible basic solution and its associated spanning tree B.* ■

From Proposition 4.1 and Theorem 3.7 we get the final result of this section:

Theorem 4.2 *If S is a Monge sequence, then for feasible problems, algorithm GREEDY*(S) terminates with an optimal basic solution and an optimal spanning tree B.* ■

As mentioned in section 2, given the optimal tree B, one can also construct the corresponding dual basic solution, by solving a system of linear equations. Note that the converse of Proposition 4.1 and Theorem 4.2 are already known to be true, by [1] and [19], respectively.

5 Gale Certificates

In this section we discuss the consequences of terminating the algorithm of Section 4 with the output 'infeasible'. For a set of vertices $S \in V$, let $e(S) = \sum_{v \in S} e_v$ be its total excess. If there exists a set of sources (or destinations) whose total supply (resp., demand) exceeds the total demand (resp., supply) of all their neighbors then the problem is obviously infeasible. Gale [9] has shown that the converse is also true, i.e., whenever the transportation problem is infeasible, there exists a set $S \in I$ (or $S \in J$) such that $e(S) > e(N(S))$. (This is a generalization of the celebrated Hall's theorem for the assignment problem[14].) Therefore, finding such a set S can be viewed as a proof of infeasibility. Let us call such a set S a *Gale certificate* of infeasibility. We shall show that whenever the greedy algorithm is guided by a feasibility sequence and terminates with the answer "infeasible", a Gale certificate can be pointed out.

For the following proposition we need some notation which refers to intermediate steps of the algorithm. Let $G^k = (V^k, E^k)$ be the graph after step k of the algorithm, and let e_v^k and d_v^k be the excess and the degree, respectively, of vertex v in that stage. $G = G^0$ is the original graph. We can view the problem $P^k = ((G^k, C), e^k)$ as a new transportation problem starting with the modified data after step k. Let B^k be the final set B produced by an algorithm operating on problem P^k as the initial problem. Note that for each k, step i of the algorithm on P^k is identical to step $k+i$ on the original problem. In particular, if a^k is the edge introduced into B in step k by the algorithm operating on the original problem P^0, then $B^k = B^{k+1} \cup \{a^k\}$. Moreover, infeasibility is detected at the same vertex in all these nested problems. Suppose the algorithm terminated after determining infeasibility at vertex w, in step l of the original problem. That is, $d_w^l = 0$ and $e_w^l > 0$. Without loss of generality assume $w \in I$. Define

$$R^k = \{u \in I \mid \exists\, \text{a path in } B^k \text{ from } w \text{ to } u\}$$
$$L^k = \{u \in J \mid \exists v \in R^k \text{ such that } vu \in E^k\}$$

In other words, R^k is the set of vertices in I which are in the same connected component of B^k with w, and L^k is the set of their neighbors in G^k. Finally, let $D^k = R^k \cup L^k$.

Proposition 5.1 *Suppose S is a feasibility sequence for G. Then with the definitions above, for $k = 0, \ldots, l$:*
(i) R^k is a Gale certificate for P^k.
(ii) For every $v \in D^k$, there exists a path from v to w containing only edges from B^k.

Proof. The proof is by backwards induction on $k = l, l-1, \ldots, 0$. For the induction basis, note that after the last step l, $d_w^l = 0$ and $e_w^l > 0$. Hence $R^l = \{w\}$ is a Gale certificate, and $D^l = \{w\}$, so (ii) is trivially true.

We now show the inductive step. Suppose (i) and (ii) hold for k, and let us prove them for $k-1$. Denote the edge introduced into B in step k by $a^k = rt$, where r is the vertex eliminated in step k. Recall that we assumed that $w \in I$. We distinguish four cases (compare figure 3):

Case I: $r \in I$, $t \in L^k$. Since $t \in L^k$, by the induction hypothesis there exists $p \in R^k$ such that $pt \in B^k$, so rt belongs to the same component as w in B^{k-1}. Is it possible that in G^{k-1}, vertex r has a neighbor which is not in L^k? Suppose $rj \in E^{k-1}$.

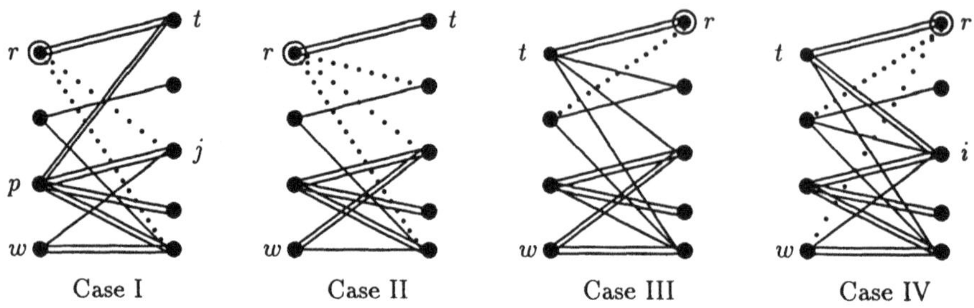

Figure 3: diagrams for the proof of the existence of a Gale certificate. In all cases r is the eliminated vertex, and dotted edges are eliminated with it. Edges included in the final set B are marked by double lines. Only vertices which are used in the proof are labeled.

Since j is active in step $k-1$, we get $rj \succ rt \prec pt$, so by the Z property of the sequence S, $pj \in E^{k-1}$. But then also $pj \in E^k$, so $j \in L^k$. Hence all neighbors of r in G^{k-1} are also present in G^k, which implies $D^{k-1} = D^k \cup \{rt\}$. This immediately implies (ii). (i) follows since

$$e^{k-1}(R^{k-1}) = e^k(R^k) + e_r^{k-1} > e^k(L^k) + e_r^{k-1} = e^{k-1}(L^{k-1}), \tag{5.1}$$

where the inequality follows from the inductive hypothesis.

Case II: $r \in I$, $t \notin L^k$. Here we want to show that $R^k = R^{k-1}$, i.e., rt is not in the connected component of w in B^{k-1}. This will imply (i) and (ii). Suppose there exists a path in B^{k-1} from r to w. Such path must contain an edge $it \in B^{k-1}$ such that i is connected to w in B^{k-1}, since r is eliminated in step $k-1$. But then $it \in B^k$ and $i \in R^k$, so $t \in L^k$, a contradiction.

Case III: $r \in J$, $t \notin R^k$. In that case $R^k = R^{k-1}$, and $r \notin L^k$, since $r \notin V^k$. (i) and (ii) are thus immediate by the inductive hypothesis.

Case IV: $r \in J$, $t \in R^k$. By the inductive hypothesis, t belongs to the same component as w in B^k. Hence in B^{k-1}, both r and t belong to that component, which implies that there exists an edge ti on the path from t to w in B^{k-1}, and that edge is also present in B^k. Since every other neighbor of t (except r) is also in V^k, by the inductive hypothesis it is also in L^k. Hence (ii) is true. (i) now follows by an argument identical to the chain of inequalities (5.1) in case I. ■

Applying the proposition with k=0 we get the desired result:

Theorem 5.2 *If S is a feasibility sequence and algorithm GREEDY* detects infeasibility at vertex w, then the vertices in w's connected component of B which are on the same side as w form a Gale certificate of infeasibility.* ■

6 Signatures

Let B be a subset of the edges in the bipartite graph $G = (I, J; E)$. In this section it will be more convenient to denote edges by ordered pairs of vertices, i.e., for the edge $(r, s) \in E$, $r \in I$ and $s \in J$. The *(row) signature* of B is defined as an integer vector $\boldsymbol{\sigma}^B = (\sigma_1^B, \ldots, \sigma_m^B)$ such that

$$\sigma_i^B = |\{k|(i,k) \in B\}| \qquad i = 1, \ldots, m. \tag{6.1}$$

The column signature is defined analogously. We shall discuss row signatures here, but the same results apply to column signatures, by interchanging the roles of I and J.

For a bipartite graph an integer vector $\boldsymbol{\sigma} = (\sigma_1, \ldots, \sigma_m)$ is called a *valid signature* if there exists a set B which forms a spanning tree and $\boldsymbol{\sigma} = \boldsymbol{\sigma}^B$, i.e., for each vertex $i \in I$ its degree in the subgraph (V, B) is σ_i. In particular, only connected graphs have valid signatures, and every valid signature vector satisfies $\boldsymbol{\sigma} > 0$ and $\sum_{i=1}^m \sigma_i = m + n - 1$. (For complete bipartite graphs, these conditions are sufficient for validity.) Balinsky and Russakoff have shown that in unrestricted problems, for every valid signature vector $\boldsymbol{\sigma}$, there exists at least one dual feasible spanning tree B with that signature [6] (see also [4, 5]). Indeed, under dual non-degeneracy, they showed that there exists one to one correspondence between dual feasible bases and valid signatures. In this section, we shall show how to use vertex elimination and feasibility sequences in order to determine quickly if a given signature is valid. Subsequently, we shall show that with a Monge sequence, one can also find a dual feasible spanning tree with that signature, if one exists. We shall also provide new characterizations based on signatures for feasibility and optimality sequences.

A preliminary question is whether one can efficiently decide if a given signature is valid. More precisely, suppose we are given a bipartite graph and a signature $\boldsymbol{\sigma}$, and asked whether there exist a spanning tree with that signature. We now show that this problem is polynomial, and also give a polynomial algorithm which constructs a spanning tree with the given signature if it is valid. Clearly, the conditions $\boldsymbol{\sigma} > 0$ and $\sum_{i=1}^m \sigma_i = m + n - 1$ are necessary. For complete bipartite graphs these conditions are also sufficient, so the question is trivial. For an arbitrary bipartite graph, having verified the necessary conditions, we can rephrase the question:

> **Problem 1:**
> Given $\boldsymbol{\sigma}$, does there exist an independent (cycle-free) set of edges D with $\boldsymbol{\sigma}^D = \boldsymbol{\sigma}$?

Let us first discuss a slightly weaker question:

> **Problem 2:**
> Given $\boldsymbol{\sigma}$, find a set of edges D of maximum cardinality satisfying (a) D is independent, and (b) $\boldsymbol{\sigma}^D \leq \boldsymbol{\sigma}$.

It is easy to verify that the collection of edge-sets which satisfies (a) forms a matroid, and so does the collections of edge-sets which satisfies (b). Hence problem 2 is that of finding a maximum cardinality set in the intersection of two matroids. This problem is known to be polynomially solvable (see, e.g., [16]). A solution D to problem 2 gives also the answer to problem 1: If $\boldsymbol{\sigma}^D = \boldsymbol{\sigma}$, the answer is positive. Otherwise, it is impossible to achieve a valid signature, since its edge-set must also satisfy (a) and (b), and have a larger cardinality than D.

We first describe an algorithm which receives as input a signature vector $\boldsymbol{\sigma}$ in addition to the graph and a permutation S of the edges, and tries to construct a spanning tree B with that signature. We assume that the input signature is positive, otherwise it is trivially invalid.

```
algorithm SIGNATURE-B(σ, S);
begin  B ← ∅; σ_i^B ← 0, i = 1,...,m;
     For i = 1,...,p do :
     begin  pick the next edge S_i = (r, s).
          if (r, s) is active then
          begin  B ← B ∪ {(r, s)};   σ_r^B ← σ_r^B + 1;
               if d_s = 1 then ELIMINATE(s)
               else if d_r = 1 or σ_r^B = σ_r then ELIMINATE(r)
               else (d_r > 1, d_s > 1, σ_r^B < σ_r) ELIMINATE(s)
          end
     end
     if σ ≠ σ^B then output "failure" else output B.
end
```

The algorithm terminates either with a set of edges B corresponding to the given signature, or with the answer "failure". Note that termination with failure can be done earlier, by stopping the algorithm when a vertex r is eliminated because $d_r = 1$ but still $\sigma_r^B < \sigma_r$.

Observe that the algorithm eliminates one vertex in every step, and it chooses to eliminate an end-vertex if such a vertex is incident on the currently examined edge. This immediately implies:

Lemma 6.1 *Algorithm SIGNATURE-B satisfies the vertex elimination properties VE.1 and VE.2.* ∎

Theorem 6.2 *If S is a feasibility sequence then algorithm SIGNATURE-B($\boldsymbol{\sigma}, S$) determines if the given vector $\boldsymbol{\sigma}$ is a valid signature, and constructs a spanning tree B with that signature if one exists.*

Proof. The fact that the algorithm constructs a spanning tree follows from Lemma 6.1 and Proposition 3.5. If the algorithm terminates successfully, it follows from the algorithm that B has the correct signature. It remains to show that the algorithm never terminates with failure for a valid signature. To prove this, it suffices to show that by performing the first step of the algorithm we do not lose the validity of the signature. Suppose G has a spanning tree $T$with signature $\boldsymbol{\sigma}$, and $a = (r, s)$ is the first edge in S. In the first step of the algorithm, one of the endpoints of a was eliminated. Let G' be the graph after the first step. We claim that there exists a set of edges B' in G' so that $B' \cup \{a\}$ is a spanning tree on G with the required signature. To show that, let $\boldsymbol{\sigma'}_r = \sigma_r - 1$, $\boldsymbol{\sigma'}_i = \sigma_i, i \neq r$ be the modified signature prescribed for G'. We shall prove that $\boldsymbol{\sigma'}$ is indeed valid for G'.

If $d_r = 1$ or $d_s = 1$ then clearly $T - \{a\}$ is a spanning tree on G' with signature $\boldsymbol{\sigma'}$, so the claim follows. Suppose $r \in I$ was the eliminated vertex in that step, $d_r > 1$ and $d_s > 1$. This can happen only if $\sigma_r = 1$, and in that case again $T - \{a\}$ has the desired properties. The only remaining case is when $s \in J$ was the eliminated vertex, $\sigma_r > 1$, $d_r > 1$ and $d_s > 1$.

Let us first show that there exists a spanning tree T' in G with signature $\boldsymbol{\sigma}$ which includes (r, s). If $(r, s) \notin T$ then $T \cup \{(r, s)\}$ contains a unique cycle. In that cycle there exists an edge (r, t), $t \neq s$. The set $T' = T \cup \{(r, s)\} - \{(r, t)\}$ has the same signature as T, it contains (r, s) and is also a spanning tree for G.

Next, suppose $(r, s) \in T$ and the spanning tree T contains additional edges incident on the vertex s. In that case, since the algorithm removes all those edges during the first step, we should show that there is another spanning tree with the same signature but without those edges. Let $(j, s) \in T$ be such an edge. Since $\sigma_r > 1$, there is some other edge $(r, k) \in T, r \neq s$. Since $(r, s) \prec (r, k)$ and $(r, s) \prec (j, s)$, by property Z of the sequence S, $(j, k) \in E$. Moreover, $(j, k) \notin T$, or else the four edges would have formed a cycle in T. Let $T^* = T \cup \{(j, k)\} - \{(j, s)\}$. The set T^* has the same signature as T, is still a spanning tree for G and has one less edge incident on s. By repeating the argument (with T^* replacing T) if necessary, we conclude that there exists a spanning tree T with signature $\boldsymbol{\sigma}$ such that (r, s) is the only edge it contains which is incident on s. Therefore, $T - \{(r, s)\}$ is a spanning tree in G' with signature $\boldsymbol{\sigma'}$. ∎

To prove the converse, we need the following lemma which shows that it is possible to extend a valid signature by "going backwards in time" and adding rather than eliminating vertices. Define the *residual signature* at any stage of the algorithm to be $\boldsymbol{\sigma} - \boldsymbol{\sigma}^B$. Namely, this is the signature that must be completed on the remaining subgraph at that stage in order to satisfy the original signature requirement.

Lemma 6.3 *Let G be a bipartite graph, and let S be any permutation of its edges where $S_1 = (i, j)$. Let G' be the subgraph induced by eliminating either i or j. If G' admits a valid signature $\boldsymbol{\sigma'}$, then there exists a valid signature $\boldsymbol{\sigma}$ on G such that after the first step of SIGNATURE-B($\boldsymbol{\sigma}, S$), the residual signature on G' is $\boldsymbol{\sigma'}$.*

Proof. By the assumptions of the lemma there exists a spanning tree T' in G' with the valid signature $\boldsymbol{\sigma'}$. If i is eliminated, define $\sigma_i = 1$ and $\sigma_k = \boldsymbol{\sigma'}_k$ for $k \neq i$. If

j is eliminated, define $\sigma_i = \boldsymbol{\sigma}'_i + 1$ and $\sigma_k = \boldsymbol{\sigma}'_k$ for $k \neq i$. In both cases, the set $T = T' \cup \{(i,j)\}$ forms a spanning tree in G, which satisfies the requirements of the lemma. ■

Theorem 6.4 *Let S be such that for every valid signature $\boldsymbol{\sigma}$, SIGNATURE-B($\boldsymbol{\sigma}, S$) generates a spanning tree with that signature. Then S is a feasibility sequence.*

Proof. Suppose S is not a feasibility sequence. Let $(i,j) \in E$ be the first edge in the sequence S which violates the Z property. In particular, there are vertices k,l such that $(i,l) \in E$, $(k,j) \in E$, $(i,j) \overset{S}{\prec} (i,l)$ and $(i,j) \overset{S}{\prec} (k,j)$ and $(k,l) \notin E$.

If (i,j) is the first edge in S, define $R_k = \{(k,s)|s \in N(k)\}$, $\delta_k = |R_k|$, $R_i = \{(i,s)|s \in N(i) - N(k)\}$, $\delta_i = |R_i| + 1$. The set $R = R_k \cup R_i$ is cycle free, hence it can be extended to a spanning tree with a valid signature $\boldsymbol{\sigma}$, satisfying $\sigma_i = \delta_i$, $\sigma_k = \delta_k$. Moreover, any such spanning tree T must contain all the edges in R_k. If in addition $(i,j) \in T$, then no other edge (i,s) with $s \in N(k)$ can be included in T, since it would form a cycle. Hence every spanning tree T which satisfies $\sigma_i = \delta_i$, $\sigma_k = \delta_k$, and $(i,j) \in T$ must contain all edges in R_k and R_i. Since in the first step the algorithm introduces (i,j) into the spanning tree and eliminates one of its endpoints, it is impossible that both $(i,l) \in R_i$ and $(k,j) \in R_k$ will be included later in T, a contradiction.

Assume now that (i,j) is the (t+1)-st edge in the sequence S, $t \geq 1$. Suppose we use the algorithm GENERATE-B on the subsequence $S_1, \ldots, S_t$, temporarily ignoring signature requirements, but making sure that none of the vertices i,j,k,l is eliminated. This can be guaranteed since none of those vertices is an end vertex incident on an edge in S_r for $r \leq t$. Let E^t be the set of active edges produced by the algorithm after step t. Since the subsequence $S_1, \ldots, S_t$ satisfies property Z, by Lemma 3.4, E^t is connected. Hence the same argument as in the first part of this proof implies that in the active graph after step t, there is a valid signature $\boldsymbol{\sigma}^t$ for which the algorithm fails. Lemma 6.3 implies that $\boldsymbol{\sigma}'$ can be extended to a valid signature $\boldsymbol{\sigma}$ in the original graph, so that algorithm SIGNATURE-B($\boldsymbol{\sigma}, S$) leads to the residual signature $\boldsymbol{\sigma}^t$ after stage t. Hence the algorithm fails for the valid signature $\boldsymbol{\sigma}$. ■

Theorem 6.5 *If S is a Monge sequence and $\boldsymbol{\sigma}$ is a valid signature, then the set B generated by SIGNATURE-B($\boldsymbol{\sigma}, S$) is a dual feasible spanning tree with signature $\boldsymbol{\sigma}$.*

Proof. By Lemma 6.1, SIGNATURE-B is a special case of algorithm GENERATE-B. The result thus follows from Theorem 6.2 and Theorem 3.7. ■

Let us now show that the converse of Theorem 6.5 is also true:

Theorem 6.6 *Let S be such that for every valid signature $\boldsymbol{\sigma}$, SIGNATURE-B($\boldsymbol{\sigma}, S$) generates a dual feasible spanning tree with that signature. Then S is a Monge sequence.*

Proof. By Theorem 6.4, S must be a feasibility sequence. Suppose it does not have the Hoffman property. Then there exist edges $(i,j),(i,l),(k,j),(k,l)$ such that

$$(i,j) \prec (i,l), \quad (i,j) \prec (k,l), \quad (i,j) \prec (k,j) \tag{6.2}$$

and

$$C_{ij} + C_{kl} > C_{il} + C_{kj}. \tag{6.3}$$

Let (i,j) be the first edge in the order S which satisfies conditions (6.2) and (6.3). Define $\overline{G}$ to be the same graph without edge (k,l). Note that S does not satisfy the Z property on $\overline{G}$, by (6.2). Let $\boldsymbol{\sigma}$ be the signature generated on $\overline{G}$ by the process described in the proof of Theorem 6.4. Since $\boldsymbol{\sigma}$ is valid for $\overline{G}$, it is also a valid signature for the original graph G. By the proof of Theorem 6.4, algorithm SIGNATURE-B($\boldsymbol{\sigma}, S$) must fail on $\overline{G}$.

Consider the operation of algorithm SIGNATURE-B($\boldsymbol{\sigma}, S$) on G. If the algorithm fails, then we get the desired contradiction. Suppose the algorithm succeeds. When edge (i,j) is examined by the algorithm, it is still active and thus is included in B. Moreover, edge (k,l) must be also included in B, or else the final set B will form a spanning tree to $\overline{G}$ with signature $\boldsymbol{\sigma}$, contradicting Theorem 6.4. Let $(\overline{u},\overline{v})$ be the dual solution vector corresponding to B. Since $(i,j) \in B$ and $(k,l) \in B$, this solution must satisfy:

$$\overline{u}_i + \overline{v}_j = C_{ij} \qquad \overline{u}_k + \overline{v}_l = C_{kl}$$

Substituting the above equations into (6.3), we obtain:

$$\overline{u}_i + \overline{v}_j + \overline{u}_k + \overline{v}_l > C_{il} + C_{kj}$$

or, equivalently,

$$(\overline{u}_i + \overline{v}_l - C_{il}) + (\overline{u}_k + \overline{v}_j - C_{kj}) > 0 \ .$$

Hence, at least one of the values in parentheses must be strictly positive, in contradiction to the dual feasibility of $(\overline{u},\overline{v})$. ∎

Theorems 6.5 and 6.6 together give another characterization to Monge sequences, analogous to the primal one: Call S a *dual feasibility sequence* for C if for every feasible signature $\boldsymbol{\sigma}$, SIGNATURE-B($\boldsymbol{\sigma}, S$) generates a dual feasible basis with that signature.

Corollary 6.7 *S is a Monge sequence for C if and only if it is a dual feasibility sequence for C.* ∎

7 Summary

This paper brings together two previously studied topics: vertex elimination, and greedily solvable bipartite flow problems. A concept which, somewhat surprisingly, turns out to be relevant to both, is the feasibility sequence. We have defined a simple vertex elimination algorithm guided by an edge order, and have shown that with a feasibility sequence, the algorithm generates a spanning tree. Moreover, only a feasibility sequence guarantees that the algorithm will generate a spanning tree for every induced subgraph. When we consider also the edge costs, then with a Monge sequence the algorithm generates a dual feasible spanning tree, and again, Monge sequences are the only sequences that guarantee this for every induces subgraph.

When supplies and demands are also introduced, we have a complete transportation problem. By combining the properties of the greedy algorithm and the vertex elimination algorithm, we have proved that with a Monge sequence, the hybrid algorithm generates an optimal basic solution and its corresponding optimal spanning tree, for every feasible supply and demand function. When that function is infeasible, the algorithm produces a Hall-type certificate of infeasibility.

Finally, we have considered spanning trees with given signatures. By adjusting the vertex elimination algorithm to take into consideration a given signature, we have shown that with a feasibility sequence, the algorithm generates a spanning tree with the input signature whenever that signature is valid (i.e., corresponds to a spanning tree). Moreover, with any other sequence, this is not the case. As we noted, building a spanning tree with a given signature can be done polynomially even without a feasibility sequence, using matroid intersection.

Several interesting problems arise from our discussions. First, given a signature, how can one find a dual feasible spanning tree which corresponds to it, or determine that none exist? We have shown how to do this only in case a Monge sequence exists. Second, if the signature algorithm terminates with infeasibility, can one give a certificate that shows that the signature is invalid, in a fashion similar to what was done in section 5? Finally, can the results of this paper be generalized to non-bipartite problems?

References

[1] I. Adler, A. J. Hoffman, and R. Shamir, Monge and Feasibility Sequences in General Flow Problems, Technical report, RUTCOR, Rutgers University, (1990). To appear in *Discrete Mathematics.*

[2] R. K. Ahuja, T. L. Magnanti, and J. B. Orlin, Network Flows, *Handbooks in Operations Research and Management Science*, Vol. I, G. L. Nemhauser, A. H. G. Rinnooy Kan, and M. S. Todd (editors), (Elsevier, Amsterdam, 1989) p. 211–369.

[3] N. Alon, S. Cosares, D. S. Hochbaum and R. Shamir, An algorithm for the detection and construction of Monge sequences, *Linear Algebra and Its Applications*, **114/115** (1989) 669–680.

[4] M. L. Balinsky, The Hirsch conjecture for dual transportation polyhedra, *Mathematics of Operations Research*, **9(4)** (1984) 629–633.

[5] M. L. Balinsky, Signature methods for the assignment problem, *Operations Research*, **33(3)** (1985) 527–536.

[6] M. L. Balinsky and A. Russakoff, Faces of dual transportation polyhedra, *Mathematical Programming Study*, **22** (1984) 1–8.

[7] R. G. Bland, New finite pivoting rules for the simplex method, *Mathematics of Operations Research* **2(2)**, (1977) 103–107.

[8] B.L. Dietrich, Monge sequences, antimatroids, and the transportation problem with forbidden arcs, *Linear Algebra and Its Applications*, **139** (1990) 133–145.

[9] D. Gale, A theorem on flows in networks, *Pacific Journal of Mathematics*, **7** (1957) 1073–1082.

[10] M. Grötschel, L. Lovász and A. Schrijver, *Geometric Algorithms and Combinatorial Optimization* (Springer, Berlin, 1988).

[11] A. V. Goldberg, É. Tardos, and R. E. Tarjan, Network Flow Algorithms, *Paths, Flows and VLSI-Layout*, B. Korte, L. Lovász and A. Schrijver (editors), (Springer-Verlag, Berlin 1990) p. 101–164.

[12] M. C. Golumbic, *Algorithmic Graph Theory and Perfect Graphs.* (Academic Press, New York, 1980).

[13] G. Hadley, *Linear Programming.* (Addison-Wesley, Reading, Massachusetts, 1962).

[14] P. Hall, On representatives of subsets, *J. London Math. Soc.*, **10** (1935) 26–30.

[15] A.J. Hoffman, On simple linear programming problems, in *Convexity: Proceedings of symposia in Pure Mathematics*, **Vol 7**, (V. Klee, Ed.), (American Mathematical Society, 1963).

[16] E. L. Lawler, *Combinatorial Optimization: Networks and Matroids.* (Holt, Reinhart and Winston, New York, 1976).

[17] D. J. Rose, Triangulated graphs and the elimination process, *J. Math. Anal. Appl.*, **32** (1970) 597–609.

[18] R. Shamir, A fast algorithm for constructing Monge sequences in transportation problems with forbidden arcs, Report 136/89 (1989), Tel Aviv University. To appear in *Discrete Mathematics.*

[19] R. Shamir and B. Dietrich, Characterization and algorithms for greedily solvable transportation problems, *Proceedings of the first ACM/SIAM Symposium on Discrete Algorithms*, (SIAM, Philadelphia, 1990) p. 358–366.

Network Optimization Problems, pp. 23-36
Eds. D.-Z. Du and P.M. Pardalos

Networks Minimizing Length Plus the Number of Steiner Points

Thomas Colthurst, Chris Cox, Joel Foisy, Hugh Howards,
Kathryn Kollett, Holly Lowy, and Stephen Root
c/o Prof. Frank Morgan, Department of Mathematics, Williams College, Williamstown, MA 01267, Internet: Frank.Morgan@williams.edu

1 Introduction

The Steiner problem seeks to connect a given set of vertices with the shortest network, generally with added vertices called Steiner points (cf. [CR] pp. 354-361, [BG]). We consider networks minimizing a cost function of length plus the number of Steiner points, modeling, for example, the expense of each intersection in a road network. Unlike solutions to the Steiner problem, these cost-minimizing networks are not scale invariant, and in general there will be different optimal solutions for similar but scaled point sets. The existence of networks minimizing length plus the number of Steiner points follows from standard compactness arguments.

This paper establishes four as the sharp bound on the number of edges that can meet at a Steiner point in a cost-minimizing network in $\Re^2$ (Theorem 3.5). We have learned recently that this was originally conjectured by W. Boyce in 1970 [B], and proved independently Rubinstein, Thomas, and Weng [RTW] using a variational approach to the Steiner ratio. In addition to our proof of Boyce's conjecture we show that in general $2n$ edges may meet in $\Re^n$, as along the coordinate axes (Theorem 2.6), and that the maximum number of edges which can meet in $\Re^n$ grows exponentially (3.2, 3.7). The higher-dimensional problem of surfaces minimizing area plus the length of singular curves models interfacial energies materials (see [MT], [M]). Our results suggest that such interfaces would meet only in threes or fours.

This paper is the work of the Geometry Group of the Williams College SMALL Undergraduate Research Project, Summer 1991, a National Science Foundation site for Research Experiences for Undergraduates (REU). For a period of ten weeks, each of twenty- four students worked in one or two of the seven groups that comprised the project. Professor Frank Morgan advised the Geometry Group.

Support for the project was provided by grants from the National Science Foundation Research Experiences for Undergraduates Program, NECUSE, Shell, and the Bronfman Science Center at Williams College.

2 Examples of Networks Minimizing Length Plus the Number of Steiner Points

Examples 2.1 shows that cost-minimizing networks connecting the vertices of a square can meet in threes and fours, depending on the size of the square. Theorem 2.6 shows that $2n$ unit vectors from the origin along the $2n$ axis directions in $\Re^n$ are cost-minimizing.

Examples 2.1 *The Square*

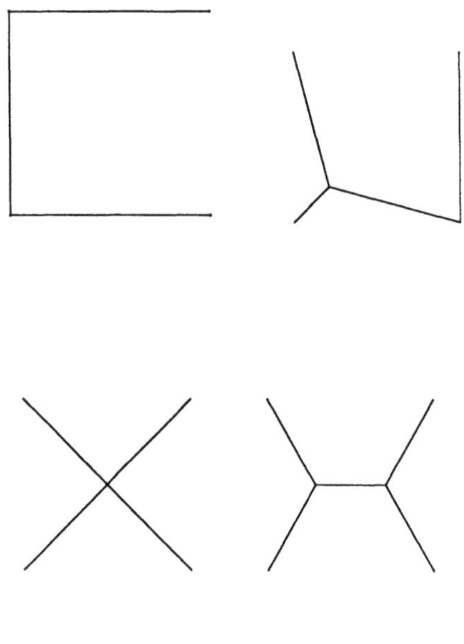

Figure 1: The four possible topological types of networks minimizing length plus number of Steiner points on the square. The network in the upper right is never the minimizer, but the other three are minimizing at some scale.

We list the cost-minimizing networks on a square to show that three and four edges can meet at a Steiner point in such a network. We do this by considering all topological types which do not have non-adjacent vertices connected by an edge, in which case an edge connecting adjacent vertices would clearly decrease cost. On the square there are four such types, as is shown in Figure 1:

1. Two Steiner points connected to each other and each connecting adjacent vertices. This is the length minimizing tree, and has cost $C(L) = 2 + (1+\sqrt{3})L \approx 2 + 2.732L$.
2. One Steiner point connecting three of the vertices and a side of the square connecting the fourth. This has cost $C(L) = 1 + (1 + \frac{\sqrt{2}}{2} + \frac{\sqrt{6}}{2})L \approx 1 + 2.932L$.
3. One Steiner point at the center of the square connecting all four vertices. This has cost $C(L) = 1 + 2\sqrt{2}L \approx 1 + 2.828L$.
4. The perimeter method, with no Steiner points and three sides of the square. Here the cost $C(L) = 3L$.

The network with two Steiner points is the unique network with two Steiner points and four vertices [C2]. We see here the scale dependence: when $L \geq \frac{1}{2\sqrt{2}-\sqrt{3}-1}$, the length minimizing network is cost-minimizing. For $\frac{1}{3-2\sqrt{2}} \leq L \leq \frac{1}{2\sqrt{2}-\sqrt{3}-1}$, the third network with four edges meeting at the center is minimizing. The perimeter method is best for relatively small values of L, when $L \leq \frac{1}{3-2\sqrt{2}}$.

Lemma 2.2 *In a network minimizing length plus the number of Steiner points, the sum of the unit vectors of segments meeting at a Steiner point must be* 0.

PROOF: For a given vector v, the derivative of $|v|$ is $\frac{v}{|v|}$. The sum of the derivatives, i.e., of the unit vectors meeting at a point must add to zero, or else we could move the Steiner point to a position that does allow the unit vectors to sum to 0, and the network would have reduced total length. □

In particular, any Steiner point of degree three must have 120° angles between its edges.

Lemma 2.3 *Suppose a network $N(0)$ is continuously transformed to a new network $N(1)$ of the same topological type. Then the function F(t) describing the length of the intermediate networks is strictly convex.*

PROOF: Consider any two networks $N(0)$ and $N(1)$ of a given topological type, with nodes $n_1(0), ..., n_m(0)$ and $n_1(1), ..., n_m(1)$. Then a new network $N(t)$ of the same type can be created by interpolating between each pair of nodes $n_i(0)$ and $n_i(1)$ to form the node $n_i(t)$. Let $F(t)$ equal the length of $N(t)$, which will clearly be a function of t. Then

$$F(t) = \sum_{n_i,n_j} f_{i,j}(t)$$

where $f_{i,j}(t) = d(n_i(t), n_j(t))$, the distance between two connected nodes n_i, n_j. Consider two pairs of corresponding points $n_i(0), n_j(0)$ and $n_i(1), n_j(1)$. Then

$$f_{i,j}(\frac{1}{2}) = d(\frac{n_i(0) + n_i(1)}{2}, \frac{n_j(0) + n_j(1)}{2})$$

so

$$f_{i,j}(\frac{1}{2}) = \frac{1}{2}\|n_i(0) + n_i(1), n_j(0) + n_j(1)\|$$

Considering the four points as vectors, it follows from the triangle inequality that

$$\frac{1}{2}\|n_i(0) + n_i(1), n_j(0) + n_j(1)\| \leq \frac{1}{2}(\|n_i(0) + n_j(0)\| + \|n_i(1) + n_j(1)\|)$$

$$= \frac{f_{i,j}(0) + f_{i,j}(1)}{2}.$$

Hence $f_{i,j}(\frac{1}{2}) \leq \frac{f_{i,j}(0)+f_{i,j}(1)}{2}$ which implies that $f(t)$ is an upward convex function, so $F(t)$ is also convex. Observe that equality holds and $F(t)$ is not strictly convex only when equality holds in the triangle inequality above. This implies that all edges are parallel. It can be shown by a simple induction argument that, assuming there are no cycles, no two distinct networks of the same topological type have all edges parallel. Hence, $F(t)$ is strictly convex. □

Lemma 2.4 *For $n \geq 2$, the cost-minimizing network connecting the $2n$ vertices $(0, ..., \pm a, ..., 0)$ of a regular cross-polytope with at most one Steiner point connects every vertex directly to the origin for $a > \frac{1}{(2\sqrt{2n} - 2n - \sqrt{2})}$.*

PROOF: Consider the cross-polytope of edge length $l = a\sqrt{2}$. Since it has $2n$ vertices, connecting them by a tree on the perimeter has cost $l(2n-1)$. Connecting its vertices to a Steiner point at the origin has cost $1 + \sqrt{2}nl$. Thus, the center network is cheaper for $l > 1/(2n - \sqrt{2}n - 1)$ and for such an l has cost greater than $1 + \frac{\sqrt{2}n}{2n - \sqrt{2}n - 1}$.

Any cost-minimizing network N with one Steiner point may be classified by the number of vertices the Steiner point connects; call this number t. Since the Steiner tree portion of N has length at least $\frac{1}{\sqrt{3}}$ that of the minimum spanning tree [GH], N has cost at least $1 + l(2n-t) + \frac{1}{\sqrt{3}}(t-1)l$. Comparing this to the center cost of $1 + \sqrt{2}nl$ one concludes that it is cheaper to connect a vertex to the Steiner point if $t < \frac{n(2-\sqrt{2}) - \frac{1}{\sqrt{3}}}{1 - \frac{1}{\sqrt{3}}}$. For t higher than this, we claim that there is a vertex not directly connected to the Steiner point which is closer to it than to the nearest vertex.

We may assume that the $2n - t$ unconnected vertices are mutually adjacent, containing no two vertices on the same axis, and form a $2n - t - 1$ dimensional simplex. For if neither of two points on the same axis are connected to the Steiner point, symmetry dictates that the Steiner point must lie in the hyperplane orthogonal to their axis. Since it must be contained in the convex hull of the vertices in the hyperplane, the Steiner point will actually be closer to both vertices than they are to any other vertices. On the other hand axis directions with both vertices connected directly to the Steiner point need not be considered either, as the Steiner point will always lie in the $\Re^{n-1}$ dimensional orthogonal hyperplane, so the problem can be considered in $\Re^{2n-t}$.

Call the distance between the origin and the Steiner point x. Then the coordinates of the Steiner point look like $(\frac{x}{\sqrt{2n}}, \frac{x}{\sqrt{2n}}, \frac{x}{\sqrt{2n}}, ..., \frac{x}{\sqrt{2n}})$, and the coordinates of a given unconnected vertex are $(\frac{-l}{\sqrt{2n-t}}, 0, 0, ...0)$. Then the distance between them is just

$$\sqrt{(\frac{x}{\sqrt{2}} + \frac{l}{\sqrt{2}})^2 + \frac{(2n-t)x^2}{2n-t}}$$

Hence, a vertex is closer to the Steiner point than to other vertices if

$$l^2 > \frac{2\sqrt{2}xl}{\sqrt{2n-t}} + 2x^2$$

or

$$x < l(2\sqrt{2 - \frac{2}{2n-t}} - \frac{1}{\sqrt{2(2n-t)}})$$

Clearly x is less than the distance from a vertex to the origin, so we know that

$$x < \frac{l}{\sqrt{2n-t}}$$

Combining the two inequalities and simplifying we get

$$\frac{3}{2} < 4n - 2t$$

Now we just use the limit on t relative to n established above and we get

$$0.6141n + 1.3660 > 1.5$$

which is true for all $n \geq 2$. Thus, the network connecting all the vertices at the center is the cheaper than networks of all other topological types with one Steiner point. It remains to show that the Steiner point should be placed at the origin in the minimal network of this type.

Consider how the length of the central network changes as the Steiner point is moved away from the center along some vector u. The change in length with respect to time is given by $\frac{dl}{dt} = -(v_1 + v_2... + v_6) \cdot u$ where the v_i are the six unit vectors along the axes. A quick calculation show that $\frac{dl}{dt} = 0$ for all u, so the origin is an equilibrium for this configuration. By Lemma 2.3 length is a strictly convex function of the position of the Steiner point, so it follows that the network connecting all the vertices to the center must be minimal. □

Lemma 2.5 *The unit vectors in the six axis directions in $\Re^3$ constitute a cost-minimizing network for the octahedron.*

PROOF:
By Lemma 2.4 we know that the unit vectors are cost-minimizing over all networks with (at most) one Steiner point. We will show that there exists a scaling (a range of l) over which the unit vectors are of lesser cost than the minimal spanning tree and all possible trees with two or more Steiner points.

The minimal spanning tree has cost $5\sqrt{2}l$ and the unit vectors have cost $6l+1$. It can be easily calculated that if $l > 0.934 > \frac{1}{5\sqrt{2}-6}$, then the unit vectors will be minimizing over the minimal spanning tree.

Now consider networks having at least two Steiner points. The cost of such a network is at least two plus the length of the shortest network connecting all six points. A group of Williams students in 1989 proved that the length-minimizing network for the octahedron had four Steiner points. They also showed the eight combinatorial choices for the minimizer [C1], by classifying the topological types according to how the four Steiner points connect pairs of vertices on the same axis. Here we examine each case individually. All we have to do is find a lower bound for each candidate, then show that for some $l > 0.934$, the network with one Steiner point is cheaper. Note that in these representations, x, y, and z represent the axis that each segment hits at distance 1 from the origin.
Case 1:

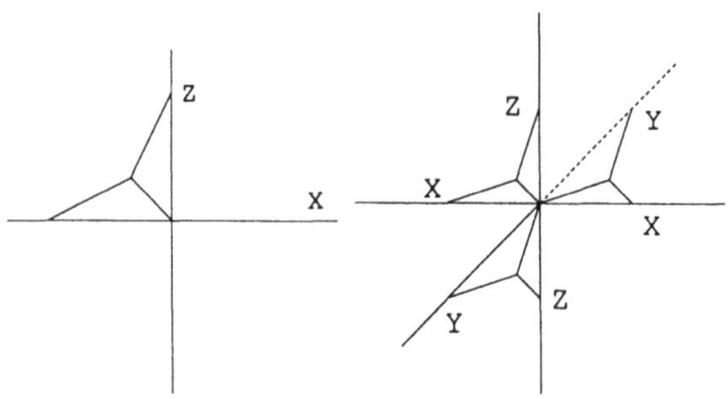

Figure 2: Case 1. The network at right has length three times that of the piece of the network component on the left, which has length $> 1.93l$.

This case uses a Steiner point at the origin which is connected to the vertices in pairs using three additional Steiner points. Hence every pair of vertices sharing an

axis are connected through three Steiner points. Here a straightforward computation shows that the length is greater than $5.7l$, which means that when $6l + 1 < 5.7l + 2$, the one Steiner point network will be cheaper. Thus for $0.934 < l < 3.333$, the one Steiner point network will be better.
Cases 2, 3, 4:

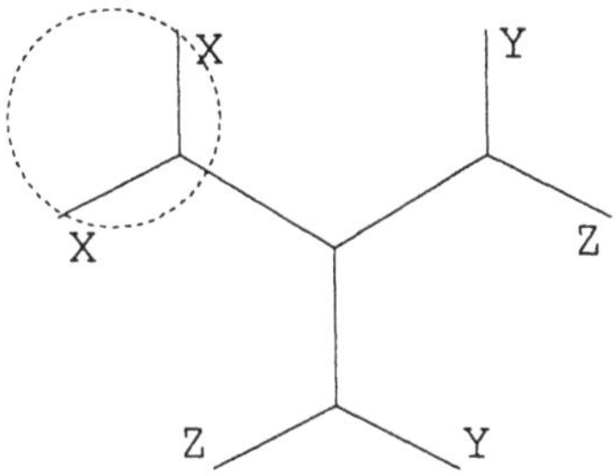

Figure 3: Case 2. The portion of the network in the dotted circle has length at least $2l$.

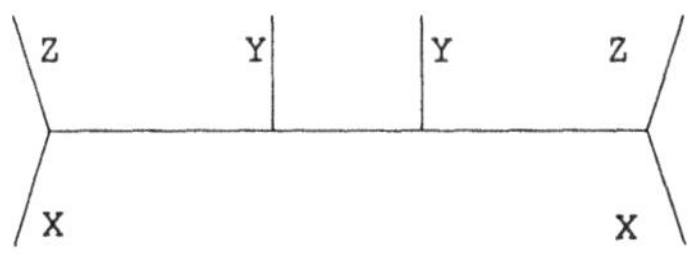

Figure 4: Case 3. The y-legs have length $\geq 2l$ and the remaining portion of the network is the minimizing network on a square with side length $\sqrt{2}l$.

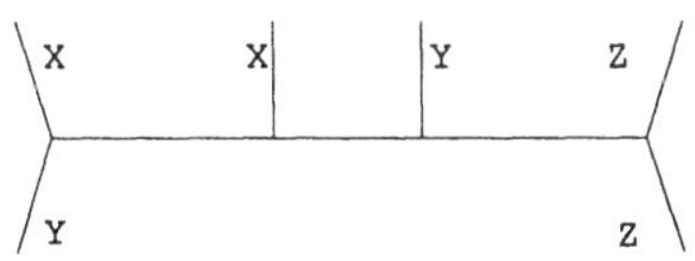

Figure 5: Case 4. The z-legs have length $\geq 2l$ and the remaining portion of the network is the minimizing network on a square with side length $\sqrt{2}l$.

Cases 2, 3, and 4 consider the possible topological types in which exactly one pair of boundary points on the same axis are connected through a single Steiner point. The minimum distance between this pair of points is $2l$. The cost of these three cases is bounded below by $2l + L$, where L is the length of the minimizing network connecting four points on the same plane forming a square with side length $\sqrt{2}l$. More specifically, the overall length is bounded by $(4\sqrt{\frac{2}{3}} + \sqrt{2} - \frac{2}{\sqrt{6}} + 2)l$. Thus, it is

cheaper to use one Steiner point when $6l + 1 < (4\sqrt{\frac{2}{3}} + \sqrt{2} - \frac{2}{\sqrt{6}} + 2)l + 2$, i.e. when $0.934 < l < \frac{1}{6-(4(0.81)+1.78)} \approx 1.04$.
Cases 5, 6:

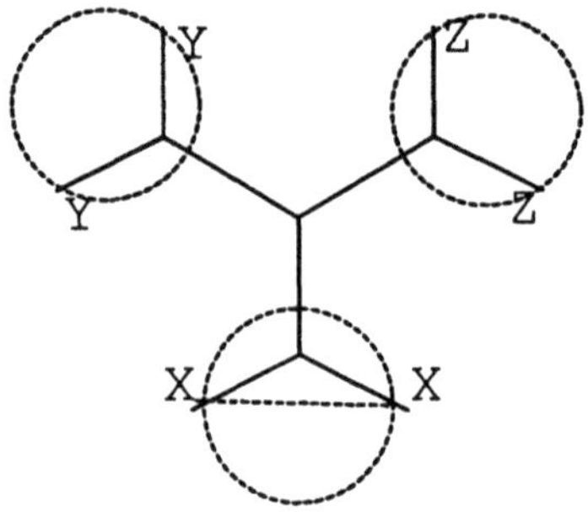

Figure 6: Case 5. The length of the portion of the network in each dotted circle is $\geq 2l$.

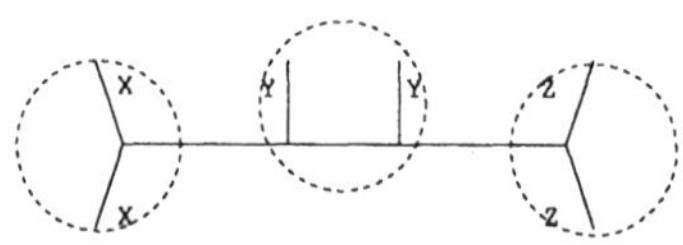

Figure 7: Case 6. The length of the portion of the network in each dotted circle is $\geq 2l$.

Cases 5 and 6 are the two topological types in which all three pairs of vertices on the same axis are connected through a single Steiner point. In these cases, it is clear that cost is bounded by $6l$. Clearly, the one Steiner point network is cheaper.
Case 7:

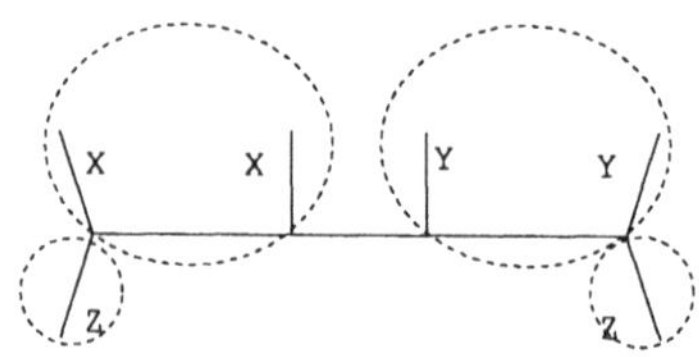

Figure 8: Case 7. The parts of the network in the dotted ellipses each have length $\geq 2l$, and the parts of the network in the dotted circles each have length $\geq \sqrt{\frac{2}{3}}$.

In this case two pairs are connected through two Steiner points and the other through four. Its cost is bounded by $(2 + 2 + 2\sqrt{\frac{2}{3}})l$. Thus, the Steiner network is

better when $6l + 1 < (2 + 2 + 2\sqrt{\frac{2}{3}})l$. Therefore the center network is cheaper when $0.934 < l < \frac{1}{6-(5.62)} \approx 2.63$.
Case 8:

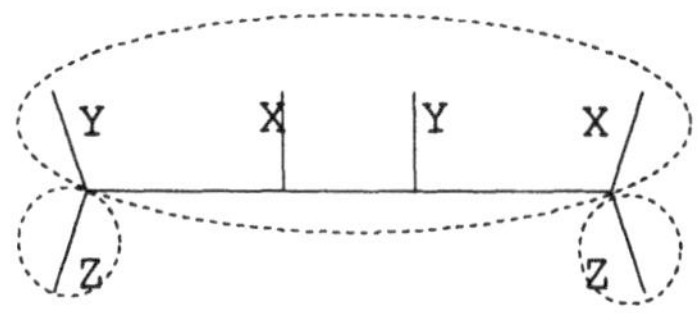

Figure 9: Case 8. The parts of the network in the dotted circles each have length $\geq \sqrt{\frac{2}{3}}l$, and the parts in the large ellipse have length $\geq (4\sqrt{\frac{2}{3}} + \sqrt{2} - \frac{2}{\sqrt{6}})l$.

In this case two pairs of vertices are connected through two Steiner points and the other through four. Its length is bounded by the shortest network connecting x, x, y and $y + 2\sqrt{\frac{2}{3}}l$. In other words, by $6\sqrt{\frac{2}{3}} + \sqrt{2} - \frac{2}{\sqrt{6}}$. Thus, the 1 point network is better when $0.934 < l < \frac{1}{6-(6(0.81)+0.58)} < 1.78$. □

We have seen that four edges of a network that minimizes length plus the number of Steiner points may meet in $\Re^2$, and that six edges may meet in $\Re^3$. We will now complete the proof that for $n \geq 2$, there is a network in $\Re^n$ that minimizes length plus the number of Steiner points and has $2n$ edges meeting at a point.

Theorem 2.6 *The $2n$ vectors of length a in the $2n$ axis directions constitute a cost-minimizing network for some value of a. In fact, if $n > 2$, then the unit vectors are cost-minimizing.*

PROOF: By Lemma 2.4, we need only consider networks with two or more Steiner points. A network with two or more Steiner points must again have length of at least $\frac{1}{\sqrt{3}}$ that of the minimum spanning tree [CH], and thus cost of at least $2+\frac{1}{\sqrt{3}}(2n-1)l$, where $l = \sqrt{2}a$ is the distance between adjacent points. But for $l < 1/(\sqrt{2}n - \frac{1}{\sqrt{3}}(2n-1))$ this is greater than the cost for the center network, and for $n \geq 5$, $1/(\sqrt{2}n - \frac{1}{\sqrt{3}}(2n-1)) > 1/(2n - \sqrt{2}n - 1)$, the value for l at which the center network becomes cheaper than the minimal spanning tree. Therefore there exists a range of values for a over which the vectors along the axes form a minimal network for $n > 5$.

For n=2 and n=3, Examples 2.1 and Lemma 2.5 provide examples in which the vectors along the axes form cost-minimizing networks.

For n = 4, the cross-polytope has tetrahedra as 3-faces, and any network on it must have length twice that of the shortest network on a tetrahedron, or $2(2.445)l$ [C1], since we may project it onto a tetrahedron in four orthogonal directions. Thus any network with two or more Steiner points has cost $\geq 2(1 + 2.445)$, which is more expensive than the center network for $1/(2n - \sqrt{2}n - 1) = 0.744 < l < 1.319$.

Thus for $n \geq 4$ there is a range of l for which $2n$ edges of length l meeting at a point is cost-minimizing. □

3 Bounds on the Number of Edges Meeting at Steiner Points

Theorem 3.5 establishes four as the sharp upper bound on the number of edges in a cost-minimizing network that can meet at a Steiner point in the plane. Proposition 3.6 and Corollary 3.7 establish lower bounds on the number of edges that can meet in $\Re^n$. First Lemma 3.1 gives a lower bound on the angle between two edges in a cost minimizing network.

Lemma 3.1 *If AB and AC are two edges meeting at a point A in a cost-minimizing network, then $\angle BAC \geq 60°$.*

PROOF: If $\angle BAC < 60°$, then $\angle ABC$ or $\angle BCA > 60°$. Without loss of generality, take $\angle ABC > 60°$. Then $\|AC\| > \|BC\|$, so $AB + BC$ is a cheaper network than $AB + AC$ which was, by assumption, minimal. □

Remark 3.2 *An exponential upper bound on the number of edges that can meet at a Steiner point.*

Lemma 3.1 yields as an upper bound on the number of segments that may meet at a point the number of balls of radius $\frac{\pi}{6}$ that can be packed on the unit sphere. This upper bound is 6 in $\Re^2$, between 12 and 14 in $\Re^3$, and grows exponentially as a function of n.

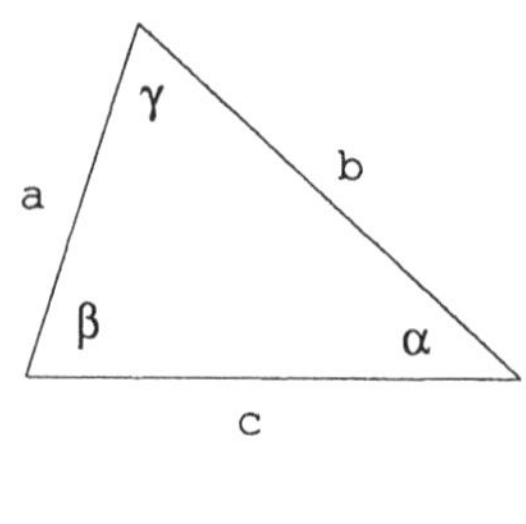

Figure 10: In a triangle with largest angle γ, $\|a\| + \|b\| \geq \|c\|(2\cos\gamma + 1)$.

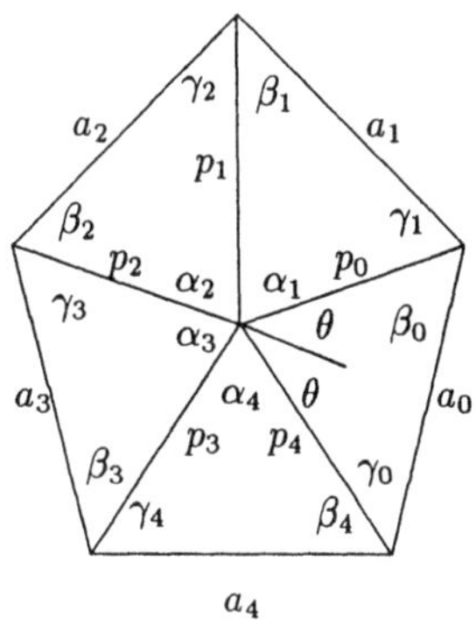

Figure 11: Five edges cannot meet at a Steiner point in a cost minimizing network.

Lemma 3.3 *Given a* $\triangle abc$ *such that* $90° \geq \gamma > \alpha$ *and* $\gamma > \beta$ *(see Figure 10), then* $\|a\| + \|b\| \geq \|c\|(2\cos\gamma + 1)$.

PROOF: Without loss of generality, fix γ and take $\beta \geq \alpha$. By the law of sines,

$$\frac{\|a\|}{\sin\alpha} = \frac{\|b\|}{\sin\beta} = \frac{\|c\|}{\sin\gamma} = \frac{\|a\|}{\sin(\beta+\gamma)}.$$

Therefore,

$$\frac{\|a\|+\|b\|}{\|c\|} = \frac{\sin(\beta+\gamma)+\sin\beta}{\sin\gamma}.$$

The minimum value of this expression when varying β occurs when $\beta = \alpha$, so

$$\frac{\|a\|+\|b\|}{\|c\|} = 1 + \frac{\sin(2\gamma)}{\sin\gamma} = 2\cos\gamma + 1.$$

□

Lemma 3.4 *If five edges,* $p_0, p_1, \ldots, p_4$*, meet at a Steiner point in a cost-minimizing network in the plane, as in Figure 11, then* $\alpha_i < 2\tan^{-1}\frac{5}{3\sqrt{3}}$*, for all* $i \in \{0, 1, \ldots, 4\}$.

PROOF: Without loss of generality, let $\angle\alpha_0 = 2\theta$ be the largest angle. By Lemma 2.2 the unit vectors must add to zero. Consider the components in the p_0 and p_4

directions: $2\cos\theta + \cos(\theta+\alpha_1) + \cos(\theta+\alpha_1+\alpha_2) + \cos(\alpha_4+\theta) = 0$. θ is maximized if $\alpha_1 = \alpha_2 = \alpha_4 = 60°$ by a convexity argument, so $2\cos\theta + \cos(\theta+60°) + \cos(\theta+120°) = 0$, or $\theta = \tan^{-1}\frac{5}{3\sqrt{3}}$, so $\alpha_0 = 2\theta < 87.795°$. □

Theorem 3.5 *Four is a sharp bound on the number of edges meeting at a Steiner point in a cost-minimizing network.*

PROOF: We have shown examples of four edges meeting (See Examples 2.1). By Lemma 3.1 it follows that seven or more edges cannot meet, and if six meet all angles must be exactly 60°. In this case, however, a regular hexagon can be substituted in some small region about the Steiner point, and if one edge is removed the length is decreased. This creates six new Steiner points, but they can be eliminated by expanding the hexagon until the new Steiner points reach the vertices at the end of the original six edges. For this reason, six edges cannot meet. Hence it remains to determine if five can meet.

Consider five edges $p_0, \ldots, p_4$ meeting at a Steiner point in a cost-minimizing network. $\alpha_0 \geq \beta_0$ and $\alpha_0 \geq \gamma_0$, since otherwise we could switch a_0 for p_0 or p_4 and decrease the overall length. Now apply Lemma 3.4 to obtain $\|p_i\| + \|p_{i+1}\| \geq \|a_i\|(2\cos\alpha_i + 1)$. Then $2\sum_{i=0}^{4}\|p_i\| \geq \sum_{i=0}^{4}\|a_i\|(2\cos\alpha_i+1)$. Without loss of generality, let a_0 be the longest side. The cost of the perimeter is $\sum_{i=1}^{4}\|a_i\|$, and the cost of five edges meeting at a Steiner point is $\sum_{i=0}^{4}\|p_i\| + 1$.

Since $\sum_{i=0}^{4}\alpha_i = 2\pi$, it follows from a convexity argument that $\sum_{i=0}^{4}\cos\alpha_i$ is minimized when the angles are at the extremum of the domain $(60°, 87.796°)$. This means that $\alpha_i = 60°, 60°, 64.41°, 87.796°, 87.796°$, and $\sum_{i=0}^{4}\cos\alpha_i > 1.5087 > \frac{3}{2}$. Therefore,

$$\|a_0\|(\cos\alpha_0 + \frac{1}{2}) + \sum_{i=1}^{4}\|a_i\|(\cos\alpha_i - \frac{1}{2}) > 0,$$

so

$$\frac{1}{2}\|a_0\|(2\cos\alpha_0 + 1) + \frac{1}{2}\sum_{i=1}^{4}\|a_i\|(2\cos\alpha_i + 1 - 2) > 0,$$

and

$$\frac{1}{2}\sum_{i=0}^{4}\|a_i\|(2\cos\alpha_i + 1) > \sum_{i=1}^{4}\|a_i\|.$$

Recall that $2\sum_{i=0}^{4}\|p_i\| \geq \sum_{i=0}^{4}\|a_i\|(2\cos\alpha_i + 1)$, so

$$\sum_{i=0}^{4}\|p_i\| > \sum_{i=1}^{4}\|a_i\|,$$

and

$$\sum_{i=0}^{4}\|p_i\| + 1 > \sum_{i=1}^{4}\|a_i\|.$$

But then the cost of five edges meeting at a Steiner point is greater than the cost of the perimeter.

Therefore, five edges cannot meet at a Steiner point in a cost minimizing network. □

Proposition 3.6 *If t points are distributed on a unit n-sphere such that the minimum angle α between any two points is greater than* 78.687947° *and if* $t > 1/(1-\frac{2}{((1+1/\sqrt{3})\beta)})$, *where* $\beta = \sqrt{2-2\cos\alpha}$ *is the minimum distance between two points, then for some scaling to a sphere of radius r, there is a cost-minimizing network in which at least a third of the points meet at a Steiner point.*

PROOF: If the n-sphere has radius r then connecting the points through the center has cost $1+rt$. Since the minimum distance between any two points is $\sqrt{2-2\cos\alpha r} = \beta r$, connecting the points through a perimeter tree has cost at least $\beta r(t-1)$. Thus the center network has lower cost for $r > 1/(\beta(t-1)-t)$, which is positive for $t > 4$ and $\alpha \geq 78.68°$. All other one Steiner point networks can be determined by the number of points they connect; call this number c. Since the Steiner tree portion of the network must have cost at least $\frac{1}{\sqrt{3}}$ times the minimal spanning tree [GH], the one Steiner point network has minimum cost $1+1/\sqrt{3}\beta r(c-1)+\beta r(t-c)$, which is greater than $1+rt$ for $c < \frac{t(\frac{1}{\beta}-1)+\frac{1}{\sqrt{3}}}{\frac{1}{\sqrt{3}}-1} \approx 0.5t - 1.36$ when $\alpha \geq 78.687947°$. Hence the optimal network with one Steiner point must have at least half of its vertices meeting at a Steiner point as t becomes large, and in fact calculations reveal that in any case $c > \frac{1}{3}t$.

Now consider networks with two or more Steiner points. Again use the fact that any Steiner network has length at least $\frac{1}{\sqrt{3}}$ that of the minimal spanning tree, which implies that the cost of such networks is at least $2+\frac{1}{\sqrt{3}}\beta r(t-1)$. A one Steiner point network is then cheaper for $r < 1/(t-\frac{1}{\sqrt{3}}\beta r(t-1))$. Thus there is a range of r for which a one Steiner point network is cost minimizing if $1/(t-\frac{1}{\sqrt{3}}\beta r(t-1)) > 1/(\beta(t-1)-t)$ or $t > 1/(1-2/((1+\frac{1}{\sqrt{3}})\beta))$. This is positive for $\alpha > 78.687947°$. □

Corollary 3.7 *The number of edges which can meet in a cost-minimizing network in* $\Re^n$ *grows exponentially in* n.

PROOF: By [EF], there is a set of points of exponential cardinality on an n-sphere such that the angle between any pair exceeds any given angle less than 90°, including the angle $\alpha \approx 78.687947°$ above. □

References

[BG] Marshall W. Bern and Ronald L. Graham, The shortest-network problem, *Scientific American*, January, 1989, pp. 84-89.

[B] W. M. Boyce, Can degree five Steiner points reduce network cost for planar sets?, 1970, preprint.

[C1] M. Conger, M. Christian, D. Cooper, J. Goodell, Length-minimizing networks in $\Re^3$: The tetrahedron and new examples, Geometry Group, Winter Study, 1989, Advisor: F. Morgan.

[C2] M. A. Conger, Energy-minimizing networks in $\Re^n$, Honors Thesis, Williams College, 1989, revised 1989.

[CR] R. Courant and H. Robbins, *What is Mathematics?*, Oxford University Press, 1941.

[EF] P. Erdős and Z. Füredi, The greatest angle among n points in the d-dimensional Euclidean space, *Annals of Discrete Mathematics* 17(1983) 275–283.

[GH] R. L. Graham and F. K. Hwang, Remarks on Steiner minimal trees, *I. Bull. Inst. Math. Acad. Sinica* 4(1976) 177- 182.

[M] Frank Morgan, Surfaces minimizing area plus length of singular curves, 1991, preprint.

[MT] Frank Morgan and Jean Taylor, Destabilization of the tetrahedral point junction by positive triple junction line energy, *Scripta Met. et Mat* 25(1991) 1907-1910.

[RTW] J. H. Rubinstein, D. A. Thomas, and J. F. Weng, Degree five Steiner points cannot reduce network costs for planar sets, *Networks*, to appear.

Network Optimization Problems, pp. 37-52
Eds. D.-Z. Du and P.M. Pardalos

Practical Experiences Using an Interactive Optimization Procedure for Vehicle Scheduling

Joachim R. Daduna
Dornier GmbH / Management Consulting
Büro Berlin, Nürnberger Straße 68 / 69, D - 1000 Berlin 30

Miodrag Mojsilovic
Peter Schütze
HanseCom Gesellschaft für angewandte Informatik und Informationsverarbeitung mbH, Spohrstraße 4, D - 2000 Hamburg 60

Abstract

During the last twenty years operations research methods became more and more important in urban mass transit planning. One of the most essential subjects in this area is the preparation of solutions for the vehicle scheduling problem.

For the HOT program system two different modules have been developed for this scheduling problem. As well as the usual scheduling procedure the system offers a module, that connects timetable preparation and vehicle scheduling. The usage of the module "sensitivity analysis" undoubtedly lead to additional savings in fleet size and operational cost. Practical experiences are described.

1 Introduction

A line of software systems has been developed since the 70's, with which efficient planning in public mass transit is attained (see Wren (1981), Bodin et al. (1983), Rousseau (1985), Daduna / Wren (1988), Desrochers / Rousseau (1992)). Early successful solutions were especially evident in vehicle scheduling (cp. Mojsilovic (1983), Luedtke (1985)), since it involved structures, which can be relatively easily described through models and then can be worked out using efficient algorithms.

One of the first of such systems available in this market was the HOT program system, by Hamburg Hochbahn Aktiengesellschaft, and used in bus, tram, and light rail operational planning, for which it was developed. The intense technological progress in the field of data processing was used to overhaul the fundamental hardware and software during the 80's (cp. Daduna / Mojsilovic (1988)). This development is made by HanseCom, a subsidiary of Hamburger Hochbahn Aktiengesellschaft.

By overhauling the software an additional module was developed, that connects two spheres of planning, timetable preparation and vehicle scheduling, which were, as a rule, treated separately in the past. The goal was to analyze with the help of an interactive optimization procedure how stable a solution is, in particular to which degree better results in scheduling could be achieved through slight alteration to the input. This module, which is called "sensitivity analysis", is now in use for several years by various urban mass transit authorities. The results achieved through this module make it evident that efficient algorithms, in conjunction with a highly developed dàta processing technology, form an im- portant basis for operational planning.

First of all the problem, as well as the formal foundations, will be described, based on Daduna (1988). Next the procedure will be presented and explained through an example. New results from urban mass transit authorities of varying size and structure will show the benefits of the sensitivity analysis.

2 Problem Formulation

The sensitivity analysis is a module that gives the user assistance in operational planning. The applicability which is described in this paper shows, in this particular form, only the advanced version HOT II. The background to this development was the consideration of how the schedule aspects could be included in timetable construction to achieve a higher efficiency rate of operation. Since a simultaneous treatment of the planning problems is not possible due to its extreme complexity, the sensitivity analysis is applied to check the entered timetable data from the point of view of scheduling.

It is systematically applied to the critical time periods of operation, i.e. rush hour. These time periods of peak capacity cause an unproportional increase in costs for the transit authorities. This implies that the maximum number of blocks in this time period determines the vehicle demand and therefore, to a certain extent, also the number of required personnel (see Daduna 1987). From this the basic objective results, namely to reduce the number of blocks during rush hour not only through vehicle scheduling with the help of an optimization procedure, as is the case with the usual planning of operation, but also by changing some timetable data. Herewith possible changes of the timetable data are determined with an interactive optimization procedure.

The utilization of such a module in the operational planning is further demonstrated by the example of Fig. 2.1. Carrying out eight trips on passenger service

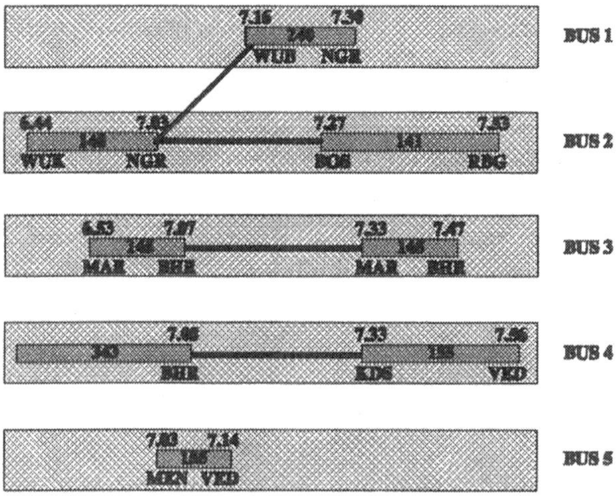

Layover:

Regular trip to deadhead trip	1min
Regular trip to regular trip on the same line	2 min
Regular trip to regular trip on different lines	5 min

Running times for deadhead trips:

	BOS	KDS	MAR	WUB
BHR	11	11	8	28
NGR	11	28	25	8
VED	21	11	20	38

Figure 2.1: Example of a vehicle schedule before an alteration to the input data

we need, in the present case, five vehicles, because only three links are feasible. The crucial point is the trip of line 140, ending in NGR at 7:03 a.m.. This trip cannot be linked with the one on line 240, starting at 7:16 a.m.. It is only one minute short. The necessary deadhead trip takes eight minutes with an additional pre-layover of five minutes after the deadhead trip and a post-layover of one minute before it.

This result will certainly be unsatisfactory to the planning staff. Therefore we must look for alterations to specific input data in view of the following essential limitations:

- It is not permitted to reduce the offered service level. The total number of trips on service, as well as routes, and running times have to remain unchanged.
- In-company and contractual agreements are not allowed to be violated, e.g. layover times and the duration of deadhead trips.

To a certain extent, however, slight alterations of departure times will be possible, because this does not mean a reduction, but only a marginal alteration without any effects on the service level. The possible range of alteration is determined by the following aspects:

- In general, published trips with even headways have to remain unchanged.
- Additional trips during peak hours are not allowed to come after a regular trip. Furthermore, the difference between the departure times of these trips should not exceed a given range, because otherwise the objective of relieving the respective regular trip will not be achieved.
- Connections to transfer stops on other lines or with other means of transport must be considered.

School trips and trips on contract hire, which, in general, are not in the timetable, yield a good scope for such alterations. In these cases, it only must be guaranteed that the vehicle arrives at the corresponding point of destination in time before the start of lesson or work. In addition, special delay buffers may be reduced, because they are fixed in the course of planning and are not restricted by in-company or contractual agreements.

The possible alterations offered by sensitivity analysis are:

- Altering the departure time of one of either the trips to be linked or of both trips
- Reducing the delay buffers (determined by the company)

In the example described above, we can shift the departure time either of the trip on line 140 or 240. The necessary alteration amounts to only one minute. The adaption of input data produces a solution, which yields a saving of one vehicle (see Fig. 2.2). This example has, at first sight, a relatively simple structure. However one should not come to the conclusion that these links can be realized in manual planning, because the example represents an isolated part of a complex structure.

3 Mathematical Formulation

Vehicle scheduling planning pursues two objectives, which are hierarchically structured, at the same time. The main objective is minimizing the number of blocks in peak hours (see Mojsilovic (1983), Carraresi / Gallo (1984), Luedtke (1985)). The other, subordinate objective pursues the reduction of non-productive times between two trips on service (deadhead trips, etc.).To a certain extent, this emerges, however, from the main objective, because considerable cuts in pull-out / pull-in trips will be achieved by reducing the number of blocks.

To solve such problems using optimization procedures, the complex structure of real world data must be reflected in a model. In doing so, a lot of different data will

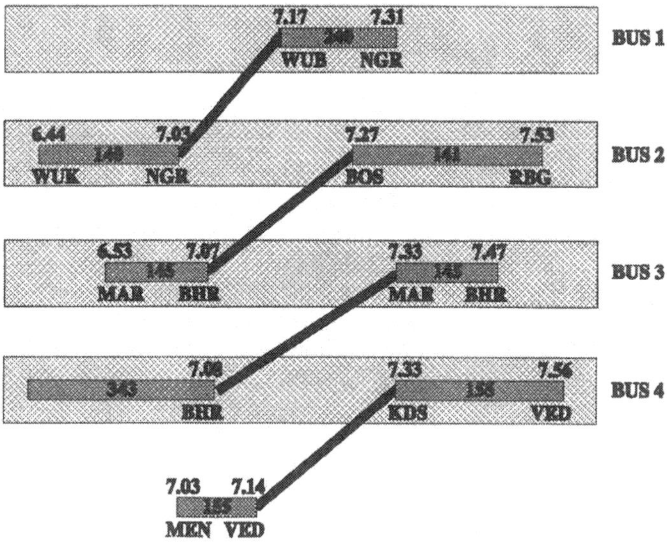

Figure 2.2: Extract from a vehicle schedule after shifting the departure time of one trip.

be qualitatively assessed and aggregated in a weight. This process represents one of the major steps in using computer-aided systems, because the result very much depends on this model form.

The basic data for vehicle scheduling are the given trips of one timetable period. In addition technical and in-company restrictions must be considered when linking two trips. For two trips i and j with i preceding j, we must take into account the following data

δ_{ij} := duration of the deadhead trip

β_i := delay buffer at the end of trip i

$\Delta^{a_i(j)}$:= layover at the end of trip i depending on trip j (post-layover)

$\Delta^{d_j(i)}$:= layover before the start of trip j depending on trip i (pre-layover)

These Δ-values resulting from in-company restrictions depend on which trips are to be linked.

A linking of two trips i and j will be admissible, if the following conditions are satisfied:

$$s_i + d_i + \Delta^{a_i(j)} + \delta_{ij} + \beta_i + \Delta^{d_j(i)} \leq s_j \tag{1}$$

and

$$s_j - \{s_i + d_i\} \leq T_{Max} \tag{2}$$

with s_i as the departure time and d_i as the running time of trip i while T_{max} defines an interval which allows the two trips to be linked. Furthermore, the trips to be linked must be served by vehicles of the same type (standard bus, articulated bus) and must be assigned to the same depot or group of depots.

The conditions (1) and (2) only yield admissible links without a qualitative assessment. This part of the model forming for vehicle scheduling refers to the first objective, minimizing the number of blocks. In regard to the second objective, i. e. minimizing the sum of non-productive times, the above mentioned qualitative weights for each link will be needed in addition to the quantitative constraints.

By defining a value for each pair of trips (i, j) we get a matrix C of elements c_{ij} with:

$$c_{ij} \begin{cases} \in [0; c* \ll \infty] & \text{if it is possible to link trip } i \text{ and } j \\ = \infty & \text{otherwise} \end{cases} \tag{3}$$

This problem may be solved as a (classical) assignment problem (see Mojsilovic (1983)).

For a sensitivity analysis a modified assignment matrix $\tilde{C}^R$ will be needed with R defining a status resulting from the number of cycles of the current initial solution. The elements $\tilde{c}_{ij}$ of the matrix have the following structure:

$$\tilde{c}_{ij} \begin{cases} = 0 & \text{if it is possible to link trip } i \text{ and } j \text{ without any dviation} \\ \in [0; \tilde{c}* \ll \infty] & \text{if it is possible to link trip } i \text{ and } j \text{ in the range of a given} \\ & \text{maximal deviation} \\ = \infty & \text{otherwise} \end{cases} \tag{4}$$

This matrix $\tilde{C}^R$ has to be updated in each iteration in accordance with the users' decision. This combinatorial problem may be solved as an assignment problem with an additional constraint. This problem results in the following formulation:

$$Minimize \sum_{i=1}^{n} \sum_{j=1}^{n} \tilde{c}_{ij} xij \tag{5}$$

with the constraints

$$\sum_{i=1}^{n} x_{ij} = 1 \quad \forall j = 1, 2, \ldots, n \tag{6}$$

$$\sum_{j=1}^{n} x_{ij} = 1 \quad \forall i = 1, 2, \ldots, n \tag{7}$$

$$x_{ij} \in \{0; 1\} \quad \forall i, j = 1, 2, \ldots, n \tag{8}$$

$$r \leq R \tag{9}$$

with n defining the number of trips and r the number of calculated cycles of the solution of the assignment problem for the given R.

4 Operational Process

The operational consists of four main steps which will be described separatedly. Fig. 4.1 shows the flow chart for these steps.

Step 1: In the first step, the user must define which type of day and which time interval will be processed and which alterations at the most will be permitted. On this basis, the necessary data structure with the matrix $\tilde{C}^R$ will be computed within the system.

Step 2: In the second step, a first optimal solution will be calculated for $R = n$, the initial solution. This produces the value r, the maximum number of blocks in the time interval under examination, without altering any data.

Step 3: This step represents the interactive part of the sensitivity analysis. The number of the blocks R will be set to $r - 1$. Starting from the existing optimal basic solution with r blocks, a re-optimization procedure will be effected to obtain a saving. The necessary alterations will be shown successively on screen. The user must decide in each case whether the suggested alteration will be accepted or not. If not, then the next suggestion follows. After an finite number of possible alterations, a saving will be achieved or the procedure will be terminated because no feasible solution for R can be found.

If the procedure is continued, the re-optimization procedure starts again with the value $R = r - 1$.

Because the sensitivity analysis works with parameters, the procedure may be repeated with different values, starting with step 1. However, repetition procedures may also be started in order to modify individual decisions in the interactive operational process.

Step 4: In the last step, the user must choose that solution which is the most suitable one in his opinion. Finally, the input data is altered accordingly in the data base.

5 Example

In the following, we will show in an example with real data, how the interactive part of the operation works from the users point of view. This data is taken from the summer timetable 1992 of Verkehrsbetriebe Hamburg-Holstein AG.

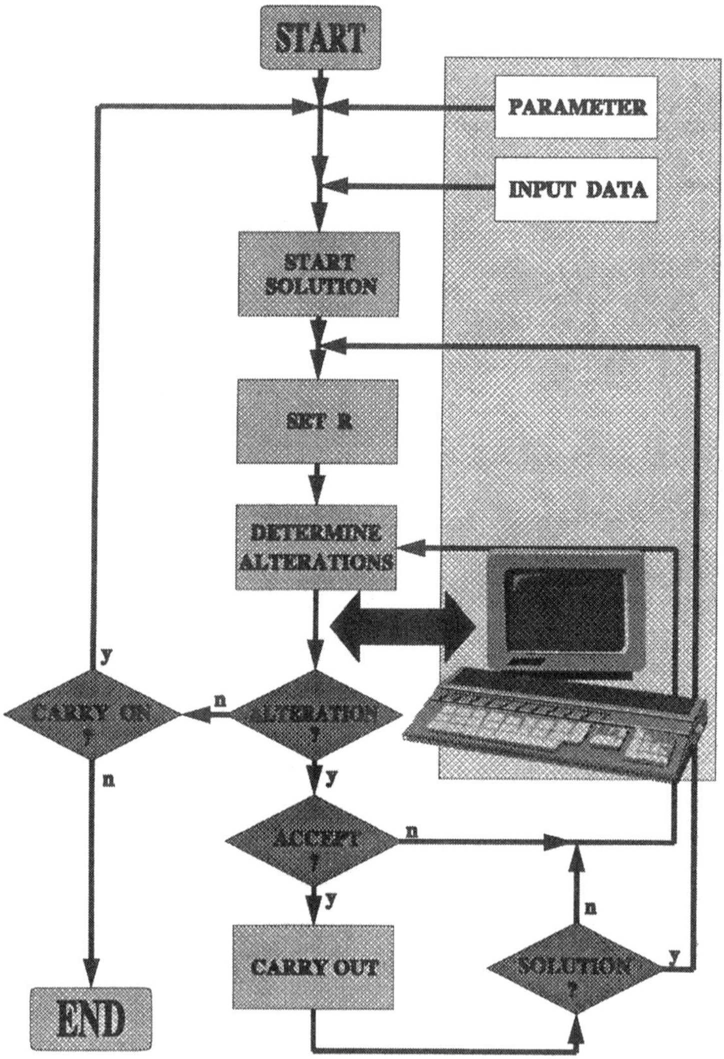

Figure 4.1: Flow chart of the sensitivity analysis

Fahrzeuge: Anfangswert 350 gegenwaertige Anzahl 350 Einsparung 0

	Linie	Abfahrt	von ->	nach	Ankunft	MWZ	F-Art	Zeit-puffer	Ver-aend.
davor	136	6:33	MAX	BBF	6:43	0	N	1	---
Fahrt 1	131	_< 6:59	SPE	BBF	7:11	0	V	-3	---
Fahrt 2	136	7:24>_	HID	SRE	7:36	0	N	8	---
danach	136	7:46	SRE	BBF	8:13	0	N	---	---

Vorschlaege seit der letzten Einsparung: 1

Koennen Sie eine Veraenderung um 3 Minuten akzeptieren ?

Figure 5.1: Suggestion for alteration shown on the scree

In the first step the following suggestion for alteration appears on screen (see Fig. 5.1): The trips shown, in this case four, represent a segment from one block. Lines 5 and 6 contain the data of those two trips which may be linked after altering the respective departure times. Lines 4 and 7 show in addition the trips before the first and after the second trip to be linked. Such an additional information helps the user to assess more easily the impacts of an alteration.

The following will be indicated: the number of the line on which the trip will be carried out, departure and arrival time, first and last stop of the trip, (in-company) delay buffer (MWZ), the type of trip (F-Art), and time range (Zeitpuffer) for alterations or the missing time for a link. This would mean in the example above that trip 1 starts at 6:59 a.m. on line 131 from stop SPE and ends at BBF at 7:11 a.m. V (the type of trip) indicates that it is an additional trip. (The other types of trips, mentioned in the following, are: N = regular trip, Y = school trip.)

The value -3 in the field "Zeitpuffer" of trip 1 at 6:59 indicates that a link with trip 2 is not possible, because there is a shortage of three minutes between the end of the first trip and the start of the second one. True, we have thirteen minutes between the two trips, but there is not enough time for a deadhead trip from BBF to HID, plus layovers.

However, subsequent to trip 2, there is still a time range of eight minutes in SRE (after subtracting two minutes for layover). True, we have the possibility to start trip 2 the necessary three minutes later. But trip 2 is a trip with an even headway, published in the timetable, and therefore in general not permitted to be shifted.

```
    Fahrzeuge: Anfangswert 350  gegenwaertige Anzahl 349   Einsparung  1

                                                                Zeit-  Ver-
          Linie  Abfahrt  von -> nach  Ankunft  MWZ  F-Art  puffer  aend.

 davor     ---   -----    ---    ---   -----     -     -      ---    ---

 Fahrt 1   275  _< 6:43   SKA    RAD    7:07     0     N       -1    ---
 Fahrt 2   275     7:08>_ RAD    MEO    7:29     0     N        0    ---

 danach    275     7:32   MEO    RAD    7:54     0     N      ---    ---

    Vorschlaege seit der letzten Einsparung:   1

      Koennen Sie eine Veraenderung um  1 Minuten akzeptieren ?
```

Figure 5.2: Second suggestion for alteration

Furthermore, there is a connection to City Rail in BBF which has to be adhered to. Trip 1, on the contrary, is an additional trip and therefore not published in the timetable. From an in-company point of view, this makes it possible to start this trip the required three minutes earlier. Should the suggestion be accepted, then already one vehicle has been saved in this example.

Often, however, further alterations will have to be made until a saving is attained, because a restructuring of the basis solution usually requires several steps.

The suggestion for alteration shown in Fig. 5.2 will be indicated in the next iteration. In this case, the line above trip one contains no data. This means that trip 1 represents the start of the block. Correspondingly, trip 2 forms the end of a block, if no information after this trip is given. This suggestion for alteration cannot be accepted for in-company reasons, because it is a regular trip with a fixed headway. Only one alteration to the departure time is suggested in Fig. 5.1. This was the only possibility in the earlier versions of the sensitivity analysis. By expanding the programme system, we can now shift the two trips which have to be linked. This is described by taking the above example one step further.

After not having accepted the suggestion in Fig. 5.2 nor three other suggestions for alteration, the user will be presented with the suggestion shown in Fig. 5.3.

For linking the school trips at 7:02 a.m. and 7:17 a.m., there is a shortage of four minutes. Due to the data structure, the departure time of one of the two trips might be altered by the required four minutes. However, it is appropriate in this situation to shift both trips by two minutes.

```
   Fahrzeuge: Anfangswert 350  gegenwaertige Anzahl 349   Einsparung  1

                                                          Zeit-  Ver-
           Linie  Abfahrt  von -> nach  Ankunft  MWZ  F-Art  puffer  aend.

davor       ---    -----    ---    ---   -----     -    -      ---    ---

Fahrt 1     293  _< 7:02    WSS    SZU    7:12     0    Y       -4    ---
Fahrt 2     293     7:17>_  WSS    SZU    7:30     0    Y        4    ---

danach      700     7:45    USU    ULS    7:53     0    Y      ---    ---

   Vorschlaege seit der letzten Einsparung:   5
------------------------------------------------------------------------
      Koennen Sie eine Veraenderung um  4 Minuten akzeptieren ?
```

Figure 5.3: Fifth suggestion for alteration in the second iteration.

At first, the user must enter whether he accepts the suggestion. If so, he will be asked to indicate a shift for one of the two trips. This is shown in Fig. 5.4 by the number "2" before the sign "_<" in the first trip. The necessary shifting of the second trip will be automatically calculated within the system. By accepting this fifth suggestion for alteration, a saving will have been made again. All alterations being made are listed in Tab. 5.1.

Line	Departure time	From	To	Arrival time	Departure time (changed)	Arrival time (changed)
131	6.59	SPE	BBF	7:11	6:56	7:08
293	7:02	WSS	SZU	7:12	7:00	7:10
293	7:17	WSS	SZU	7:30	7:19	7:32

Table 5.1: Alterations in the departure times

```
     Fahrzeuge: Anfangswert 350  gegenwaertige Anzahl 349  Einsparung 1

                                                              Zeit-  Ver-
           Linie  Abfahrt  von -> nach  Ankunft  MWZ  F-Art  puffer  aend.

  davor    ----   -----    ---    ---   -----     -    -      ---    ---

  Fahrt 1   293  2< 7:02   WSS    SZU    7:12     0    Y       -4    ---
  Fahrt 2   293     7:17>_ WSS    SZU    7:30     0    Y        4    ---

  danach    700     7:45   USU    ULS    7:53     0    Y      ---    ---

     Vorschlaege seit der letzten Einsparung:   5
  Bitte bei einer Fahrt die Verschiebung angeben
```

Figure 5.4: Alteration in the departure times of both trips by two minutes each

6 Results

The programme system HOT is being used by several urban mass transit companies in their operational planning. The experiences and results of four of these companies from using the sensitivity analysis in the bus networks are explained.

Hamburger Hochbahn AG (HHA)

HHA operates 142 mainly inner-city bus lines. In the summer timetable 1992 the maximum number of trips to be served at the same time is 634 in the morning peak. This requires a total of 735 buses. By using the sensitivity analysis, eight vehicles could be saved (ca. 1.1

Verkehrsbetriebe Hamburg-Holstein AG (VHH)

The VHH bus network consists of 121 lines in total in a city area as well as on the outskirts. Out of the original 350 blocks in the summer timetable 1992 (with 260 trips as maximum) 10 could be saved (ca. 2.9 which operate at the same time, amounts to 327 (see fig. 7.1). This difference results from specific in-company conditions, which, in the scheduling procedure, lead to short blocks ending before the maximum peak.

Dresdner Verkehrsbetriebe AG (DVB)

Dresdner Verkehrsbetriebe AG mainly serves the municipal area. In the summer timetable 1992 it was possible to save in total four busses out of 143 (ca. 2,8

Compagnie des Transport de Besancon (CTB)

CTB operates 20 bus lines in the municipal area. For the winter timetable 1992/93, in which several lines were restructured and at the same time the service was increased, it was possible to reduce the number of vehicles in the morning peak hours by eleven vehicles from 137 to 126 (ca. 8.0

As the results demonstrate, the savings differ obviously from each other. We are not able to define the exact causes, but some factors can be indicated. With municipal lines, even headways and connections to rail systems, only a small choice of possible suggestions for alterations will be available to the sensitivity analysis (e.g. with HHA). This applies especially when no or not enough school trips are offered, because this transit is part of the regular timetable.

If the service run mainly in the outskirts, the number of school trips and trips on contractional hire increases sharply (e.g. with VHH), which offers a greater choice of alterations. The considerable improvements with CTB divert clearly from the other results. They are probably due to the restructuring of the network and the transitional problems involved.

However, we cannot exclude the considerable individual influence of the user in the process. Because the alterations are not fixed to rigid rules and therefore not automated, the user holds a central position. By accepting or refusing, he strongly influences the result by his assessment and, of course, in-company specialized knowledge. However, how long the sensitivity analysis has already been used by the company might also be an important question. More experience might result in impacts on the decisions in the interactive part.

The attained percentage savings might, at first sight, appear to be small, especially in the case of HHA. A more detailed view yields, however, another interpretation. First, we must consider that the savings through the sensitivity analysis represent an additional effect to the actual vehicle scheduling. When calculating the basic solution, an optimization step will be made which, in comparison with manual planning, offers considerable savings.

In any case, a proportional value is often only partially meaningful. Rather the absolute results of the savings made in the operating cost must be looked at for an interpretation, because these may, even with small percentages, show a considerable amount. Take the operating cost, (including payroll cost) claimed by vehicle producers per vehicle and year, which rank between US170.000*andUS* 270.000 for middle-European countries, and the additional positive effects will become more clear.

7 Conclusions

The results described show that the strongest effects are obviously made in restructured networks when using the sensitivity analysis. However, this does not limit the usage to such cases.

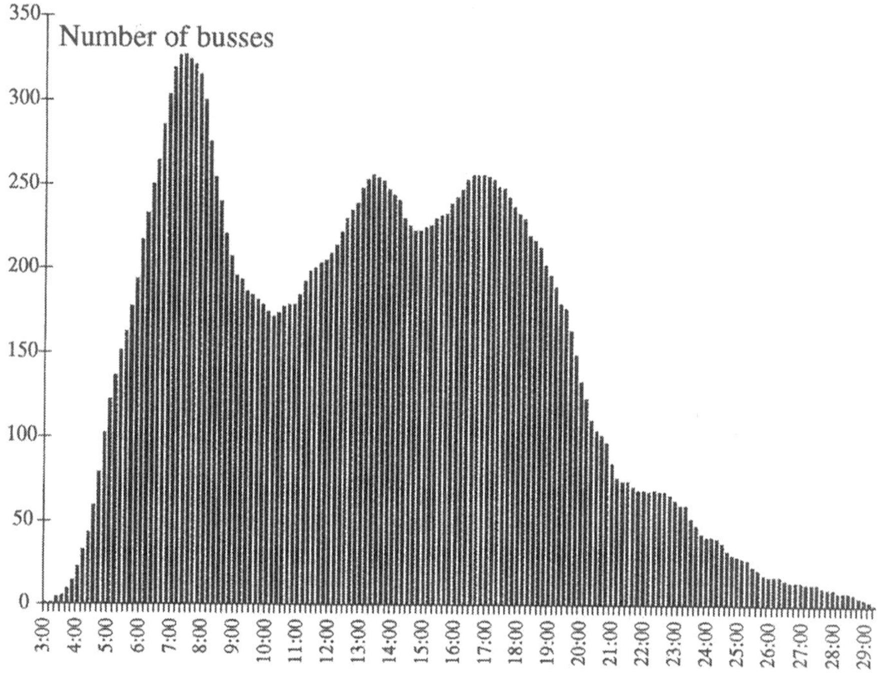

Figure 7.1: Distribution curve of the number of vehicles (VHH)

The examples of HHA and VHH, two companies which have used the sensitivity analysis for several years, show that cost-saving potentials result even from, to some extent, small alterations to the timetable in successive schedule periods. The cause can be found in the complexity of the problem structure. Even small alterations may allow entirely new combinations.

The main objective for the sensitivity analysis is, also because of its incorporation in the planning process, the reduction of the maximum fleet size. This, however, does not mean that the usage has to be limited to the time before and after the absolute peak, which in the case of VHH (see Fig. 7.1) lies between 5:30 a.m. and 8:40 a.m. Because the number of vehicles required at certain times determines the number of duties to some extent, it might be of interest with regard to duty scheduling to analyse also the local maxima. With a distribution curve of the number of vehicles like the one of CTB (see Fig. 7.2), which in total has four almost equal maxima, all peaks must be analysed with regard to the maximum number of vehicles. By altering an originally absolute maximum, it is possible with such a structure that due to a

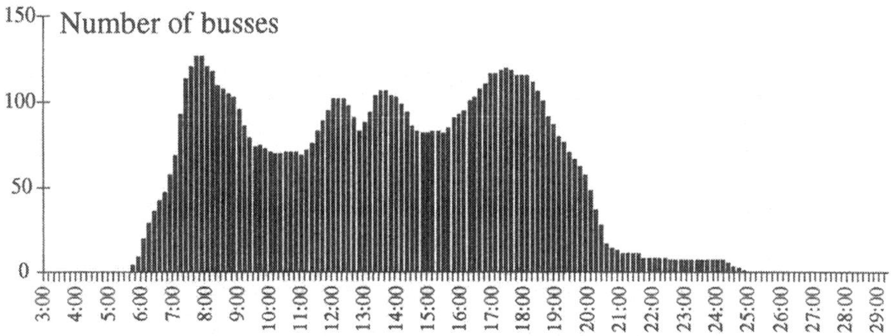

Figure 7.2: Distribution curve of the number of vehicles (CTB)

reduction a new absolute maximum occurs in another time period.

8 Outlook

With the sensitivity analysis the program system HOT II offers an efficient module. Its efficiency results from an effective interaction of algorithm and in-company experience. Naturally, this module will be further developed. The aim is to take advantage of technological progress in the field of hardware and software and furthermore to include the experiences from in-company usage. This is made by HanseCom, the developer and distributor of the HOT II system.

References:

[1] L. Bodin, B. Golden, A. Assad, and M. Ball (1983): Routing and Scheduling of Vehicles and Crews: The State of the Art, in: *Computer and Operations Research* 10, 63-211

[2] P. Carraresi and G. Gallo (1984): Network Models for Vehicle and Crew Scheduling, in: *European Journal of Operations Research* 16, 139-151

[3] J.R. Daduna (1987): DV-gestützte Dienstplanbildung f¡129¿r das Fahrpersonal in Verkehrsbetrieben - Problematik und Lösungsansätze, in: Operations Research Proceedings 1986, (Springer, Berlin, Heidelberg) 76-85

[4] J.R. Daduna (1988), A Decision Support System for Vehicle Scheduling in Public Transport, in: W. Gaul and M. Schader (eds.) *Data, Expert Knowledge and*

Decisions (Springer, Berlin, Heidelberg, New York, London, Paris, Tokyo) 93-102

[5] J.R. Daduna and M. Mojsilovic (1988): Computer-Aided Vehicle and Duty Scheduling Using the HOT Programme System, in: J.R. Daduna and A. Wren (eds.) 133-146

[6] J.R. Daduna and A. Wren (eds.) (1988): *Computer-Aided Transit Scheduling* (Springer, Berlin, Heidelberg, New York, London, Paris, Tokyo)

[7] M. Desrochers and J.-M. Rousseau (eds.) (1992): *Computer-Aided Transit Scheduling* (Springer, Berlin, Heidelberg, New York, London, Paris, Tokyo)

[8] Luedtke, L.K. (1985): RUCUS II: A Review of System Capcilities, in: J.-M. Rousseau (ed.), 61-115

[9] M. Mojsilovic (1983): Verfahren für die Bildung von Fahrzeugumläufen, Dienstplänen und Dienstreihenfolgenplänen in Verkehr und Transport, in: HEUREKA '83 - Optimierung in Transport und Verkehr, Karlsruhe, 178-191

[10] J.-M. Rousseau (ed.) (1981): *Computer Scheduling of Public Transport 2* (North Holland, Amsterdam, New York, Oxford)

[11] A. Wren (1981): *Computer Scheduling of Public Transport 1* (North Holland, Amsterdam, New York, Oxford)

Network Optimization Problems, pp. 53-62
Eds. D.-Z. Du and P.M. Pardalos

Subset Interconnection Designs: Generalizations of Spanning Trees and Steiner Trees

Ding-Zhu Du
Computer Science Department, University of Minnesota, Minneapolis, MN 55455, USA and Institute of Applied Mathematics, Academia Sinica, Beijing 100080, PRC

Panos M. Pardalos
Department of Industrial and Systems Engineering, University of Florida, Gainesville, FL 32611 USA

Abstract

Given an edge-weighed graph with vertex set X and a family of subsets of X, we consider problems of finding a subgraph G with the minimum total weight such that G gives some necessary connection between vertices in each subset in the family. The problems are generalizations of the spanning tree problem and the Steiner tree problem. Recently, many results for such problems have appeared. We give a survey of these together with some new results.

1 Introduction

Consider the following two problems:

(1) Communication Network Design: There is a set of employees and a set of projects. In each project there is a subset of employees. Each employee may work in several projects. Now, we want to find a communication network between employees such that any two members of a project are able to communicate through a sequence of other members of that project. We also wish to find the less expensive network.

(2) Vacuum System Design: A vacuum usually contains some valves. The function of valves is to give different interconnection states between components of the system.

The system must have specific interconnection states for each job. Now, given a particular job, how do we design a vacuum system to do the job? Here, we want the number of valves to be as small as possible.

Both problems ask for a network to have multi-phase interconnections. In the first problem, a phase corresponds to a project; for each project, the network needs to give a spanning tree interconnecting all members in the project, where a *spanning tree* for a point set is just a tree with the point set as its vertex set. In the second problem, a phase corresponds to a state; for each state, the network needs to give a Steiner tree interconnecting a subset of the components with some restrictions (see Section 3 for details), where a *Steiner tree* for a point set is a tree with the vertex set containing the point set and leaf set contained by the point set.

Problem (2) was first proposed by Du and Chen [4] in 1976. It was a very practical problem. In fact, they were in a bulb-making factory and wanted to promote new products. For each new product, they had to design a new vacuum system for the new job, which raised the problem. In [4] the authors gave a mathematical model for problem (2) and several examples in the real world as well as an approach to adjust the design when valves of more complicated types are used.

Since problem (2) appeared, a special case has been studied extensively in which, the restricted Steiner trees are spanning trees. Prisner [18] [19] generalized this case and proposed problem (1). In (1), all edges are weighed; different edges may have different weights. In (2), however, only the number of edges is counted; i.e., all edges have a uniform weight of one (called the unit-weight case).

Both problems have been known to be NP-hard [6]; special cases and heuristics have gained attention. In this paper we survey existing results, extend some results from the unit-weight case to the general case, and point out some open questions.

Before doing so, let us give some notations and terminologies. For a graph G, we denoted by $\|G\|$ the number of edges in G. For a set Y, we denote by $|Y|$ the number of elements in Y. Let $G = (V, E)$ and $G' = (V', E')$. Then the union, the intersection and the difference of graphs G and G' are defined to be $G \cup G' = (V \cup V', E \cup E')$, $G \cap G' = (V \cap V', E \cap E')$ and $G \setminus G' = (V, E \setminus E')$, respectively. The *complement* of a graph G is the graph $\bar{G}$ with vertex set $V(G)$ such that an edge is in $\bar{G}$ if, and only if, it is not in G.

2 Multi-Phase Spanning Networks

Given an edge-weighed complete graph with vertex set X ($|X| = n$) and subsets $X_1, \cdots, X_m$ of vertices, the problem considered in this section is to find a minimum weighed subgraph G such that for every $i = 1, \cdots, m$, G contains a spanning tree for X_i. We will call this problem the *Multi-Phase Spanning Network Problem* (MPSpN). A graph is called a *feasible graph* for $(X_1, \cdots, X_m)$ if for every X_i it contains a spanning tree for X_i. A solution of the MPSpN is called a *minimum feasible graph* for $(w\ ;\ X_1, \cdots, X_m)$ where w is the weight function for the edge set. When $w \equiv 1$, the solution of the MPSpN is said to be a minimum feasible graph for $(1;\ X_1, \cdots, X_m)$.

When the MPSpN is studied, the following are usually assumed.

(a) $X = \cup_{i=1}^m X_i$. (In fact, every minimum feasible graph has no edge incident to a point in $X \setminus \cup_{i=1}^m X_i$.)

(b) $|X_i| \geq 2$ for all i. ($|X_i|$ with $|X_i| = 1$ can always be deleted.)

(c) Every feasible graph G satisfies $G = \cup_{i=1}^m G_i$. In fact, an edge not in $\cup_{i=1}^m G_i$ can be deleted without changing the feasibility.

To see an example, consider five subsets $X_1 = \{v_1, v_2\}, X_2 = \{v_1, v_2, v_3\}$, $X_3 = \{v_3, v_4, v_5\}$, $X_4 = \{v_1, v_2, v_4\}$, and $X_5 = \{v_2, v_4, v_5\}$. These subsets and the weight function w are as shown in Fig.1. Some feasible graphs for $(X_1, X_2, X_3, X_4, X_5)$ are in Fig.2. Among them, G^* is a minimum feasible graph for $(w\ ;\ X_1, X_2, X_3, X_4, X_5)$.

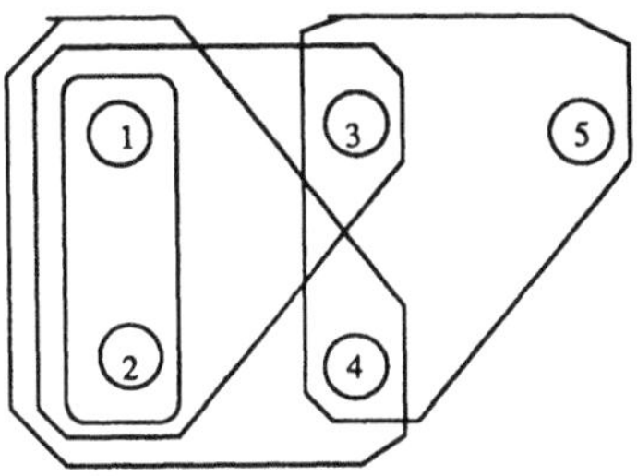

w	1	2	3	4	5
1		5	6	7	8
2			5	6	7
3				5	6
4					5
5					

Figure 1: An input of the MPSpN

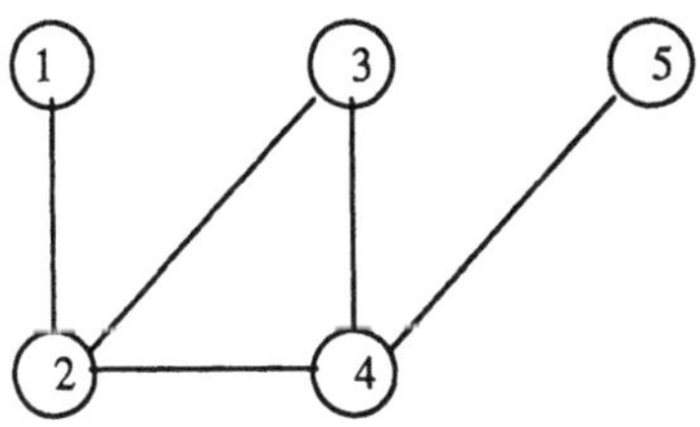

Figure 2: Feasible graphs for the input of Fig.1

The MPSpN is an NP-hard problem. It is closely related to several classical problems. For $m = 1$, the MPSpN is exactly the minimum spanning tree problem. Graham [13] gave an interesting historical survey. The most popular polynomial-time algorithm for the minimum spanning tree problem was discovered by Kruskal [15].

A set system $(X_1, \cdots, X_m)$ that has a feasible graph to be a forest is called *subtree hypergraph.* Such a system has various applications in computer science [1] and statistics [16]. It is also related to chordal graphs [9][10]. How can we tell whether a set system is a subtree hypergraph? Tarjan and Yannakakis [21] found an $O(m+n)$-time algorithm.

If $(X_1, \cdots, X_m)$ is a subtree hypergraph, then we can find a minimum feasible graph for $(w \ ; \ X_1, \cdots, X_m)$ in polynomial time. In fact, a subtree hypergraph $(X_1, \cdots, X_m)$ must satisfy the following condition.

(A) There exists a feasible graph G such that for any $i, j \in \{1, \cdots, m\}$, $G[X_i] \cap G[X_j]$ is a tree, where $G[X_i]$ is the subgraph of G induced by X_i.

The proof is easy. Suppose that G is the feasible forest. Consider two vertices u and v in $G[X_i] \cap G[X_j]$. Since G is feasible, there exists a path in G_i from u to v and there also exists a path in $G[X_j]$ from u to v. However, G is a forest. So, the two paths are identical, which is a path in $G[X_i] \cap G[X_j]$. Thus, $G[X_i] \cap G[X_j]$ is connected. A connected subgraph of a forest must be a tree. Therefore, (A) holds. The following theorem has been proved in [6] for the unit-weight case. We will show that it is also true in general.

Theorem 2.1 *If $(X_1, \cdots, X_m)$ satisfies condition (A), then computing a minimum feasible graph for $(w \ ; \ X_1, \cdots, X_m)$ can be done in polynomial time.*

Before doing so, let us first look at an efficient special case given by Du and Miller [6]. Consider the following condition.

(B) For any two subsets X_i and X_j, if $|X_i \cap X_j| \geq 2$, then there exists a subset X_k such that $X_i \cap X_j = X_k$.

This condition implies that for any $I \subseteq \{1, \cdots, m\}$ either $|\cap_{i \in I} X_i| \leq 1$ or there exists $k \in \{1, \cdots, m\}$ such that $X_k = \cap_{i \in I} X_i$. The following theorem was shown in [6].

Theorem 2.2 *If (B) holds, then all complements of feasible graphs form a family of independent subsets of a matroid.*

¿From this theorem and the theory of matroids [12], we see that the MPSpN with (B) can be solved by the following two greedy algorithms.

ALGORITHM 1.

$G :=$ the complete graph of X;

Sort edges of G in weight-increasing order, $e_1, \cdots, e_{n(n-1)/2}$;

for $i := 1$ to $n(n-1)/2$ *do*

if $G \setminus e_i$ is feasible *then* $G := G \setminus e_i$.

ALGORITHM 2.

Sort all edges of the complete graph of X in weight-decreasing order, $e_1, \cdots, e_{n(n-1)/2}$;

$G := \emptyset$;

for $i := 1$ to $n(n-1)/2$ *do*

if there exists j such that e_i connects two connected components of $G[X_j]$ *then* $G := G \cup e_i$.

Algorithm 2 is a variation of a greedy algorithm given by Prinsner [19]. It was proved in [6] that if (A) holds, then the feasible graph in (A) is a minimum feasible graph for $(1;\ X_1, \cdots, X_m)$ and every minimum feasible graph for $(1;\ X_1, \cdots, X_m)$ possesses the property in (A). In addition, all minimum feasible graphs for $(1;\ X_1, \cdots, X_m)$ are minimum feasible graphs for $(1;\ \cap_{i \in I} X_i, I \subseteq \{1, \cdots, m\})$. It follows that:

Theorem 2.3 *If (A) holds, then all minimum feasible graphs for* $(1; X_1, \cdots, X_m)$ *form a family of bases of a matroid.*

This theorem means that the MPSpN with (A) can also be solved by greedy algorithms like Algorithms 1 and 2. Thus, Theorem 2.1 is proved.

An application of Theorem 2.1 is to study the MPSpN with small m. It was proved by Tang [11] that for $m = 1, 2, 3$, (A) holds. So, we have:

Corollary 2.4 *The MPSpN is polynomial-time computable for* $m = 1, 2, 3$.

Since the MPSpN is equivalent to a problem about intersection of m matroids, the case for $m = 3$ is a little surprising. In fact, a problem about three matroid intersection is usually NP-hard [11]. Tang [11] also constructed a counterexample which shows that (A) does not hold for $m \geq 4$. The following question, therefore is still open.

Open Problem 1 *With a fixed* $m \geq 4$, *is the MPSpN polynomial time computable?*

Note that in the unit-weight case, it is easy to show that for m fixed, MPSpN is polynomial-time solvable. So, the problem is significant only for general weight function.

An extension of bounding m is the following condition.

(α, β) for any α distinct points $x_1, x_2, \cdots, x_\alpha \in X$, there are at most β X_i's containing all $x_1, x_2, \cdots, x_\alpha$

where α and β are two natural numbers. Clearly, (α, β) implies (α', β') whenever $\alpha' \geq \alpha$ and $\beta' \geq \beta$. Under the assumption that all weights equal one, Du [8] proved that for $\alpha \leq 2$ and $\beta \leq 2$, the problem is polynomial-time solvable while for $\alpha \geq 3$ or $\beta \geq 6$ or $(\alpha \geq 2, \beta \geq 3)$ the problem is NP-hard. The following is an open problem left in [8].

Open Problem 2 *What is the computational complexity of computing a minimum feasible graph for* $(1; X_1, \cdots, X_m)$ *in the case of* $\alpha = 1$ *and* $3 \leq \beta \leq 5$*?*

We believe that it is polynomial-time solvable in case ($\alpha = 1, \beta = 3$), but is NP-hard in cases ($\alpha = 1, \beta = 4$) and ($\alpha = 1, \beta = 5$).

The interesting work in [8] is about the case of $\alpha = \beta = 2$. The problem is transfered to a maximum matching problem on a graph. It is not known whether this work could be extended to arbitrary edge-weights.

An application of the results on $\alpha = \beta = 2$ is to construct approximate solutions for the MPSpN. There are several ways. The first is to divide the collection of subsets $(X_1, \cdots, X_m)$ into several small collections satisfying condition ($\alpha = 2, \beta = 2$), construct a minimum feasible graph for each small collection, and then union all of them. The second is as follows: When a pair of points appears in more than two subsets X_i's, we stick them together. In this way, we can reduce the original collection of subsets to a new one, satisfying condition ($\alpha = 2, \beta = 2$). After a minimum feasible graph for the new collection of subsets is found, we break stuck pairs by adding some edges.

Although several heuristics have appeared in the literature, none of them has been proved to be a bounded heuristics; i.e., a heuristic that produces an approximation solution with total weight within a constant factor from optimal. Thus, the following is an important question for the MPSpN.

Open Problem 3 *Does the MPSpN have a bounded heuristic?*

Du and Miller [6] proved the following.

Theorem 2.5 *Any set system* $(X_1, \cdots, X_m)$ *can be partitioned in time* $O(mn)$ *into less than* $\sqrt{2m}$ *subsystems such that for each* $(X_{i_1}, \cdots, X_{i_j})$ *of them,* $X_{i_1}, \cdots, X_{i_j}$ *have an element in common.*

When all weights equal one, a minimum feasible graph for $(1;\ X_1, \cdots, X_m)$ with $\cap_{i=1}^m X_i \neq \emptyset$ is the star with vertex set X and the center chosen from $\cap_{i=1}^m X_i$. This fact with Proposition 2. together yields a $\sqrt{2m}$-heuristic for the MPSpN in the unit-weight case. However, the same idea does not work for the MPSpN in general. To see this, we point out the following.

Theorem 2.6 *Computing a minimum feasible graph for* $(w\ ;\ X_1, \cdots, X_m)$ *with* $\cap_{i=1}^m X_i \neq \emptyset$ *is NP-hard.*

To prove this, let us mention an NP-hard problem, the vertex covering: Given a graph G, find a smallest subset of vertices that cover all edges [11]. To do polynomial-time reduction, choose a point s different from all vertices and for each edge uv, give a set $X_{uv} = \{v, u, s\}$. Moreover, we assign weights in the following way: the distance between any two vertices of G is one; the distance from s to any vertex of G is n^2

where n is the number of vertices of G. Now, we claim that there is a vertex covering of size at most k if and only if there is a feasible graph of total weight at most $\|G\| + kn^2$.

In fact, if a vertex covering of size at most k exists, then connecting s to every vertex in the vertex covering, we obtain a feasible graph of weight at most $\|G\| + kn^2$. For the other direction, suppose that a feasible graph of weight at most $\|G\| + kn^2$ exists. In such a graph the degree of s must be at most k. Otherwise, the total weight would exceed $\|G\| + kn^2$. Now, all vertices adjacent to s in the feasible graph form a vertex covering of size at most k. Thus, we proved that the vertex covering can be polynomial-time reduced to the considered problem in Theorem 2.

If the endpoints of an edge belong to more X_i's, then they are more likely to appear in a minimum feasible graph. From this thought, Prisner [19] gave the concept of "benefit". For a graph G and a set system $(X_1, \cdots, X_m)$, the benefit $b(uv, G)$ of an edge uv is the number of $G[X_i]$'s such that uv connects two connected components of $G[X_i]$. The benefit-cost ratio of uv is $b(uv, G)/w(uv)$ where $w(uv)$ is weight of the edge uv. Using this concept, Prisner [19] discovered a $\sum_{i=2}^{K} \frac{1}{i}$-heuristic for the MPSpN where K is the maximum number of X_i's, which have two elements in common. His heuristic runs in time $O(n^4 + mn^2)$ as follows.

ALGORITHM 3.

$G := \emptyset$;

while there is an edge of positive benefit *do*

choose an edge uv with largest benefit-cost ratio and set $G := G \cup uv$.

Prisner's heuristic is the best known heuristic for the MPSpN.

3 Multi-Phase Steiner Networks

Given an edge-weighed graph B with vertex set X and subsets $X_1, Y_1, \cdots, X_m, Y_m$ of X with $X_i \cap Y_i = \emptyset$, the problem considered in this section is to find a minimum weighed subgraph G such that for every $i = 1, \cdots, m$, G contains a Steiner tree for X_i without using vertices not in Y_i. We will call this problem the *Multi-Phase Steiner Network Problem* (MPStN) and its solution a *minimum feasible graph* for $[w\ ;\ X_1, Y_1, \cdots, X_m, Y_m]$. For $m = 1$, the MPStN is the minimum Steiner tree problem [22] which is already NP-hard. Thus, the MPStN is much harder than the MPStN.

It is a well-known fact [14] that the minimum spanning tree is a 2-approximation of the minimum Steiner tree; i.e., the total length of a minimum spanning tree is not bigger than twice the total length of a minimum Steiner tree. However, there does not exist a constant c such that the total weight of a minimum feasible graph for $(w\ ;\ X_1, \cdots, X_m)$ is not bigger than c times the total weight of a minimum feasible graph for $[w\ ;\ X_1, Y_1, \cdots, X_m, Y_m]$. To see this, we look at the following example.

Let $X_{ij} = \{i, j\}$ and $Y_{ij} = \{i, j, 0\}$ for $1 \leq i < j \leq n$. A minimum feasible graph for $(1;\ X_{ij}, 1 \leq i < j \leq n)$ then has weight $n(n-1)/2$. However, a minimum feasible graph for $[1;\ X_{ij}, Y_{ij}, 1 \leq i < j \leq n]$ has weight n. Thus, the ratio between them is $(n-1)/2$, which cannot be bounded by a constant.

What is the relation between minimum feasible graphs for $(w\ ;\ Y_1, \cdots, Y_m)$ and $[w\ ;\ X_1, Y_1, \cdots, X_m, Y_m]$? In the unit-weight case, Chao and Du [3] proved the following.

Theorem 3.1 *If G is a minimum feasible graph for $(1\ ;\ Y_1, \cdots, Y_m)$ satisfying condition (A), and for every edge uv of G there exists an index i such that $u, v \in X_i$, then G is a minimum feasible graph for $[1\ ;\ X_1, Y_1, \cdots, X_m, Y_m]$.*

So far, the following question is still open.

Open Problem 4 *Could Theorem 3.1 hold for general weight function?*

Concerning the ratio between minimum feasible graphs for $(w\ ;\ Y_1, \cdots, Y_m)$ and $[w\ ;\ X_1, Y_1, \cdots, X_m, Y_m]$, we look at the following example. Let $X_1 = \{1, 2\}$ and $Y_1 = \{1, 2, \cdots, n\}$. Then a minimum feasible graph for $(1; Y_1)$ must have weight $n-1$ while a minimum feasible graph for $[1; X_1, Y_1]$ has weight 1. Such a ratio cannot have a constant bound.

In the above example, the minimum feasible graph for $(1; Y_1)$ contains many unnecessary edges for $[1; X_1, Y_1]$. Such edges are easily deleted. So, the following question arises:

Open Problem 5 *Starting from a minimum feasible graph for $(w\ ;\ Y_1, \cdots, Y_m)$, could we modify it in polynomial time to obtain a bounded approximation solution of minimum feasible graph for $[w\ ;\ X_1, Y_1, \cdots, X_m, Y_m]$?*

In [6], two operatuions for simplifying a feasible graph for $(1;\ X_1, \cdots, X_m)$ were described. Such operations can be extended to simplify a feasible graph G for $[w\ ;\ X_1, Y_1, \cdots, X_m, Y_m]$ as follows.

(O1) Check each edge e of G. If the removal of edge e preserves the feasibility of the graph G, then delete e.

(O2) Check each edge e of $\bar{G}$, the complement of G. If we can use operation (O1) to remove some edges of $e \cup G$ with total weight more than the weight of e, then remove such edges and add e.

Clearly, the two operations work in polynomial time. An open question left in [6] is as follows.

Open Problem 6 *Can we obtain a bounded approximation of the minimum feasible graph for $(1;\ X_1, \cdots, X_m)$ from the minimum feasible graph for $(1;\ \cap_{i \in I} X_i, I \subseteq \{1, \cdots, m\})$ by using operations (O1) and (O2)?*

We can ask a similar question between $[w\ ;\ X_1, Y_1, \cdots, X_m, Y_m]$ and $[w\ ;\ \cap_{i \in I} Y_i, I \subseteq \{1, \cdots, m\}]$.

An interesting special case for the MPStN is $\cap_{i=1}^m Y_i \neq \emptyset$. In this case, there exists a star which is a feasible graph. Many problems in the real world can be decomposed into several subproblems in such a special case. So, a feasible graph can be obtained by taking the union of solutions of subproblems. Unfortunately, it was proved in [7] that:

Theorem 3.2 *Computing a minimum feasible graph for* $[1;\ X_1, Y_1, \cdots, X_m, Y_m]$ *with* $\cap_{i=1}^m Y_i \neq \emptyset$ *is still NP-hard.*

For $[1;\ X_1, Y_1, \cdots, X_m, Y_m]$ with $\cap_{i=1}^m Y_i \neq \emptyset$, the star feasible graph has, at most, twice the weight of a minimum feasible graph. However, it is not known whether or not for $[w\ ;\ X_1, Y_1, \cdots, X_m, Y_m]$ with $\cap_{i=1}^m Y_i \neq \emptyset$ there is a bounded approximation with a factor two from optimal.

Bern and Plassmann [2] proved that the minimum Steiner tree problem in graphs is MAX SNP-hard; i.e., it is unlikely to have a polynomial-time approximation scheme. The MPStN clearly inherits this property.

References

[1] C. Beeri, R. Fagin, D. Maier, M. Yannakakis, On the desidability of acyclic database schemes, *J. ACM* 30 (1983) 479-513.

[2] M. Bern and P. Plassmann, The Steiner problem with edge lengths 1 and 2, *Information Processing Letters* 32 (1989) 171-176.

[3] S.-C. Chao and D.-Z. Du, A sufficient optimality condition for the valve-placement problem, *J. North-East Heavy Industry Institute*, 4 (1983) (in Chinese).

[4] D.-Z. Du and Y.-M. Chen, Placement of valves in vacuum systems, *J. Electric Light Sources*, 4 (1976) (in Chinese).

[5] D.-Z. Du, An optimization problem, *Discrete Appl. Math.*, 14 (1986) 101-104.

[6] D.-Z. Du and Z. Miller, Matroids and subset interconnection design, *SIAM J. Disc. Math.*, 1 (1988) 416-424.

[7] D.-Z. Du and X.-F. Du, A special case of valve-placement problem, *Acta Mathematics Applicatae Sinica*, 4 (1991) (in Chinese).

[8] D.-Z. Du, On complexity of subset interconnection designs, *DIMACS TR* 91-23.

[9] P. Duchet, Propriete de Helly et problemes de representation, in: *Colloque International Paris-Orsay* 260 (1978) 117-118.

[10] C. Flament, Hypergraphes arbores, *Discrete Mathematics* 21 (1978) 223-227.

[11] M.R. Garey and D.S. Johnson, *Computers and Intractability, A Guide to the Theory of NP-Completeness*, Freeman, San Francisco, 1979.

[12] M. Gondran and M. Minoux, *Graphs and Algorithms*, John Wiley, New York, 1984.

[13] R.L. Graham, P. Hell, On the history of the minimum spanning tree problem, *Ann. History Computing* 7 (1985) 43-57.

[14] R.M. Karp, Reducibility among combinatorial problem, in R.E. Miller and J.W. Thatcher (ed.), *Complexity of Computer Computation* (Plenum Press, New York, 1972) 85-103.

[15] J.B. Kruskal, On the shortest spanning subtree of a graph and the traveling saleman problem, *Proc. Amer. Math. Soc.* 71 (1956) 48-57.

[16] S.L. Lauritzen, T.P. Speed, K. Vijayan, Decomposable graphs and hypergraphs, *J. Austral. Math. Soc.* A 36 (1984) 12-29.

[17] E. Prisner, Intersection-representation of graphs in n-cyclomatic graphs, *Ars Combinatoria*, to appear.

[18] E. Prisner, Familien zusammenhangender Teilgraphen eines Graphen und ihre Durchschnittsgraphen, Dissertation in Universitat Hamburg 1988.

[19] E. Prisner, Two algorithms for the subset interconnection design, *Networks*, to appear.

[20] R.E. Tarjan, M. Yannakakis, Simple linear-time algorithms to test chordality of graphs, test acyclicity of hypergraphs, and selectively reduce acyclic hypergraphs, *SIAM J. Comput.* 13 (1984) 566-579.

[21] T.-Z. Tang, An optimality condition for minimum feasible graphs, *Applied Mathematics*, 2 (1989) 21-24 (in Chinese).

[22] S. Voß, *Steiner Probleme in Graphen*, (Hain, Frankfurt, 1990).

Network Optimization Problems, pp. 63-92
Eds. D.-Z. Du and P.M. Pardalos

Polynomial and Strongly Polynomial Algorithms for Convex Network Optimization

Dorit S. Hochbaum[1]
School of Business Administration and Industrial Engineering and Operations Research Department, University of California, Berkeley

Abstract

Numerous recent developments in complexity theory and the design and analysis of algorithms have contributed to new insights into nolinear network flows problem. In particular, it is now known that the separable convex costs network flow problem is polynomial, in either integers or continuous variables, the latter for output with specified number of significant digits. Moreover, assuming a reasonable computation model consisting of the arithmetic operations and the floor operation and comparisons, it was demonstrated that strongly polynomial algorithms are impossible for the problem. It follows that the available polynomial algorithms are very close in their running time to optimal algorithms.

Several properties of convex separable network problems are investigated here. These include so-called proximity theorems that establish the proximity of integer and continuous solutions, and also that of the closeness of an optimal solution to a scaled piecewise linear approximation of the nonlinear problem to the optimial solution to the nonlinear problem. Such proximity theorems make it possible to use scaled piecewise linear approximations in algorithms of polynomial complexity and obtain integer solutions from continuous and vice versa. Classes of algorithms that use the proximity properties are described.

Quadratic network problems have some special properties differentiating them from other nonlinear instances of the problem. It is shown how these properties could be exploited for faster and more efficient algorithms. Issues related to quadratic nonseparable network problems are discussed as well.

Complexity issues and NP-hardness of classes of nonlinear network flow problems are presented and important directions for future work are delineated.

1 Introduction

The general nonlinear costs network flow problem is an instance of nonlinear optimization subject to linear constraints. Given a network $G = (N, A)$, let T be the $n \times m$ node-arc adjacency matrix of the network, $n = |N|$ and $m = |A|$, $\boldsymbol{x}$ the vector of flows on the arcs, $\boldsymbol{x} = \{x_{ij}\}_{(i,j)\in A}$, b_i the supply (demand) of node i, and u_{ij} the capacity upper bounds. The integer formulation of the nonlinear flow problem is,

$$\begin{aligned} \min \quad & F(\boldsymbol{x}) \\ s.t. \quad & T\boldsymbol{x} = \boldsymbol{b} \\ & \boldsymbol{0} \leq \boldsymbol{x} \leq \boldsymbol{u} \\ & \boldsymbol{x} \text{ integer} \end{aligned}$$

For $F(\boldsymbol{x})$ a linear function, the problem of minimum cost flows have been studied since the early 60's, and has been studied extensively in the past decade. Until recently, results for the nonlinear problem were limited to iteratively convergent algorithms, rather than algorithms with finite established complexity. An extensive review of such iterative algorithms for the continuous nonlinear constrained problem is given in [Min86a]. The reason for the lack of finite algorithms is that in general the continuous version of the problem (when the integrality requirement is omitted) may posses solutions that cannot be expressed as finite output. Such solutions are irrational numbers or even nonalgrbraic numbers that cannot be represented as a solution to a certain polynomial equation. This property renders complexity theory inapplicable, as it requires finite length input and output.

The nonconvex continuous version of the problem is NP-hard [Sah74] even for the quadratic case and so are many simplified versions of the nonseparable problem even if convex. The corresponding concave minimization problem is in general NP-hard. The complexity of the concave network problem as well as some algorithms (exponential) are discussed in [GP90], [GP91a] and [GP91b]. An excellent unifying presentation of some polynomial instances of the concave flow problem is given in [EMV87]. There the problem is proved polynomial when the arcs are uncapacitated and number of demand nodes is fixed. It is also proved polynomial for certain classes of planar graphs. Yet, even with the assumption of convexity, a quadratic nonseparable problem when the set of constraints on the adjacency matrix is empty, is NP-complete. (An example is provided in section 4). It is therefore the case, that for general purpose polynomial algorithms we focus primarily on the class of problems with $F(\boldsymbol{x}) = \sum f_{ij}(x_{ij})$ where f_{ij} are convex.

Throughout we discuss convex network flow problems with the classification:

1. Separable or nonseparable
2. Quadratic or nonquadratic and nonlinear
3. Integer or continuous.

Hochbaum and Shanthikumar proved several proximity results for convex separable nonlinear optimization ([HS90]). For the nonlinear separable network flow problem these demonstrate closeness properties between an integer optimal solution to the problem and a continuous optimal solution, between two scaled solutions to the problem, and between a scaled solution and an optimal (integer or continuous) solution. A scaled solution is a solution to the problem when the objective is given on a piecewise linear grid determined by the scaling constant. The closeness in all these case is that the two vectors differ in the maximum norm by at most m or $2m$ units.

Edmonds and Karp's [EK72] approach of capacity scaling could also be viewed as a proximity result, except that for the minimum cost network flow problem, the solution at each phase is not necessarily a feasible solution to the scaled problem. We shall discuss this approach in the context of proximity.

Defining what it means to obtain a solution in the continuous case is not straightforward due to the potential irrationality of the output for nonlinear problems. We therefore use the definition of ϵ-*accurate* solution, where $\log \frac{1}{\epsilon}$ is the number of significant digits desired in the output. A solution, $\boldsymbol{x}^{(\epsilon)}$ is ϵ-*accurate* if there exists an optimal solution $\boldsymbol{x}^*$ such that $\|\boldsymbol{x}^{(\epsilon)} - \boldsymbol{x}^*\|_\infty \leq \epsilon$. That is, ϵ is the accuracy required in the solution space.

When the functions f_{ij} are convex nonlinear functions, both the continuous and integer versions of the problem are solvable by algorithms running in a number of steps that is a polynomial function of m and n, but also of $\log_2 \|\boldsymbol{u}\|_\infty$ or $\log_2 \|\boldsymbol{b}\|_\infty$. Such an algorithm for the integer case is described by Minoux [Min86] using the capacity scaling approach, and for the integer and continuous cases in [HS90], using the proximity results, where the running time for the continuous case depends polynomially on the number of digits required to describe the solution. These algorithms are polynomial but not strongly polynomial, as they depend on the magnitudes of the numbers appearing in the problem instance, as well as the problem size parameters. On the practice side, a heuristic implementation of this algorithm, using the interior point method, indicates that it is very efficient, [HS91].

The topic of strong polynomiality became an important issue once a polynomial algorithm, the Ellipsoid method, was devised for solving linear programming problems. The ellipsoid method, as well as all other polynomial algorithms known for linear programming, is polynomial but not strongly polynomial. That is, the running time depends on the data coefficients, rather then only on the number of variables and constraints. Consequently, solving linear programming problems with different degrees of accuracy in the cost coefficients results in different running times. That is, the actual number of arithmetic operations grows as the accuracy of the data, and hence the length of the numbers in the input, increases. Such behavior of an algorithm is undesirable as it requires careful monitoring of the size of the numbers appearing in the data describing the problem instance, and hence limits the efficient applicablity of the algorithm.

Whether linear programming is solvable in strongly polynomial time is still an open problem, yet much progress has been reported. The most important work in this context is that of Tardos, [Tard86], that established that "combinatorial" linear programming problems, those with a constraint matrix having small coefficients, are solvable in strongly polynomial time. In particular, minimum cost network flow problems, which have coefficients that are either 0, 1 or -1 in the constraint matrix, are solvable in strongly polynomial time, [Tard85].

For non-quadratic objective functions, it was proved in [Hoc90] that there is no algorithm solving the single source separable convex Transportation problem in strongly polynomial time. This lower bound result holds either in the comparison model or in the algebraic computation tree model with the four arithmetic operations $+, -, \cdot, :$, and comparisons, and it holds even if the floor operation is added. Since the single source Transportation problem is a special case of separable convex minimization problems over totally unimodular constraint matrices or over polymatroidal constraints, then in particular all nonlinear convex network flow problems cannot be solved in strongly polynomial time.

The question of whether there exists a strongly polynomial algorithm for quadratic separable network flow problems remains open. The lower bound result in [Hoc90] does not apply to this case. Moreover, for quadratic problems there is a polynomial bound on the length of the output (this is because the optimality conditions are linear). Partial results have been derived recently. Tamir, [Tam89], devised a strongly polynomial algorithm for minimum convex quadratic separable cost flow when the underlying network is series-parallel, in the presence of a single source-sink pair. In [CH90] a strongly polynomial algorithm for the Quadratic Transportation problem with fixed number of sources or sinks is described. The strongly polynomial algorithm described in that paper delivers optimal continuous solutions using the operations mentioned above, without the floor operation. For the integer version, a strongly polynomial algorithm follows from a technique proposed in [HS90] that relies on the proximity result. The technique is a strongly polynomial algorithm for deriving an integer optimal solution from a continuous optimal solution to nonlinear convex problems over totally unimodular constraint matrix. This technique requires the floor operation. In [HH92], Hochbaum and Hong describe a strongly polynomial algorithm for the quadratic separable convex network Allocation problem with complexity equal to that of a single maximum flow algorithm. This problem is not a general network flow problem as only some of the edges have costs associated with them. All other edges only have capacity limits. Again, the idea here is to solve the continuous problem optimally, and then use the technique for converting a continuous optimal solution to a integer optimal solution.

There is no hope of solving the integer version in strongly polynomial time *without* the use of the floor operation. This follows from an observation by Tamir, [Tam89], illustrating that the floor operation is essential for solving integer quadratic problems. Tamir has demonstrated this via the following quadratic Allocation problem which may be viewed as a quadratic Transportation problem with a single source node:

$$\begin{array}{ll} \min & [\frac{1}{2}x_1^2 + \frac{1}{2}(a-1)x_2^2] \\ s.t. & x_1 + x_2 = b \\ & x_1,\ x_2 \geq 0, \text{ integer} \end{array}$$

The optimal value of x_2 is $\lfloor \frac{b}{a} \rfloor$. Therefore the floor operation can be executed via a routine that solves a quadratic Transportation problem. This, along with the impossibility result of strongly polynomial algorithms for non-quadratic problems mentioned above, implies that the floor operation, if helpful in devising strongly polynomial algorithms, is limited in its power to quadratic problems only.

Although separable quadratic problems may be solvable in strongly polynomial time, the question of strong polynomiality of *nonseparable* quadratic continuous optimization problems subject only to nonnegativity constraints, is at least as hard as the question of strong polynomiality of linear programming (this insight is due to I. Adler). Therefore, we do not investigate this issue here, as it should be treated in the framework of the strong polynomiality of linear programming. Nonseparable quadratic *integer* problems are NP-hard, [MK87], so in that sense also these problems are beyond the scope of our current investigation.

Nonseparable convex continuous problems, as well as nonseparable quadratic convex problems are solvable as follows. A solution approximating the optimal objective value to the convex continuous problem is in principle obtainable in polynomial time, provided that the gradient of the objective functions are available and that the value of the optimal solution is bounded in a certain interval. Such work, based on the Ellipsoid method, is described by Yudin and Nemirovsky ([NY83]). In the quadratic case, exact solutions are possible. Indeed, the polynomial solvability of continuous convex quadratic programming problems over linear constraints was established as a byproduct of the ellipsoid algorithm for Linear Programming (see Kozlov, Tarasov and Khachian, [KTK79]). The best running time reported to date is by Monteiro and Adler [MA89], $O(m^3L)$, where L represents the total length of the input coefficients and m the number of variables. Similar results were also given by Kapoor and Vaidya [KV86].

The case for the integer problems that are nonseparable is harder. We give in Section 5 a condition for such problems to be solvable in polynomial time for the convex nonseparable problem. We then describe special cases of integer nonseparable network flow problems that are polynomial, all of which with quadratic objective. [HSS92] gave a polynomial algorithm for a specific quadratic nonseparable Transportation problem in integers. That algorithm is strongly polynomial with some restrictions on the size of the linear coefficients in the objective function. Granot and Skorin-Kapov [GSK90] consider also a case of Transportation problem in integers solvable in polynomial time.

Cases where quadratic convex nonseparable optimization over box constraints (which are a special case of the quadratic network flow with an empty set of flow balance constraints), have been shown to have strongly polynomial algorithms are: by Barahona [Bar86], by Baldick and Wu [BW90] and by Hochbaum [Hoc89]. All these are special cases in which the quadratic matrix in the objective can be made separable by using a totally unimodular transformation of the variables. When this is the case there is a polynomial time algorithm for separable convex optimization over totally unimodular constraints, that delivers optimal integer solutions in polynomial time ([HS90]). Baldick ([Bal91]) has also presented several classes of matrices that can be diagonalized (i.e. the quadratic problems are made separable) via the use of totally unimodular matrices.

The plan of this paper is as follows. In section 1 we describe the scaled piecewise linear approximation and the proximity theorems for the separable flow problem and a general purpose polynomial algorithm. In Section 2 we propose a framework for obtaining several classes of algorithms that rely on these properties and on scaling. Section 3 provides the description of the lower bound proof that results in the impossibility of strongly polynomial algorithm for convex network flow problems. In Section 4 we demonstrate several results concerning the case of quadratic separable network flow, and Section 5 gives some limited results for nonseparable problems.

The notation in this paper, includes bold letters for denoting vectors, and the notation $\mathbf{e}$ is used for the vector $(1, 1, \ldots, 1)$.

2 Proximity Theorems and Piecewise Linear Approximations

2.1 The Scaled Piecewise Linear Approximation

Let $f_{ij} : R \to R$, $(i,j) \in A$, be m convex functions and define

$$F(\boldsymbol{x}) := \sum_{(i,j)\in A} f_{ij}(x_{ij}).$$

We are interested in the solutions to the nonlinear convex integer network flow, (INF), problem:

$$\begin{array}{lll} (INF) & \min & F(\boldsymbol{x}) \\ & \text{s.t.} & T\boldsymbol{x} = \boldsymbol{b} \\ & & 0 \le \boldsymbol{x} \le \boldsymbol{u} \\ & & \boldsymbol{x} \text{ integer} \end{array}$$

and to its continuous (real) relaxation

$$\begin{array}{lll} (RNF) & \min & F(\boldsymbol{x}) \\ & \text{s.t.} & T\boldsymbol{x} = \boldsymbol{b} \\ & & 0 \le \boldsymbol{x} \le \boldsymbol{u} \end{array}$$

Here T is an $n \times m$ adjacency matrix of the network $G = (V, A)$, $\boldsymbol{b}$ is a demand-supply n -vector, and $\boldsymbol{u}$ the capacity upper bounds vector on the arcs.

The idea of linearizing a nonlinear function in order to obtain solutions is not new. In a 1959 book [Den59], Dennis writes regarding quadratic cost networks: *"The electrical model for network flow problems can be extended to include flow branches for which the total cost contains terms depending on the square of the individual branch flows.... It appears that the algorithms presented in this chapter could be generalized.... These methods however are combinatorial in character and could require prohibitive calculation time, even on relatively simple networks. Certainly the simplicity and elegance of the diode-source algorithms would be absent. It would seem that the most practical means of attacking flow problems with quadratic costs would be to approximate the cost curve with piece-wise linear curve and substitute an appropriate number of linear cost branches connected in parallel."* Whereas it is clear that solving the problem on a piecewise linear approximation yields a feasible solution, the quality of such solution and its closeness to an optimal solution was not evaluated. [HS90] analyzed and described these proximity properties between the piecewise linearized problem's optimal solution and an optimal solution for the convex separable minimzation over linear constraints. We focus here only on the special case of network flow problems.

We now describe formally the piecewise linear approximation. For this we introduce the following class of problems. For any scaling constant $s \in R_+$ let the scaled problem $(RNF - s)$ be defined by

$$\begin{aligned} (RNF - s) \quad \min \quad & F(s\boldsymbol{y}) \\ & T\boldsymbol{y} = \boldsymbol{b}/s \\ & 0 \leq \boldsymbol{y} \leq \boldsymbol{u}/s \end{aligned}$$

By setting $\boldsymbol{x} = s\boldsymbol{y}$ in $(RNF - s)$ it is easily seen that $(RNF - s)$ is the same as (RNF). If we define the integer scaled problem to be $(RNF - s)$ with an integerality requirement that $\boldsymbol{y}$ is integer, then there is a feasible solution only if $\boldsymbol{b}/s$ is integer. In [HS90], the constraints are given as inequality constraints. The scaled problem is then,

$$\begin{aligned} (INF - s) \quad \min \quad & F(s\boldsymbol{y}) \\ & T\boldsymbol{y} \geq \boldsymbol{b}/s \\ & -T\boldsymbol{y} \geq -\boldsymbol{b}/s \\ & 0 \leq \boldsymbol{y} \leq \boldsymbol{u}/s \\ & \boldsymbol{y} \text{ integer} \end{aligned}$$

Note that a solution to $(INF - s)$ is not necessarily feasible for the original problem. The amount of unsatisfied supply (demand) is bounded by ns units. This problem can be solved as a linear network flow problem by rounding the scaled supplies down to the nearest integer, and the scaled demands (that are negative numbers) up to the nearest integer and adding a dummy node that absorbes the difference between the total supply and total demand at cost 0. When $s = 1$, the difference betweent the scaled supply and demand is 0, and hence there is no need for using the dummy node. Edmonds and Karp [EK72] used such idea of *capacity scaling* for the maximum flow problem that can be formulated as a minimum cost problem with $\boldsymbol{b} = \mathbf{0}$. Since the network flow problem readily provides integer solutions for integer right hand sides, and 0 is integer, feasibility was not an issue.

Instead of the scaled version of (RNF) we now use a linearized version of (RNF) as defined below. For any $s > 0$ let $f_{ij}^{L:s} : R \rightarrow R$ be the linearized version of f_{ij} such that $f_{ij}^{L:s}$ takes the same value as f_{ij} at all integer multiples of s: that is, $f_{ij}^{L:s}(sy) = f_{ij}(sy)$, for y integer and

$$f_{ij}^{L:s}(x_{ij}) = (\lfloor \frac{x_{ij}}{s+1} \rfloor - \frac{x_{ij}}{s}) f_{ij}(s\lfloor \frac{x_{ij}}{s} \rfloor) + (\frac{x_{ij}}{s} - \lfloor \frac{x_{ij}}{s} \rfloor) f_{ij}(s\lfloor \frac{x_{ij}}{s} + 1 \rfloor),$$

where $\lfloor \frac{x_{ij}}{s} \rfloor$ (the floor of $\frac{x_{ij}}{s}$) is the largest integer value smaller than or equal to $\frac{x_{ij}}{s}$. Clearly $f_{ij}^{L:s}$ is a piecewise linear function which is convex if f_{ij} is convex. $f_{ij}^{L:s}$ is depicted in Figure 1. Now define

$$\begin{array}{lll}(LNF-s) & \min & F^{L:s}(\boldsymbol{x}) \\ & & T\boldsymbol{x} \geq \boldsymbol{b} \\ & & 0 \leq \boldsymbol{x} \leq \boldsymbol{u}\end{array}$$

where

$$F^{L:s}(\boldsymbol{x}) := \sum_{(i,j)\in A} f_{ij}^{L:s}(x_{ij}), \boldsymbol{x} \in R^m.$$

Note that the optimal solution to the integer program

$$\begin{array}{lll}(INF'-s) & \min & F^{L:s}(s\boldsymbol{y}) \\ & & T\boldsymbol{y} \geq \boldsymbol{b}/s \\ & & -T\boldsymbol{y} \geq -\boldsymbol{b}/s \\ & & 0 \leq \boldsymbol{y} \leq \boldsymbol{u}/s \\ & & \boldsymbol{y} \text{ integer}\end{array}$$

is also an optimal solution to $(INF-s)$ because $F^{L:s}$ and F take the same value at integer multiples of s : i.e., $F^{L:s}(s\boldsymbol{y}) = F(s\boldsymbol{y})$ for all $\boldsymbol{y}$ integers. Hence, to solve $(INF-s)$ one can solve instead the piecewise linear version, $(INF'-s)$.

The set of feasible solutions $\{\boldsymbol{x} \mid T\boldsymbol{x} = \boldsymbol{b}, 0 \leq \boldsymbol{x} \leq \boldsymbol{u}\}$ is bounded in a box of length $B = \min\{\|\boldsymbol{u}\|_\infty, \|\boldsymbol{b}\|_1\}$ in each dimension (the flow on each edge cannot exceed capacity, or total sum of demands). In the initial iteration we substitute the tightest bound for the upper bound on the capacities. In the proximity-scaling algorithm, the

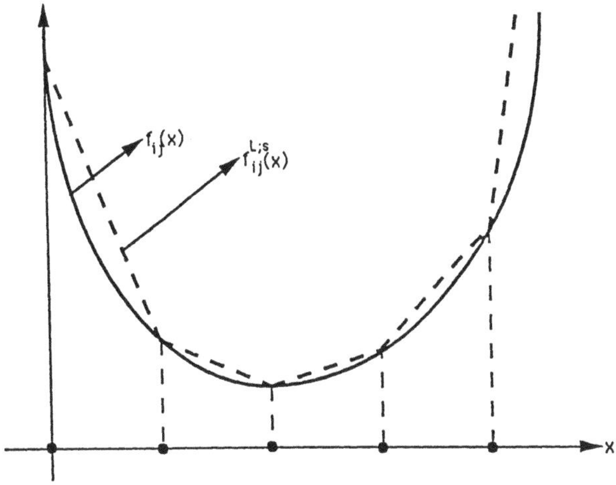

Figure 1:

upper and lower bounds on the variables will be updated at each iteration. So in a generic iteration the capacity bounds are replaced by constraints:

$$L_{ij} \le x_{ij} \le U_{ij}, \qquad (i,j) \in A.$$

These constraints will be scaled as well. In our procedure, at each iteration we shall work with the upper and lower bounds, U_{ij} and L_{ij}, and a scaling constant s, such that the length $\frac{U_{ij}-L_{ij}}{s}$, is an integer constant. We denote N=$\frac{U_{ij}-L_{ij}}{s}$. Each variable y_{ij}, for the integer case, is then substituted by a sum of N 0-1 variables:

$$y_{ij} = \lfloor \frac{L_{ij}}{s} \rfloor + \sum_{k=1}^{N} z_{ij}^{(k)}$$
$$z_{ij}^{(k)} \text{ is 0 or 1, for} k = 1, \ ..., \ N.$$

For the continuous case the substitution for x_i is:

$$x_{ij} = s\{\lfloor \frac{L_{ij}}{s} \rfloor + \sum_{k=1}^{N} z_{ij}^{(k)}\}$$
$$0 \le z_{ij}^{(k)} \le 1 \qquad \text{for } k = 1, ..., N.$$

So now $(LNF-s)$ and $(INF'-s)$ are piecewise linear convex minimization problems on the variables $z_{ij}^{(k)}$.

The modified objective function for the linear programming formulation (e.g. [D63] pp.482-486) of both $(LNF - s)$ and $(INF' - s)$ is

$$\min \sum_{(i,j)\in A} f_{ij}^{L:s}(s\lfloor \frac{L_{ij}}{s} \rfloor) + \sum_{k=1}^{N} \sum_{(i,j)\in A} [f_{ij}^{L:s}(s(\lfloor \frac{L_{ij}}{s} \rfloor + k)) - f_{ij}^{L:s}(s(\lfloor \frac{L_{ij}}{s} \rfloor + k - 1))] z_{ij}^{(k)}.$$

Let the column of T corresponding to arc (i,j) be denoted by $\boldsymbol{a}_{ij}$. Each such column is duplicated N times. Alternatively, in the graph, each arc is multiplied N times where each duplicated arc has the capacity 1. Let T^N be the matrix T in which each column is duplicated N times. The constraint set is then,

$$T^N \boldsymbol{z} = \boldsymbol{b}'.$$

where $b'_p = b'_p/s - \sum(\boldsymbol{a}_{ij})_p \lfloor L_{ij}/s \rfloor$. The linear programming version of the $(LNF - s)$ problem is then (omitting the constant from the objective function),

$$\begin{aligned} (LNF' - s) \quad \min \quad & \sum_{(i,j)\in A} \sum_{k=1}^{N} [f_{ij}^{L:s}(s(\lfloor \frac{L_{ij}}{s} \rfloor + k)) - f_{ij}^{L:s}(s(\lfloor \frac{L_{ij}}{s} \rfloor + k - 1))] z_{ij}^{(k)} \\ \text{s.t.} \quad & T^N \boldsymbol{z} = \boldsymbol{b}'. \\ & 0 \le z_{ij}^{(k)} \le 1, k = 1, ..., N, (i,j) \in A. \end{aligned}$$

One then has from well known results (e.g. [D63]),

Lemma 2.1 *Let $\hat{\boldsymbol{z}}$ be an optimal solution to $(LNF' - s)$. If f_{ij} is convex for each $(i,j) \in A$ then $\hat{\boldsymbol{x}}$ defined by $\hat{x}_{ij} = s\lfloor \frac{L_{ij}}{s} \rfloor + \sum_{k=1}^{N} \hat{z}_{ij}^{(k)}$, for all (i,j), is an optimal solution to $(LNF - s)$.*

Polynomial time algorithms for the piecewise linear problem are essentially an adaptation of the minimum cost flow problem's algorithms to graphs with multiple edges between a pair of nodes. An algorithm for this problem is given in [ABG84]. The drawback is that the number of edges in the graph depends now on the number of grid intervals, or alternatively on the scaling constant. Although a straightforward implementation works in polynomial time, it is nevertheless possible to speed up the running time. Pinto and Shamir [PS90] presented an algorithm that is more efficient than the straightforward approach.

2.2 Proximity Results

In this section we give proximity results for the optimal solutions of (INF), (RNF), $(INF - s)$ and $(LNF - s)$. Proofs for these results in more general context are given in [HS90]. The first proximity result is between the optimal solution of (INF) and (RNF). This is useful either when a continuous solution is more easily achievable than an integer one, such as in the case of quadratic network flow. Or, it is useful when an integer solution is achievable, and then this proximity is used to obtain an *eps*-accurate solution for the continuous problem. Note that integrality can be defined with respect to an arbitrary grid, in particular, with respect to an *eps* size grid.

Theorem 2.2

(i) For each optimal solution $\hat{x}$ for (RNF) there exists an optimal solution z^* for (INF) such that $\|\hat{x} - z^*\|_\infty \leq m$.

(ii) For each optimal solution $\hat{z}$ for (INF), there exists an optimal solution x^* for (RNF) such that $\|x^* - \hat{z}\|_\infty \leq m$.

Theorem 2.2 for the quadratic case was also proved in [GSK90], and an equivalent result for the general nonlinear case was independently derived by Werman [W91]. The following proximity is between the optimal solutions to (INF) and $(INF-s)$. It allows to solve the scaled problem on finer and finer grids in order to derive a solution for the integer problem (INF).

Theorem 2.3 *Let s be a positive integer.*

(i) For each optimal solution $\hat{y}$ for $(INF - s)$ there exists an optimal solution z^* for (INF) such that $\|s\hat{y} - z^*\|_\infty \leq ms$.

(ii) For each optimal solution zhat for (INF) there exists an optimal solution y^* for $(INF - s)$ such that $\|sy^* - \hat{z}\|_\infty \leq ms$.

Remark: Since (RNF) is a continuous relaxation of $(INF - s)$ with $x := sy$, from Theorem 2.2, one sees that for any optimal solution $\hat{y}$ of $(INF-s)$, there exists an optimal solution x^* to (RNF) such that $\|x^*/s - \hat{y}\|_\infty \leq m$. Also from Theorem 2.2 we know that there exists an optimal solution z^* to (INF) such that $\|x^* - z^*\|_\infty \leq m$. Combining these two proximity results using the triangular inequality of the l_∞ norm, one has $\|s\hat{y} - z^*\|_\infty \leq m(1+s)$. Similarly, it can be shown that for any optimal solution $\hat{y}$ of $(INF-s)$, there exists an optimal solution y^* for $(INF-s)$ such that $\|sy^* - \hat{z}\|_\infty \leq m(1+s)$. These bounds, however, are slightly improved in Theorem 2.3.

The following proximity result between the optimal solutions of (INF) and $(LNF-s)$ could be used to develop an algorithm for (INF) in case $(LNF-s)$ is more efficiently solvable than $(INF-s)$.

Theorem 2.4 *Let s be a fixed positive integer. For every optimal solution $\hat{x}$ for $(LNF - s)$ there exists an optimal solution z^* for (INF) such that*

$$\|\hat{x} - z^*\|_\infty \leq 2ms.$$

To develop an algorithm for the continuous case, (RNF), we use the following proximity result between the optimal solutions to (RNF) and $(LNF - s)$.

Theorem 2.5 *Let $s > 0$ be fixed. For every optimal solution $\hat{y}$ for $(INF - s)$ there exists an optimal solution x^* for (RNF) such that $\|\hat{y} - x^*\|_\infty \leq ms$.*

2.3 The Proximity-Scaling Algorithm

In this section we give an overview of the [HS90], which will be used as a framework for the implementation. The proof to all claims in this section are in [HS90].

The algorithm maintains a box that contains an optimal solution to the problem. At each iteration the volume of this box is reduced by a factor of 2^m. This is achieved by reducing each dimension of the box by a factor of 2. A direct implementation will therefore require a total of $\log_2 B$ iterations, where B is the length of a side of the initial cube for the integer problem. For the continuous problem solved to *eps*-accuracy there are $\log_2 \frac{B}{2\epsilon}$ such iterations.

The iterations continue, until, the optimal solution interval is reduced to a size at most 2ϵ. The complexity of the algorithm is summarized in the following theorem:

Theorem 2.6 *Let the complexity of solving the scaled network flow problem on m unit capacity arcs and n nodes, $(INF-s)$, be $T(m,n)$, then the complexity of solving (INF) for an integer optimal solution is $\log_2 \frac{B}{m} T(4m^2, n)$, and the complexity of solving (RNF) for an ϵ-accurate optimal solution to the nonlinear separable and convex network flow problem is $\log_2 \frac{B}{\epsilon} T(4m^2, n)$.*

The idea of the algorithm is as follows. At each iteration the optimal value of each variable is to be found in a bounded length interval, which is initially of length B. The interval is subdivided in $4m$ units, each of size s. The scaled problem is then solved for s (solved optimally or within some approximate feasibility). The optimal solution to the scaled problem is then within m units, each of size s away from an optimal solution. (The same statement can be made with regard to an optimal solution that is not feasible in that it satisfies only a portion of the supplies and demands). The optimal solution is then to be found in a new interval the length of which is only $2m$ units of size s each. We then divide s by 2 to produce a new scaling constant for the half length interval centered around the solution derived in the previous iteration. At the last iteration the length of the interval for each variable is $O(m)$. Therefore then the piecewise linear approximation on the integer grid provides an optimal solution to the problem.

Since we stop when the scaling constant is 1 and the length of the interval containing each variable $U_{ij} - L_{ij}$ is m, the total number of iterations of this algorithm is at most $\log_2 \frac{B}{m}$. In order to solve the continuous problem, we solve the integer problem on a grid of size $\frac{\epsilon}{m}$ and a final interval of length ϵ. From Theorem 2.2 this integer solution is at most $m\epsilon$ units away from the continuous optimal solution. It is therefore an ϵ -accurate solution.

Let $B = \min\{\|\boldsymbol{u}\|_\infty, \|\boldsymbol{b}\|_1\}$. The generic proximity-scaling algorithm is described below.

Procedure Proximity-Scaling

Step 0: Let $s = 2^{\lceil \log \frac{B}{4n} \rceil}$, $\boldsymbol{U}^{(1)} = B$, $\boldsymbol{L}^{(1)} = 0\boldsymbol{e}$, $i = 1$.

Let $y^{(0)} = 0e$.

Step 1: Solve, (using $2 \cdot y^{(i-1)}$ as an initial feasible solution):

$$
\begin{aligned}
(INF' - s) \quad \min \quad & F^{L:s}(sy) \\
& Ty \geq b/s \\
& -Ty \geq -b/s \\
& 0 \leq y \leq u/s \\
& L^{(i)} \leq y \leq U^{(i)} \\
& y \text{ integer}
\end{aligned}
$$

Let an optimal solution be $y^{(i)}$.
If $s = 1$, (for the continuous case $s = \epsilon/m$), the output is $y^{(i)}$. Stop, $y^{(i)}$ is an optimal solution.

Step 2: Set $U^{(i+1)} = y^{(i)} + sne$
Set $L^{(i+1)} = y^{(i)} - sne$
Set $s \leftarrow \frac{s}{2}$.
Set $i \leftarrow i + 1$.
Go to step 1.

The proximity scaling algorithm is clearly polynomial, yet its efficiency depends on the algorithm used for each scaled problem.

One way of solving the scaled problem is using the *successive shortest paths method* ([Je58], [Ir60] and [BG61] with some improvements by [EK72]). This algorithm works with a solution that is capacity feasible, dual feasible - all reduced costs are nonnegative, and iteratively progresses towards the feasibility of the flow balance constraints. The initial solution is the 0 vector and the initial reduced costs are all set to 0 as well. This algorithm can be used directly for a piecewise linear convex function with the costs being the left segment derivative. For each node with a unit of unsatisfied supply (excess), there exists at least one node with a unit of unsatisfied demand (deficit), since sum of supplies equals sum of demands. The basic iteration of the algorithm is to find all the shortest paths in the residual network, originating at a node with an excess using as costs the reduced costs. The reduced costs are then updated with the distance label - an operation that maintains their nonnegativity, and one unit of excess supply is sent to some node with a deficit. Each such iteration involves finding single source shortest paths in a graph with nonnegative costs. This can be done using Dijkstra's algorithm in $O(m + n \log n)$ operations, on a graph with n nodes and m arcs. Notice that also in the graph we process at each iteration, that has each arc duplicated $O(m)$ times, the single source shortest paths can be evaluated in the same running time, since among all the duplicated arcs, there is only one of lowest cost to be considered, and maintaining the arcs sorted is straightforward.

Since for a given problem there are initially $\frac{1}{2}\|b\|_1 = \frac{1}{2}\sum_{i=1}^{n}|b_i|$ unit of excess, the total running time of the successive shortest paths algorithm is $O(\|b\|_1(m + n\log n))$ steps.

The piecewise linear problem, $(INF' - s)$, differs from the linear problem in that the sum of supplies is not to the sum of demands. Each supply value b_i is rounded down to $s\lfloor\frac{b_i}{s}\rfloor$ whereas each demand value (which is a negative number), is rounded down in absolute value, i.e. to $s\lceil\frac{b_i}{s}\rceil$. Once all these demands and supplies are satisfied there are up to n unit multiples of s yet unsatisfied. Also, because capacity upper bounds in $(INF' - s)$ are effectively rounded down it may be impossible to find a path of capacity 1 in the residual network from an excess node to a deficit node. Since the original problem (INF) was feasible, the infeasibility per arc is at most one unit of s. It follows that each arc can prevent at most one unit of excess from getting cancelled against deficit. Hence, applying the successive shortest paths algorithm at an iteration will result in a solution satisfying all but $O(n + m)$ unit multiples of s of supply and demand. Therefore, if we start iteration i with the initial solution $2 \cdot y^{(i-1)}$, this solution will be capacity-feasible and dual feasible for the problem with $s_i = \frac{1}{2}s_{i-1}$, while at most $O(m + n)$ units of excess need to be processed. Hence the running time of each iteration for a given scaling factor is only $O((m+n)(m+n\log n))$. Since there are $\log fracBm$ iterations, the running time for the integer convex problem is $O(\log\frac{B}{m}(m+n)(m+n\log n))$, and for the ϵ-accurate solution it is $O(\log\frac{B}{\epsilon}(m+n)(m+n\log n))$.

This complexity analysis does not depend on the value of $U^{(i)} - L^{(i)}$, the interval size. Without the validity of the proximity theorems, the proximity-scaling algorithm will still produce an optimal solution on any specified ϵ-grid, with $\epsilon = 1$ for the integer case. Yet, in order to derive an ϵ-accurate solution for the continuous case, one must make use of proximity Theorem 2.5. Indeed, Minoux [Min86] was first to discover a polynomial algorithm for the integer convex flow problem. Minoux adapted the out-of -kilter method with scaling, so that at each iteration there is a factor reduction in the sum of kilter numbers. The reported running time in [Min86] is $O(\log_2\|u\|_\infty \cdot mn^2)$.

Recently, Ahuja Magnanti and Orlin [AMO92], introduced another algorithm for the integer convex network flow problem using a capacity scaling algorithm. They again use the solution at one scaling step as the initial solution in the next iteration for the next scaling step. The running time they report is $O(\log_2\|b\|_\infty \cdot m(m+n\log n))$, which is the same as the one reported here. Interestingly, this running time is also the same as that of the capacity scaling algorithm applied to the *linear* network flow problem. All these three algorithms depend in their complexity on $\log_2 B$, (recall that $B = \min\{\|u\|_\infty, \|b\|_1\}$) which is essentially the length of the right hand sides. For linear network flow there are polynomial algorithms with running times independent of the right hand sides and independent of the objective function - strongly polynomial algorithms. Since as observed for the capacity scaling algorithm, one can achieve algorithms for the convex case with same running time as linear case, it is conceivable that the strongly polynomial algorithms could also be adapted to the convex case.

This is however impossible as proved in the next section.

It follows that algorithms with the running time as the proximity-scaling algorithm are in a sense close to being optimal (with smallest complexity possible). This statement holds in the sense that in order to derive more efficient algorithms for the convex case, there must be more efficient algorithms for the linear case of type that depend on the right hand sides in their complexity. We believe that any improvement of those should be extremely challenging.

3 The Impossibility of Strongly Polynomial Algorithms

The lower bounds presented in this section are from [Hoc90]. The problem for which the lower bound is given is the *Allocation problem* which is identical to a single source Transportation problem in maximization form, $\max\{\sum_{i=1}^n f_i(x_i) | \sum_{i=1}^n x_i = B, \boldsymbol{x} \geq 0\}$. The generic presentation of this problem is as a maximization problem with a concave objective function. We first present a comparison model lower bound followed by an algebraic tree model lower bound.

3.1 A Comparison Model Lower Bound

We provide a lower bound proof that establishes that no algorithm exists that can solve SRA in less than $\log_2 B$ comparisons. In this, we rely on a result of information theory according to which there is no algorithm that finds a value in a monotonic decreasing n-array that is the first to be smaller than some specified constant in less than $\log_2 n$ time.

We first show the lower bound result for the Allocation problem in two variables. Suppose one can solve the problem:

$$\begin{aligned} \max \quad & f_1(x_1) + cx_2 \\ & x_1 + x_2 = B \\ & x_1, x_2 \geq 0, \text{ and integer.} \end{aligned}$$

Let the function $f_1(x_1)$ be given as an array of B increments at the B integer points. Namely, if $f_1(i) = a_i$ the array of increments is $\{a_0, a_1 - a_0, a_2 - a_1, \ldots, a_B - a_{B-1}\}$. Since the function is concave, the entries in the array are monotonic nonincreasing, $a_{i+1} - a_i \leq a_i - a_{i-1}$. The optimal solution to this problem is $x_1 = j$ and $x_2 = B - j$, where j is the largest index such that $a_j - a_{j-1} \geq c$.

Since the array of the differences between consecutive entries of the array is monotonic decreasing, determining in the array of differences the index j can be done using binary search in $\log_2 B$ comparisons. The information theoretic lower bound, is also $\log_2 B$ comparisons. This is because the comparison tree has B leaves so the path of comparisons leading from the root to a leaf could be as long as $\log_2 B$. (See [K73]).

Suppose the problem could be solved independently of B, then given a monotonic decreasing array and a value c, it has a corresponding concave function, f_1 such that the solution to the problem given above (name) is independent of B. Consequently, the required entry in the vector could be found independently of B, which is a contradiction.

A similar proof works for the problem with a single variable if the constraint is an inequality constraint. The problem is,

$$\begin{aligned} \max \quad & f(x) \\ & x \leq B \\ & x \text{ nonnegative integer}. \end{aligned}$$

Since f is concave it has an increment array corresponding to it as before. The array of differences is monotonic decreasing, and we shall look for the largest index such that the entry with that index is still nonnegative. In other words the role of c above is replaced here by 0. The conclusion follows precisely as above.

The same arguments can be extended to prove that in the comparison model the Allocation problem on $n+1$ variables has complexity $\Omega(n \log_2 \frac{B}{n})$. Let the problem be defined for $c > 0$,

$$\begin{aligned} & \sum_{j=1}^{n} f_j(x_j) + c \cdot x_{n+1} \\ & \sum_{j=1}^{n+1} x_j = B \\ & x_j \geq 0 \text{ and integer} \qquad j = 1, \ldots, n+1. \end{aligned}$$

Let the functions f_j be concave and monotonic increasing in the interval $[0, \lfloor \frac{B}{n} \rfloor]$, and zero in $[\lfloor \frac{B}{n} \rfloor, B]$. Solving this problem is then equivalent to determining in n arrays of length $\lfloor \frac{B}{n} \rfloor$ each, the last entry of value $\geq c$. Since the arrays are independent, the information theory lower bound is $\Omega(n \log \lfloor \frac{B}{n} \rfloor)$. Similarly, for the case of an inequality constraint the same lower bound applies for the problem on n variables,

$$\begin{aligned} \max \quad & \sum_{j=1}^{n} f_j(x_j) \\ & \sum_{j=1}^{n} x_j \leq B \\ & x_j \geq 0 \quad \text{integer} \qquad j = 1, \ldots, n \end{aligned}$$

since x_{n+1} can simply be viewed as the slack and $c = 0$.

This comparison model lower bound holds also for the quadratic case. It is therefore impossible to solve the quadratic problems in strongly polynomial time using only comparisons. Indeed, as stated in the introduction, the floor operation is essential for the quadratic integer problem.

3.2 The Algebraic-Tree Model

One might criticize the choice of the comparison model for this problem as being too restrictive. Indeed, the use of arithmetic operations may help to reduce the problems complexity. This is the case for the quadratic simple Allocation problem, which is solvable in linear time, O(n), [B84]. The lower bound here demonstrates that such success is not possible for other nonlinear functions. The computation model that is used hereafter allows the arithmetic operations $+, -, \times, \div$ as well as comparisons and branching based on any of these operations. It is demonstrated that the nature of the lower bound is unchanged even if the floor operation is permitted as well.

We rely on Renegar's lower bound proof ([R87]) in this arithmetic model of computation for finding *epsilon*-accurate roots of polynomials of fixed degree ≥ 2. In particular, the complexity of identifying an ϵ-accurate single real root in an interval $[O, R]$ is $\Omega(\log\log \frac{R}{\epsilon})$ even if the polynomial is monotone in that interval. Let $p_1(x)$, ..., $p_n(x)$ be n polynomials each with a single root to the equation $p_i(x) = c$ in the interval $[0, \frac{B}{n}]$, and each $p_i(x)$ a monotone decreasing function in this interval. Since the choice of these polynomials is arbitrary, the lower bound on finding the n roots of these n polynomials is $\Omega(n \log\log \frac{B}{n\epsilon})$. Let $f_j(x_j) = \int_0^{x_j} p_j(x)dx$. The f_js are then polynomials of degree ≥ 3. The problem,

$$
\begin{aligned}
(P_\epsilon) \qquad \max \quad & \sum_j f_j(x_j \cdot \epsilon) + c \cdot x_{n+1} \cdot \epsilon \\
& \sum_{j=1}^{n+1} x_j = \frac{B}{\epsilon} \\
& x_j \geq 0 \\
& x_j \text{ integer },
\end{aligned}
$$

has an optimal solution $\boldsymbol{x}$ such that $\boldsymbol{y} = \epsilon \cdot \boldsymbol{x}$ is the $(n\epsilon)$-accurate vector of roots solving the system

$$
\begin{cases}
p_1(y_1) = c \\
p_2(y_2) = c \\
\vdots \\
p_n(y_n) = c.
\end{cases}
$$

This follows directly from the Kuhn-Tucker conditions of optimality, and the fact that an optimal integer solution to the scaled problem with a scaling constraint, s, $\boldsymbol{x}^*$, and the optimal solution to the continuous problem $\boldsymbol{y}^*$ satisfy $\|\boldsymbol{x}^* - \boldsymbol{y}^*\|_\infty \leq ns$ (Theorem 2.3). Hence, a lower bound for the complexity of solving (P_ϵ) is $\Omega(n \log\log \frac{B}{n^2\epsilon})$. For $\epsilon = 1$, we get the desired lower bound for the integer problem.

In [MST88] there is a lower bound proof for finding *epsilon*-accurate square roots that allows also the floor, $\lfloor\rfloor$, operation. In our notation the resulting lower bound for our problem is $\Omega(\sqrt{\log\log \frac{B}{\epsilon}})$, hence even with this additional operation the problem cannot be solved in strongly polynomial time. Again, the quadratic objective is an exception and the algorithms for solving the quadratic objective simple resource Allocation problems rely on solving for the continuous solution first, then rounding down, using the floor operation, and proceeding to compute the resulting integer vector to feasibility and optimality using fewer than n greedy steps. See for instance, [IK88] for such an algorithm. Since the lower bound result applies also in the presence of the floor operation, it follows that the "ease" of solving the quadratic case is indeed due to the quadratic objective and not to this, perhaps powerful, operation.

4 Quadratic Network Flow Problems

As noted earlier, the quadratic problem takes a special place among nonlinear optimization problems over linear constraints. This is because the optimality conditions are linear (only the derivative of the objective function appears), and the solution to a system of linear inequalities is of polynomial length in the size of the coefficients. So for quadratic problems an optimal continuous solution is a polynomial function of the length of the input. In addition, the proof of impossibility for strongly polynomial algorithms using the algebraic tree computation model, is not applicable to the quadratic case (although for the comparison computation model the proof is valid and it is impossible to derive strongly polynomial algorithms using only comparisons). .PP All this raises a number of issues: are there strongly polynomial algorithms for quadratic separable convex network flow? Are there strongly polynomial algorithms for the nonseparable convex continuous case (where polynomial algorithms exist)?, and what classes of nonseparable quadratic integer problems, that are generally NP-hard, are solvable in polynomial time?. We survey here some results related to these questions.

4.1 Strongly Polynomial Algorithms for the Separable Case

Using a known transformation, any minimum cost network flow problem can be written as a Transportation problem. It is therefore sufficient in the search for efficient algorithms for the quadratic separable network flow problem to focus on the quadratic separable Transportation problem. .PP The Quadratic Transportation problem (QTP) is defined on a bipartite network, with k supply nodes and n demand nodes. The cost of transporting flow from a supply node to a demand node is a convex quadratic function of the flow quantity. The formulation of the continuous problem is as follows:

$$\min \quad \sum_{i=1}^{k}\sum_{j=1}^{n}[a_{ij}x_{ij} + \frac{1}{2}b_{ij}x_{ij}^2]$$

$$\text{s. t.} \quad \sum_j x_{ij} = s_i \qquad i = 1, ..., k$$
$$\sum_i x_{ij} = d_j \qquad j = 1, ..., n \qquad (QTP)$$
$$x_{ij} \geq 0 \qquad i = 1, ..., k, \quad j = 1, ..., n$$

where $b_{ij} > 0$, $s_i > 0$, and $d_j > 0$ are rational numbers and $\sum_i s_i = \sum_j d_j$.

It is not known whether the quadratic Transportation problem is solvable in strongly polynomial time. When the number of supply nodes k is fixed, Cosares and Hochbaum gave a strongly polynomial algorithm. The algorithms presented in [CH90] exploit the relationship between the Transportation problem and the Allocation problem. The continuous Allocation problem can be solved by identifying a Lagrangean multiplier associated with the single constraint. In the quadratic case this can be done in linear time. The algorithm for the Quadratic Transportation problem entails relaxing and aggregating source constraints, and then searching for optimal values for the Lagrangean multipliers. For the case of two supply nodes, $k = 2$, the algorithm is linear. For greater values of k, the algorithm has running time of $O(n^{k+1})$. A recent result by Megiddo and Tamir, [MT91], which invokes an alternative searching method, yields a linear running time of the algorithm for fixed k (the constant coefficient is exponential in k).

The Allocation problem may be viewed as a single source Transportation problem. We use this problem to illustrate the technique of converting a continuous solution to an integer solution. Allocation problems are all characterized by solvability using the greedy algorithm that amounts to adding one unit increment to a variable if its marginal contribution to the objective function is minimized. The greedy algorithm however is not a polynomial algorithm. For the *simple Allocation* problem, $\min\{\sum_{i=1}^n f_i(x_i) | \sum_{i=1}^n x_i = B, \boldsymbol{x} \geq 0\}$, the running time of the greedy is $O(B)$. The most general case of Allocation problem involves separable convex optimization over polymatroidal constraints: Given a *submodular* rank function $r : 2^E \rightarrow R$, for $E = \{1, ..., n\}$, i.e. $r(\phi) = 0$ and for all $A, B \subset E$,

$$r(A) + r(B) \geq r(A \cup B) + r(A \cap B).$$

The *polymatroid* defined by the rank function r, is the polytope $\{\boldsymbol{x} | \sum_{j \in A} x_j \leq r(A), \quad A \subseteq E\}$. We call the system of inequalities $\{\sum_{j \in A} x_j \leq r(A), \quad A \subseteq E\}$, the *polymatroidal constraints*. The general Allocation problem, GAP, is

$$(GAP) \qquad \min \quad \sum_{j \in E} f_j(x_j)$$
$$\sum_{j \in E} x_j = B$$
$$\sum_{j \in A} x_j \leq r(A) \qquad A \subset E$$
$$x_j \geq l_j \text{ and integer } \; j \in E.$$

The *network* allocation problem is special case of (GAP) as proved in [FG86a] and [FG86b]. As such it is solvable in pseudo polynomial time by the greedy algorithm. It is defined with respect to any network (or graph), with a single source - root, and a set of sinks.

Let $G = (V, A)$ be a directed network with node set V and arc set A. Let $s \in V$ be the source and $T \subseteq V$ be the set of sinks. Let the total supply of the root, $B > 0$ be given. Let C_{uv} be the capacity limit on each arc (u, v). Let the vector of the flow be $\phi = (\phi_{uv} : (u, v) \in A)$.

$$
\begin{array}{lll}
\min & \displaystyle\sum_{j\in T} a_j x_j + \frac{1}{2} b_j x_j^2 & \\
i) & \displaystyle\sum_{(v,u)\in A} \phi_{vu} - \sum_{(u,v)\in A} \phi_{uv} = 0 & v \in V - T - \{s\} \\
ii) & \displaystyle\sum_{(s,u)\in A} \phi_{su} - \sum_{(u,s)\in A} \phi_{us} \le B & \\
iii) & \displaystyle\sum_{(u,j)\in A} \phi_{uj} - \sum_{(j,u)\in A} \phi_{ju} = x_j & j \in T \\
iv) & 0 \le \phi_{uv} \le k_{uv} \quad (u, v) \in A & \\
 & 0 \le x_j \le u_j \quad \text{integers } j \in T. &
\end{array}
$$

The total sum of flow leaving the source B cannot exceed the minimum cut in the network. Also, so long as each variable is bounded in an interval where the derivative is negative and the sum of upper bounds is at least as large as B, the amount of flow in the network will be *equal* to B. So subject to such preprocessing the problem can be stated either with equality or inequality constraint on the source.

The quadratic network allocation problem is not the same problem as a quadratic cost flow problem. In the latter problem there is an underlying graph and a quadratic cost associated with the flow along each arc. In this problem there is also an underlying network, but costs (quadratic) are associated only with the flow arriving at each sink. For this purpose we add a new dummy sink "t", and send all flows from the set of nodes T to that node. The costs are then only associated with arcs adjacent to node t. This graph is described in Figure 2, where only the wiggly arcs, that connect the sink to the "variables" nodes have costs associated with them.

The quadratic network allocation problem is solvable in strongly polynomial time, $O(|A||V|\log(|V|^2/|A|)$, as described in [HH92]. The description of that algorithm is beyond the scope of this paper, but the algorithm for the simple allocation problem illustrates the important computational concerns that are used in the network algorithm as well as other algorithms for special cases of the *network* Allocation problem and the Quadratic Transportation Problem.

Brucker described a linear time algorithm for the quadratic continuous simple Allocation problem in [Bru84]. Our adaptation of the algorithm, and its application to the integer case is now described.

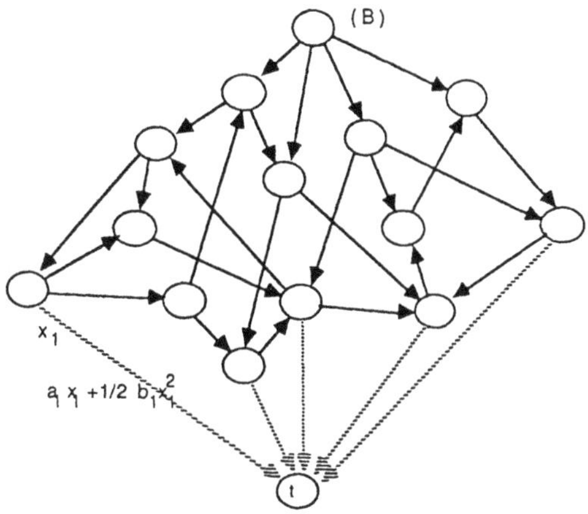

Figure 2:

The algorithm for the Quadratic Resource Allocation problem, (QRA), is based on a search for an optimal Lagrange multiplier. The continuous Quadratic Resource Allocation problem (QRA) is formulated as follows:

$$\begin{aligned} QRA: \quad \min \quad & \sum_{i=1}^{n} [a_i x_i + \frac{1}{2} b_i x_i^2] \\ s.t. \quad & \sum_{i=1} n x_i = d \\ & x_i \geq 0 \quad i = 1, \ \ldots, \ n \end{aligned}$$

where d is positive and each b_i is positive.

The convexity of the objective function guarantees that a solution satisfying the Kuhn-Tucker conditions is also optimal. In particular, we seek a non-negative solution $\boldsymbol{x}^*$ and a value δ^* such that:

$$\begin{aligned} & \sum_{i=1} n x_i^* = d \\ & x_i^* > 0 \text{ implies that } a_i + b_i x_i^* = \delta^* \end{aligned}$$

The situation is illustrated by the following figure:

The value set for δ determines associated values for x_i. For any value δ, the associated solution $\boldsymbol{x}$ is:

$$x_i = 0 \qquad \text{for } i \text{ such that } a_i > \delta$$

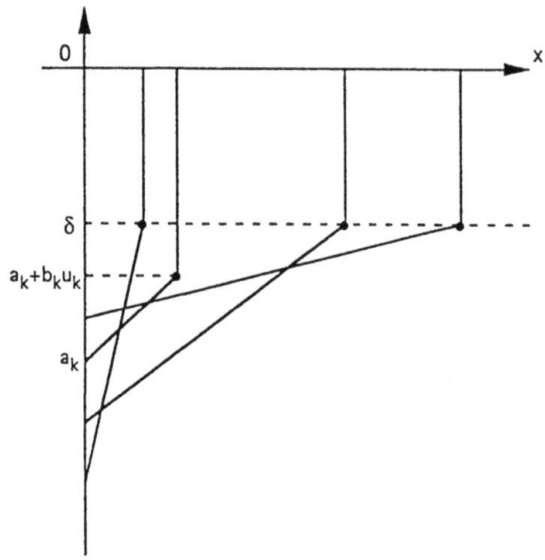

Figure 3:

$$x_i = \frac{\delta - a_i}{b_i} \qquad \text{for } i \text{ such that} \quad a_i \leq \delta$$

Finding the optimal solution to QRA is equivalent to finding a value δ^* such that the associated solution satisfies $\hat{d} = \sum_{i=1}^{n} x_i$ is equal to d. If $\hat{d} < d$, then we could conclude that δ^* is greater than δ, because any smaller value would yield an even smaller value for $\hat{d}$. Similarly, if $\hat{d} > d$, then δ^* is less than δ.

For any δ, the value of $\hat{d}$ is dependent on the coefficients in the set, $\{i \mid a_i > \delta\}$. Consequently, $\hat{d}(\delta)$, is a monotonic, piecewise linear function having breakpoints at the values a_i, $i = 1, ..., n$. Its monotonicity allows for a binary search for the optimal value, δ^* satisfying $\hat{d}(\delta) = d$.

Since the value of d is finite, it follows that there is a finite optimal δ^* for every instance of QRA. The algorithm we propose for finding δ^*, chooses "guesses", (from among the breakpoint values, a_i), until it finds two consecutive breakpoints which contain δ^* in the interval between them. In this range, $\hat{d} = \sum_i x_i$ is a linear function in δ. The problem is then solved by finding the particular value of δ for which $\hat{d} = d$, (i.e., by solving the linear equation in one variable).

In the algorithm the parameters A and B maintain partial sums necessary to evaluate $\sum_{i=1}^{n} x_i$, without computing the sum at every iteration from scratch.

Procedure quadratic-Allocation

Step 0: Let $L = \{a_1, ..., a_n\}$, $I = \{1, ..., n\}$

Let $A = \sum_{i=1}^{n} \frac{a_i}{b_i}$, $B = \sum_{i=1}^{n} \frac{1}{b_i}$,

Step 1: Set δ to a_m, the median value from the set L.
Let $\hat{A} = A - \sum_{i \in I^+} \frac{a_i}{b_i}$, $\hat{B} = B - \sum_{i \in I^+} \frac{1}{b_i}$,
where $I^+ = \{i \in I | a_i > \delta\}$

Step 2: Let $\hat{d} = \hat{B}\delta - \hat{A}$
If $\hat{d} = d$ then STOP, $\delta = \delta^*$.
If $\hat{d} > d$ then $\delta > \delta^*$.
If $\hat{d} < d$ then $\delta < \delta^*$.

Step 3: If $\delta > \delta^*$ then:
Set $I = \{i \in I | a_i < \delta\}$, $L = \{a_i | i \in I\}$
Set $A = \hat{A} - \frac{a_m}{b_m}$, $B = \hat{B} - \frac{1}{b_m}$,
If $\delta < \delta^*$ then:
Set $I = \{i \in I | a_i \geq \delta\}$, $L = \{a_i | i \in I\}$

Step 4: If $|L| \geq 2$, GOTO Step 1.
Else $\delta^* = \frac{d+A}{B}$

The algorithm outputs a value δ^*. The optimal solution $\boldsymbol{x}^*$ is then readily available, and can be determined in linear time:

$$x_i^* = \begin{cases} \frac{\delta^* - a_i}{b_i} & \text{for } i \text{ such that } a_i \leq \delta^* \\ 0 & \text{otherwise} \end{cases}$$

Theorem 4.1 *Algorithm quadratic Allocation finds δ^* and $\boldsymbol{x}^*$ in $O(n)$ time.*

Proof: For any guess δ, the values of $\hat{A}$ and $\hat{B}$ are set to assure that $\hat{d} = \sum x_i$ is set to the appropriate value. (i.e. $x_i = 0$ when $a_i > \delta$). The element a_i is removed from L if either it is known to be greater than δ^* or if it is less than an established lower bound for δ^*. When L contains only one element, say a_i , then we can conclude that δ^* is between a_i and a_j, the next largest of the a's. Furthermore, since $\hat{d}$ is a linear function of δ in this range, (i.e. $\hat{d} = \hat{B}\delta - \hat{A}$), δ^* and $\boldsymbol{x}^*$ are determined as in Step 4.

The $O(n)$ complexity of the algorithm follows from the fact that each of Steps 1, 2 and 3 can be performed in a number of arithmetic operations that is linear in the cardinality of the set L, including the selection of the median value ([BFPRT72]). Since the number of elements in the set is initially n and is cut in half after each pass, the total work is linear in $(n + n/2 + n/4 + \dots\ ) = 2n$, so the complexity of the algorithm is $O(n)$. □

When the problem is to be solved in integers we apply the proximity Theorem 2.2 (or alternatively a tighter proximity theorem developed for all Allocation problems in [Hoc90]). From the optimal continuous solution $\boldsymbol{x}^*$ we create a lower bound vector to the optimal integer solution $\boldsymbol{x}^* - n\boldsymbol{e}$ (or using the tighter proximity theorem, $\boldsymbol{x}^* - \mathbf{e}$ is a tighter lower bound). Since $\sum x_j^* = B$, there are only n^2 more units to add (or rather only n more units to add), which can be determined in such number of iterations of the greedy algorithm, each taking a constant time. The running time is therefore linear for the integer version of the problem.

The question of the strong polynomiality of the general quadratic Allocation problem is not settled yet. Still, all known special cases of the *network* Allocation problem are all solvable very efficiently. These problems include the *tree* Allocation problem, the *nested* Allocation problem and the *Generalized Upper Bounds* Allocation problem (and of course the simple allocation problem). In [HH92] we describe algorithms for these problems with complexities, $O(n \log n)$, $O(n \log n)$ and $O(n)$, respectively. All these algorithm are based on efficient search for the optimal Lagrange multipliers and in that sense generalize the procedure *quadratic-allocation.*

Finally, unrelated to the Allocation problem, Tamir ([Tam89]) devised a strongly polynomial algorithm for minimum convex quadratic separable cost flow when the underlying network is *series-parallel*, in the presence of a single source-sink pair. This algorithm is exploiting the series-parallel structure of the network in order to construct an optimal continuous solution, namely, the solution is constructed from series parallel operations applied to subproblems defined on subgraphs. An integer solution is determined using the same approach as described above.

4.2 The Complexity of Some Nonseparable Cases

We first illustrate that if there exists a "separating scheme" such that the separating matrix with T forms a totally unimodular matrix, then can still solve the integer nonseparable network flow problem in polynomial time ([HS90]). This relies on an algorithm that was developed based on proximity and scaling, which is polynomial when the largest subdeterminant of the constraint matrix has a polynomial size. The algorithm solves in polynomial time, in integers, separable convex problems over constraints for which the constraint matrix is totally unimodular. Such problems generalize the integer separable convex network flow problem. Consider the nonseparable problem,

$$\begin{array}{ll} \min & F(\boldsymbol{x}) \\ s.t. & T\boldsymbol{x} = \boldsymbol{b} \\ & 0 \le \boldsymbol{x} \le \boldsymbol{u} \\ & \boldsymbol{x} \text{ integer} \end{array}$$

Suppose there exists an *ntimesn* matrix U such that $F(U\boldsymbol{x})$ is a separable function, then the newly stated problem is,

$$\begin{array}{ll} \min & F(\boldsymbol{y}) \\ s.t. & TU^{-1}\boldsymbol{y} = \boldsymbol{b} \\ & 0 \le U^{-1}\boldsymbol{y} \le \boldsymbol{u} \\ & U^{-1}\boldsymbol{y} \text{ integer} \end{array}$$

Now, if the matrix U is totally unimodular, then the integrality requirement is preserved. If the new matrix of constraints $\begin{bmatrix} TU^{-1} \\ U^{-1} \end{bmatrix}$ is totally unimodular, then the noseparable flow problem is solvable in polynomial time and in integers.

The nonseparable *quadratic* convex continuous problem is polynomial as indicated in the introduction. With the integrality requirement, the problem becomes NP-hard as illustrated by the following known reduction from the independent set problem. Consider the maximization of the weight of an independent set in a graph: Given a graph $G = (V, E)$ with nonnegative weights W_v for each $v \in V$, find a subset of vertices $U \subseteq V$ such that for any $i, j \in U$, $\{i, j\} \notin E$, and such that the total weight $W(U) = \sum_{v \in U} W_v$ is maximum. The weighted independent set problem can be posed as the quadratic maximization problem:

$$\begin{aligned} \max \quad & \sum_{v\in V} W_v x_v - \sum_{\{u,v\}\in E} W(V)\cdot x_u \cdot x_v \\ & x_v \in \{0,1\}. \end{aligned}$$

Let $\boldsymbol{x}^*$ be the optimal solution. The maximum weight independent set is then $\{v | x_v^* = 1\}$. Note that the reduction also applies for the unweighted case. So even in the absence of the flow balance constraints, the integer problem is NP-hard. The objective function in this case is not necessarily concave. The question is then asked whether the complexity of the problem is not a result of the indefiniteness of the quadratic matrix. Baldick gives a negative answer to this question [Bal91], by constructing a reduction from the set splitting problem to 0-1 minimization of a convex quadratic function.

The formulation of a quadratic nonseparable problem on box constraints (i.e. the set of flow balance constraints is empty), is as follows:

$$\begin{aligned} \max \quad & \sum_{j=1}^{n} a_i x_i + \sum q_{ij} \cdot x_i \cdot x_j \\ & l_i \le x_i \le u_i \in \{0,1\}. \\ & x_i \quad \text{integer} . \end{aligned}$$

The idea of making the problem separable so that the [HS90] algorithm applies is translated here to finding a diagonalizing matrix U, so that for the matrix $Q = (q_{ij})$, $U^{-1}QU$ is a diagonal matrix. Baldick and Wu [BW90] used this approach for a problem of electric distribution system where only box constraints are present.

Baldick [Bal91] has further identified several classes of matrices Q where a "diagonalizing" scheme witha totally unimodular matrix exists. The two classes are:

(1) Diagonally dominant matrices, $q_{ii} \ge \sum_{i \ne j} |q_{ij}|$.

(2) Matrices with forest structure, these are matrices with a partial order on the positive coefficients inducing a forest.

For both these classes, with T empty, there are polynomial algorithms. Also if the constraint matrix, $\begin{bmatrix} TU^{-1} \\ U^{-1} \end{bmatrix}$ is totally unimodular then still integer solutions can be obtained in polynomial time. Continuous solutions can be obtained in polynomial time if the largest subdeterminant of the constraint matrix is bounded by a polynomial.

Barahona, [Bar86], discovered that the case of quadratic nonseparble 0-1 optimization is polynomial if the quadratic matrix Q has a series-parallel graph characteristic structure. That is, there exists a series-parallel graph $G = (V, E)$ with $q_{ij} \neq 0$ if and only if $(i, j) \in E$. This is again a case of network flow with an empty set of flow balance constraints.

A class of nonseparable problems over box constraints (upper and lower bound constraints) which is solvable in polynomial (and strongly polynomial) time is presented in [Hoc89]. The class is identified by the property of the multivariate polynomial objective function called *the bipartition property.* This property can be easily described on a graph for quadratic objective functions: $G = (V, E)$ with $V = V_1 \cup V_2$ so that $q_{ij} > 0$ only if both $i, j \in V_1$ or $i, j \in V_2$. This property was discovered independently, for quadratic objective functions, by Hansen and Simeone, [HS86]. In [HS86], an objective function with this property is called a *unate function.*

In [HSS92] a "high multiplicity" minimum weighted tardiness scheduling problem was discussed. This problem was formulated as a quadratic Transportation problem with a nonseparable objective function. This problem is unique among the problems discussed in this section in that the set of flow balance constraints is not empty. In that problem, the right hand sides (supplies and demands), and the linear coefficients in the objective function are large, so the aim was to find an algorithm for the integer case the running time of which is independent of these numbers. Such algorithm was found by solving a related continuous problem (not the relaxation), the solution of which could be rounded using a simple procedure, to derive an optimal integer solution.

All problems presented in this section are special classes. There is still a need to discover the extent of the polynomial solvability of nonseparable network flow problems, although this cannot be expected to be so unified as in the separable case.

References

[ABG84] R. K. Ahuja, J. L. Batra and S. K. Gupta, A Parametric Algorithm for the Convex Cost Network Flow and Related Problems, *Europ. J. Oper. Res.*, 16, (1984), 222-235.

[AMO92] R. K. Ahuja. T. L. Magnanti and J. B. Orlin, *Network Flows: Theory, Algorithms and Applications*, manuscript, to be published by Prentice Hall in 1993.

[Bal91] R. Baldick, A Unification of Polynomially Solvable Cases Of Integer "Non-separable" Quadratic Optimization, Lawrence Berkeley Laboratory manuscript, November (1991).

[BW90] R. Baldick and F.F. Wu, Efficient Integer Optimization Algorithms for Optimal Coordination of Capacitators and Regulators, *IEEE Transactions on Power Systems*, 5(3), 805-812, (1990).

[Bar86] F. Barahona, A Solvable Case of Quadratic 0-1 Programming, *Discrete Applied Mathematics*, 13 (1) (1986), 23-28.

[BS85] M. S. Bazaraa and C. M. Shetty, *Nonlinear Programming: Theory and Algorithms*, (John Wiley & Sons, 1979).

[BG61] R. G. Busacker and P. J. Gowen, A Procedure for Determining Minimal-Cost Network Flow Patterns, Operational Research Office, John Hopkins University, Baltimore, MD, (1961).

[BFPRT72] M. Blum, R.W. Floyd, V.R. Pratt, R.L. Rivest and R.E. Tarjan, Time Bounds for Selection, *J. Computer Systems Science* 7 (1972) 448-461

[Bru84] P. Brucker, An $O(n)$ Algorithm for Quadratic Knapsack Problems, *Operations Research Letters*, 3 (1984), 163-166.

[CH90] S. Cosares and D. S. Hochbaum A Strongly Polynomial Algorithm for the Quadratic Transportation Problem with Fixed Number of Suppliers, manuscript, U.C. Berkeley, August 1990, to appear *Mathematics of Operations Research.*

[D63] G.B. Dantzig, *Linear Programming and Extensions*, (Princeton University Press, 1963).

[Den59] J. B. Dennis, *Mathematical Programming and Electrical Networks*, (Technology Press Research Monographs, Technology Press and Wiley, 1959), pp.74-75.

[D67] I.I. Dikin, Iterative Solution of Problems of Linear and Quadratic Programming, *Soviet Mathematics Doklady*, 8, (1967), 674-675.

[EK72] J. Edmonds and R. M. Karp, Theoretical Improvements in Algorithmic Efficiency for Network Flow Problems, *Journal of ACM* , 19, (1972), 248-264.

[EMV87] R.E. Erickson, C.L. Monma and A.F. Veinott, Send-and-Split Method for Minimum-Concave-Cost Network Flows, *Math. of Oper. Res.*, 12 (1987), 634-664.

[FG86a] A. Federgruen and H. Groenevelt, The Greedy Procedure for Resource Allocation Problems: Necessary and Sufficient Conditions for Optimality, *Operations Research*, 34, (1986), 909-918.

[FG86b] A. Federgruen and H. Groenevelt, Optimal Flows in Networks with Multiple Sources and Sinks, with Applications to Oil and Gas Lease Investment Programs, *Operations Research*, 34, (1986), 218-225.

[F85] R. Fourer, Simplex Algorithm for Piecewise-Linear Programming I: Derivation and Proof, *Mathematical Programming*, 33 (1985) 204-233.

[GSK90] F. Granot and J. Skorin-Kapov, Some Proximity and Sensitivity Results in Quadratic Integer programming, *Mathematical programming*, 47, No. 2, (1990), 259-268.

[GSK90a] F. Granot and J. Skorin-Kapov, Strongly Polynomial Solvability of a Non-separable Quadratic Integer Program with Applications to Toxic Waste Disposal, manuscript, December (1990).

[GP90] G. Guisewite and P.M. Pardalos, Minimum Concave Cost Network Flow Problems: Applications, Complexity, and Algorithms, *Annals of Operations Research*, 25, (1990), 75-100.

[GP91a] G. Guisewite and P.M. Pardalos, Algorithms for the Single-Source Uncapacitated Minimum Concave-Cost Network Flow Problem, *Journal of Global Optimization*, 1, No. 3 (1991), 245-265.

[GP91b] G. Guisewite and P.M. Pardalos, Global Search Algorithms for Minimum Concave Cost Network Flow Problems, *Journal of Global Optimization*, 1, No. 4 (1991), 309-330.

[HS86] P. Hansen and B. Simeone, Unimodular Functions, *Discrete Applied Mathematics* 14 (1986), 269-281.

[Hoc89] D. S. Hochbaum, On a Polynomial Class of Nonlinear Optimization Problems, manuscript, U.C. Berkeley, (1989).

[Hoc90] D. S. Hochbaum, On the Impossibility of Strongly Polynomial Algorithms for the Allocation Problem and Its Extensions, *Proceedings of the 1st Integer Programming and Combinatorial Optimization Conference*, May (1990), 261-274. Also see, D. S. Hochbaum, Lower and Upper Bounds for the Allocation Problem and Other Nonlinear Optimization Problems, to appear, *Mathematics of Operations Research.*

[HH92] D. S. Hochbaum and S. P. Hong, About Strongly Polynomial Algorithms for Quadratic Network Allocation Problems, manuscript, U.C. Berkeley, (1992).

[HS91] D. S. Hochbaum and S. Seshadri, The Empirical Performance of a Polynomial Algorithm for Constrained Nonlinear Optimization, (1991), to appear, *Annals of Operations Research.*

[HSS92] . S. Hochbaum, R. Shamir and J. G. Shanthikumar, A Polynomial Algorithm for an Integer Quadratic Nonseparable Transportation Problem, *Mathematical Programming*, Vol. 55, No. 3, pp. 359-372, (1992).

[HS90] D.S. Hochbaum and J.G. Shanthikumar, Convex Separable Optimization is not Much Harder Than Linear Optimization, *Journal of ACM* Vol. 37, No 4, (1990), 843-862.

[IK88] T. Ibaraki and N. Katoh, *Resource Allocation Problems: Algorithmic Approaches*, (MIT Press, 1988).

[Ir60] M. Iri, A New Method of Solving Transportation Network Problems, *Journal of the Operations Research Society of Japan*, 3, 27-87, (1960).

[Je58] W. S. Jewell, Optimal Flow Through Networks, Technical report No. 8, Operations research Center, MIT Cambridge, MA (1958).

[KV86] S. Kapoor and P. M. Vaidya, Fast Algorithms for Convex Quadratic Programming and Multicommodity Flows, *Proc. 18th STOC*, (1986), 147-159.

[KTK79] M.K. Kozlov, S. P. Tarasov and L.G. Khachian, Polynomial Solvability of Convex Quadratic Programming, *Doklady Akad. Nauk SSR* 5 (1979), 1051-1053 [Translated in *Soviet Mathematics Doklady* 20 (1979), 1108-1111].

[MST88] Y. Mansour, B. Schieber and P. Tiwari, Lower Bounds for Computations with the Floor Operations, manuscript, December (1988).

[MT91] N. Megiddo and A Tamir, Linear Time Algorithms for Some Separable Quadratic Programming Problems, manuscript, Nov (1991).

[Min84] M. Minoux, A Polynomial Algorithm for minimum Quadratic Cost Flow Problems, *European Journal of Operational Research*, 18 (1984), 377-387.

[Min86] M. Minoux, Solving Integer Minimum Cost Flows with Separable Convex Cost Objective Polynomially, *Mathematical Programming Study*, 26 (1986), 237-239.

[Min86a] M. Minoux, *Mathematical Programming, Theory and Algorithms*, (Wiley, 1986), Chapters 5,6.

[MA89] R. D.C. Monteiro and I. Adler, Interior Path Following Primal-Dual Algorithms. Part II: Convex Quadratic Programming, *Mathematical Programming* 44 (1989) 43-66.

[MS87] K. G. Murty and S. N. Kabadi, Some NP- Complete Problems in Quadratic and Nonlinear Programming, *Mathematical programming*, 39, 117-129, (1987).

[NY83] A. S. Nemirovsky and D. B. Yudin, *Problem Complexity and Method Efficiency in Optimization*, (Wiley, 1983).

[PS90] Y. Pinto and R. Shamir, Efficient Algorithms for Minimum Cost Flow Problems with Convex Costs, Technical Report, Department of Computer Science, Tel Aviv University, Tel Aviv, Israel, (1990).

[R87] J. Renegar, On the Worst Case Arithmetic Complexity of Approximation Zeroes of Polynomials, *J. of Complexity*, 3, (1987) 9-113.

[Sah74] S. Sahni, Computationally related Problems, *SIAM J. on Computing*, 3, 262-279 (1974).

[Tam89] A. Tamir, A Strongly Polynomial Algorithm for Minimum Convex Separable Quadratic Cost Flow Problems on Series-Parallel Networks, (1989), to appear *Mathematical Programming.*

[Tard85] E. Tardos, A Strongly Polynomial Minimum Cost Circulation Algorithm, *Combinatorica* 5 (1985) 247-255

[Tard86] E. Tardos, A Strongly Polynomial Algorithm to Solve Combinatorial Linear Programs, *Operations Research* 34 (1986) 250-256

[W92] M. Werman, The relationship between Integer and Real Solutions of Constrained Convex Programming, *Mathematical Programming*, 51, (1991), 133-135.

Network Optimization Problems, pp. 93-110
Eds. D.-Z. Du and P.M. Pardalos

Hamiltonian Circuits for 2-Regular Interconnection Networks

F. K. Hwang
AT&T Bell Laboratories, Murray Hill, NJ 07974

Wen-Ching Winnie Li*
Department of Mathematics, Pennsylvania State University, University Park, PA 16802

Abstract

A 2-regular directed graph, in which each node has two inlinks and two outlinks, is a popular topology for interconnection networks due to its simple and homogeneous structure, and also due to the reduced transmission delay and the increased reliability as compared to a ring network. When the network is used as a processor network, it has been argued that a hamiltonian circuit embedded in such a network provides a maximum linear array of processors even under a fault, and also reduces the number of spare processors required. The problem of the existence of a hamiltonian circuit has been solved for the double loop network, the generalized deBruijn network, the Imase-Itoh network and the torus network. In this paper we study the same problem for extended double loop network, semi-torus network and semi-Manhattan street network. Our method enables us to unify the aforementioned known results and provides generalizations at the same time.

*This research, supported in part by the grant from NSA no. MDA904-90-H-1021, was done when the author was visiting AT&T Bell Laboratories, Murray Hill, NJ.

1 Introduction

Two important features in designing an interconnection network are a small transmission delay and some kind of fault tolerance. Represent the network by a graph with stations as nodes and wires between stations as links. Then delay is usually measured by the *diameter* (the distance of a pair of nodes farthest apart) and the fault tolerance by *connectivity* (the size of the smallest cutset). If the network is used as a processor network, then embedding a hamiltonian circuit is desirable [12] since it provides a linear array of n processors even with a link failed (or $n-1$ processors with a processor failed). It has also been argued [9] that the existence of a hamiltonian circuit reduces the number of spare processors in certain types of 2-regular interconnection networks.

In practice, simplicity and homogeneity are also important considerations. These are often built into the design by posing as constraints. Thus *regular networks* (networks having the same number ℓ of outlinks and inlinks at each node) are popular since the nodes are homogeneous. One would also like to keep ℓ small to reduce the hardware of the network as well as to keep the control operation within a node simple. The simplest regular network with n nodes is $\ell = 1$, which, if *strongly connected* (i.e., there is a path from every node to every other node), is a hamiltonian circuit. However, the 1-regular network is known to have large diameter and 1-connectivity.

The next level of simplicity is when $\ell = 2$. Not only have we shortened the diameter and increased the connectivity, we also now have an abundance of networks to choose from. Let $[i, j]$ denote a link from node i to node j. We consider two general classes.

(i) The *extended double loop class.* A network in this class has n nodes $0, 1, \ldots, n-1$ and $2n$ links $[i, a_1 i + b_1]$, $[i, a_2 i + b_2]$ for $i = 0, 1, \ldots, n-1$, where a_1, a_2, b_1, b_2 are fixed integers and addition is modulo n. With $a_1 = a_2 = 2$, $b_1 = 0$, $b_2 = 1$, we have the generalized deBruijn network [6, 14] ($n = 2^p$ being the graph corresponding to the deBruijn sequence). With $a_1 = a_2 = -2$, $b_1 = -1$, $b_2 = -2$, we have the Imase-Itoh network [7]. With $a_1 = a_2 = 1$ we have the usual double loop network. If, furthermore, $b_1 = 1$, then the network is known as FLBH (forward loop backward hop) [15].

(ii) The *mesh class.* A network in this class is defined on a grid lying on the surface of a torus. The nodes are the grid points and the links are the grid segments. All links on the same grid line have the same direction. When the horizontal and vertical grid lines are all unidirectional, the network is known as a *torus network* [3, 10]. (A variation is known as a *rectangular array* [8].) When the horizontal and vertical grid lines alternate in direction, except if the number of horizontal or vertical lines is odd, then one line does not alternate, the network is known as a *Manhattan street network* [11]. We also relax the requirements on line directions of torus networks and Manhattan street networks to only either the horizontal or the vertical lines, and call them *semi-torus networks* and *semi-Manhattan street networks*, respectively.

Necessary and sufficient conditions for the existence of hamiltonian circuits have been determined for the usual double loop network and the torus network, which are directed Cayley graphs on abelian groups with two generators [13], and also for the generalized deBruijn network and the Imase-Itoh network [1, 2]. In this paper we investigate the hamiltonian property for the extended double loop network (Theorem 3.11), the semi-torus network (Theorem 4.4), and the semi-Manhattan street network (Theorem 5.1). Our method is rather general, which enables us to unify the aforementioned known results and meanwhile provides generalizations to non-Cayley digraphs as well.

2 A General Approach

Consider a 2-regular network G with two types of links. For an extended double loop network, the two types are defined by the two equations $g_1(i) = a_1 i + b_1$ and $g_2(i) = a_2 i + b_2$. For a mesh network the two types are the horizontal and the vertical links. Let L_1 and L_2 denote the sets of type 1 and type 2 links of G, respectively. A subgraph F of G is called a *factor* if every node of G has exactly one inlink and one outlink in F. It is easily verified that L_1 and L_2 are factors for the extended double loop network and for the mesh network.

Let F be a factor of G. Suppose that F contains a type 1 link $[i_1, j_1]$. Then F cannot contain the type 2 link $[i_2, j_1]$ since otherwise the node j_1 has two inlinks in F. Consequently, F must contain the type 1 link $[i_2, j_2]$ or the node i_2 would have no outlink in F. A recursive argument shows that if F contains a link $[i, j]$, then F must contain a set of links $S([i, j])$ of links of the same type as $[i, j]$.

Suppose that the two types of links can be described by two invertible functions g_1 and g_2 on the nodes such that the two outlinks of the node i are $[i, g_1(i)]$ and $[i, g_2(i)]$. For $j = 1, 2$, represent the set $S_j(i) = S([i, g_j(i)])$ by the starting nodes of the links. Put $g = g_2^{-1} \circ g_1$ and $g^{-1} = g_1^{-1} \circ g_2$. Then $S_1(i) = \{i, g(i), g^2(i), \ldots, g^{m-1}(i)\}$ and $S_2(i) = \{i, g^{-1}(i), g^{-2}(i), \ldots, g^{-(m-1)}(i)\}$, where m is the smallest integer k such that $g^k(i) = i$. Observe that $S_1(i) = S_2(i)$, denote it simply by $S(i)$. Clearly, $i' \in S(i)$ if and only if $S(i') = S(i)$. Therefore the sets $S(i)$ partition the nodes of the network G. The network G is said to be *m-uniform* if the sets $S(i)$ have the same cardinality m. In this case we obtain a partition of nodes of G into n/m classes, each containing m nodes. A path p of length n/m from node a to node b is called *basic* if all the starting nodes on p belong to different classes and a and b belong to the same class. Therefore if a factor F of an m-uniform G contains a basic path p of length n/m, then F is completely determined by p since the outlink (of F) at a node i has the same type as the outlink of the unique starting node on p belonging to the class of i. In the factor determined by a basic path p of length n/m, there is a shortest cycle containing p of length, say, k. Call km/n the *period* of p.

Theorem 2.1 *Suppose that G is m-uniform and every path in G of length n/m is basic. Then G is hamiltonian if and only if there is a path of length n/m and period m.*

Proof. The assumptions on G imply that when we move along a path or cycle contained in a factor of G, only every n/m nodes belong to the same class, hence a hamiltonian circuit exists if and only if there is a path of length n/m and period m.

The above approach is essentially the "arc-forcing subgroup" technique introduced by Rankin [13] in dealing with a group theoretic problem which can be used to determine if certain Cayley graphs are hamiltonian. There are many papers in the literature dealing with the hamiltonian property of a Cayley graph, see [19] and the papers referred therein. Our formulation affords a broader application, including non-Cayley graphs as well.

We end this section by noting the following

Lemma 2.2 *A 2-regular digraph is connected if and only if it is strongly connected.*

Proof. Clearly, strong connectedness implies connectedness. Conversely, let X be a connected 2-regular digraph. We show that it is strongly connected. Let v be a node of X. Denote by V the set of nodes reachable from v. Since all the outlinks of nodes in V must end at nodes of V, the $2|V|$ outlinks of V must be the same $2|V|$ inlinks of V. This shows that the subgraph induced by V is a connected component of X. As X is connected, V contains all nodes of X, that is, there is a path from v to any other node of V. Thus X is strongly connected.

In view of this lemma, we shall only use the term "connected" from now on.

3 The Extended Double Loop Network

Given a positive integer n and integers a, b, e, f, denote by $G(n; a, e; b, f)$ the directed graph with nodes being the elements of $\mathbb{Z}/\kappa\mathbb{Z}$ and the outlinks from a node i being $[i, ai+e]$ and $[i, bi+f]$.

Lemma 3.1 *$G(n; a, e; b, f)$ is 2-regular if and only if either $gcd(a,n) = gcd(b,n) = 1$ or $gcd(a,n) = gcd(b,n) = 2$ and $e - f$ is odd.*

Proof. By definition, there are two outlinks from each node. There are two inlinks at a node i if and only if the two equations

$$ax + e \equiv i \pmod{n} \quad \text{and} \quad bx + f \equiv i \pmod{n}$$

have two solutions modulo n altogether. If a is prime to n, then $ax + e \equiv i \pmod{n}$ has a unique solution for all i, and hence the same must be true for $bx + f \equiv i \pmod{n}$, this implies that b is also prime to n. We have shown that $gcd(n,a) = 1$ if and only if $gcd(n,b) = 1$. Next consider the case $gcd(n,a) = d > 1$. An equation

$ax + e \equiv i \pmod{n}$ has zero or d solutions modulo N. As it is solvable when $i = e$, we conclude that $d = 2$. As $gcd(n, b)$ is also > 1 in this case, we have $gcd(a, n) = gcd(b, n) = 2$. Further, for each i, exactly one of $ax + e \equiv i \pmod{n}$ and $bx + f \equiv i \pmod{n}$ is solvable if and only if e and f have opposite parity. This proves the lemma.

Call $G(n; a, e; b, f)$ an *extended double loop network* when it is *2-regular*. The special case $a = b = 1$ gives the usual double loop network $G_n(e, f)$. In this section we study the hamiltonian property of extended double loop networks.

First of all, observe the following. The translation by an integer k defines a bijection ϕ on $\mathbb{Z}/\varkappa\mathbb{Z}$, that is, $\phi(i) = i + k$ for all i. As $\phi(ai + e) = ai + e + k = a(i + k) + e + k(1 - a)$ and $\phi(bi + f) = b(i + k) + f + k(1 - b)$, ϕ yields an isomorphism from $G(n; a, e; b, f)$ to $G(n; a, e'; b, f')$, where $e' = e + k(1 - a)$ and $f' = f + k(1 - b)$. Let $d = gcd(e - f, a - b, n)$. Then $gcd\left(\frac{e-f}{d}, \frac{a-b}{d}, \frac{n}{d}\right) = 1$ and d divides $e' - f'$. Choose k to be the product of the prime factors of n/d which do not divide $(e - f)/d$. Then for such k we have $gcd\left(\frac{e'-f'}{d}, \frac{n}{d}\right) = gcd\left(\frac{e-f}{d} + k\frac{b-a}{d}, \frac{n}{d}\right) = 1$, in other words, $gcd(e' - f', n) = d$. This shows that among the extended double loop networks of the form $G(n; a, e'; b, f')$ which are isomorphic to $G(n; a, e; b, f)$, we can choose suitable e', f' such that $gcd(e' - f', n)$ divides $a - b$. Therefore *we shall always assume that* $gcd(e - f, n)$ *divides* $a - b$ for an extended double loop network $G(n; a, e; b, f)$.

Note that a hamiltonian digraph is necessarily connected. So we begin by investigating when an extended double loop network $G(n; a, e; b, f)$ is connected.

Lemma 3.2 *Suppose $gcd(n, a) = gcd(n, b) = 2$. Then the extended double loop network $G(n; a, e; b, f)$ is the line graph of $G(n/2; a, e; b, f)$.*

Proof. Denote the nodes of $G(n/2) = G(n/2; a, e; b, f)$ by $\bar{i}$, where $i \in \mathbb{Z}/(\varkappa/\not\varkappa)\mathbb{Z}$ and call the links $[\bar{i}, \overline{ai + e}]$ of type 1 and $[\bar{i}, \overline{bi + f}]$ of type 2. We shall label the n links of $G(n/2)$ by elements in $\mathbb{Z}/\varkappa\mathbb{Z}$. More precisely, the type 1 outlink from $\bar{i}$ is labelled $ai + e$ and the type 2 outlink labelled $bi + f$. Since $ai + e \equiv ai' + e \pmod{n}$ (resp. $bi + f \equiv bi' + f \pmod{n}$) implies $i \equiv i' \pmod{n/2}$ and since $e - f$ is odd and a, b, n are even, we have $ai + e \not\equiv bj + f \pmod{n}$ for all i, j, therefore no two links are labelled the same. Further, the labels of the two inlinks at the node $\bar{i}$ are congruent to $i \pmod{n/2}$ from construction, so they are i and $i + n/2$. The nodes of the line graph of $G(n/2)$ are the links of $G(n/2)$, which we identify with the elements of $\mathbb{Z}/\varkappa\mathbb{Z}$ via labelling, and the out-neighbors of a node i of the line graph are the outlinks at the node $\bar{i}$, namely, $ai + e$ and $bi + f$. This shows that the line graph of $G(n/2; a, e; b, f)$ is precisely $G(n; a, e; b, f)$.

Theorem 3.3 *Let $G(n; a, e; b, f)$ be an extended double loop network with n even. Let n' be the largest odd factor of n. Suppose that $gcd(a, n) = gcd(b, n) = 2$. Then the following three statements are equivalent.*

(1) $G(n; a, e; b, f)$ is hamiltonian,

(2) $G(n; a, e; b, f)$ is connected.

(3) $G(n'; a, e; b, f)$ is connected.

Proof. (3) $\Longrightarrow$ (1). Since $G(n';a,e;b,f)$ is 2-regular and connected, it is Eulerian. Its line graph, which is $G(2n';a,e;b,f)$ by Lemma 3.2, is hamiltonian and hence connected. If $n > 2n'$, then an iterative argument concludes that $G(n;a,e;b,f)$ is hamiltonian.

(1) $\Longrightarrow$ (2) is obvious.

(2) $\Longrightarrow$ (3) If $G(n';a,e;b,f)$ is not connected, then its line graph $G(2n';a,e;b,f)$ is also not connected. Again, an iterative argument shows that $G(n;a,e;b,f)$ is not connected.

The reader is referred to [16] for a similar, but more general situation.

Given an extended double loop network $G(n;a,e;b,f)$, denote by

$$n' = \begin{cases} n \text{ if } gcd(n,a) = gcd(n,b) = 1 \ , \\ \text{the largest odd factor of } n \text{ if } gcd(n,a) = gcd(n,b) = 2 \text{ and } e-f \text{ is odd} \ . \end{cases}$$

Note that $gcd(n',a) = gcd(n',b) = 1$ and $gcd(e-f,n) = gcd(e-f,n')$.

Theorem 3.4 *Let $G(n;a,e;b,f)$ be an extended double loop network with $d = gcd(e-f,n)$. Assume that $G(n';a,e;b,f)$ is $\frac{n'}{d}$-uniform. Then $G(n;a,e;b,f)$ is connected if and only if $G(d;a,e)$ is hamiltonian.*

Here $G(m;k,h)$ denotes the directed graph with nodes being the elements of $\mathbb{Z}/{>}\mathbb{Z}$ and the outlink at a node i being $[i, ki+h]$.

Proof. By Theorem 3.3, $G(n;a,e;b,f)$ is connected if and only if $G(n';a,e;b,f)$ is. Hence it suffices to prove the theorem for the case a, b both prime to n. Write b' for the integer between 1 and n such that $bb' \equiv 1 \pmod{n}$, and let $c = ab'$.

Suppose nodes x and y have outlinks to the same node, say, $ax + e \equiv by + f \pmod{n}$, then x and y are in the same connected component and they are related by $y \equiv b'ax + b'(e-f) \equiv cx + b'(e-f) \pmod{n}$. Therefore the nodes in $S(x) = \{x, cx + (e-f)b', c^2x + (e-f)b'(c+1), \ldots, c^m x + (e-f)b'(c^{m-1} + \cdots + c + 1), \ldots\}$ are in the same connected component. Since $c \equiv 1 \pmod{d}$ and $e \equiv f \pmod{d}$, all nodes in $S(x)$ are congruent to x modulo d. This together with the uniformity assumption implies that $S(x)$ consists of the n/d nodes of the network which have the same residue as x when divided by d. Further, $G(n;a,e;b,f)$ is connected if and only if the new graph obtained from $G(n;a,e;b,f)$ modulo d, that is, $G(d;a,e)$, is connected (since $b \equiv a \pmod{d}$ and $f \equiv e \pmod{d}$), or equivalently, $G(d;a,e)$ is hamiltonian. This proves the theorem.

The following algebraic criterion for the uniformity of an extended double loop network is proved in [5].

Theorem 3.5 *Let $G(n;a,e;b,f)$ be an extended double loop network with $gcd(a,n) = gcd(b,n) = 1$ and $d = gcd(e-f,n)$. Then $G(n;a,e;b,f)$ is n/d-uniform if and only if every prime factor of n/d divides $(a-b)/d$ and 4 divides $a-b$ whenever it divides n/d.*

Observe that the algebraic condition is trivially satisfied when $a = b$. In particular, all the usual double loop networks $G_n(h_1, h_2) = G(n; 1, h_1; 1, h_2)$ are n/d-uniform, where $d = gcd(h_1 - h_2, n)$.

A criterion for $G(m; k, h)$ to be hamiltonian is given in [4], which we now recall. Let $\delta = gcd(k-1, m)$. Then $G(m; k, h)$ is hamiltonian if and only if (1) h is prime to m, (2) δ and m have the same prime factors, (3) 4 divides δ whenever it divides m. Hence a necessary condition for $G(n; a, e; b, f)$ to be connected is $gcd(e, d) = gcd(e, gcd(e-f, n)) = gcd(e, f, n) = 1$. When $a = b = 1$, this condition is also sufficient as $a - 1 = 0$ satisfies the other requirements. We have shown

Corollary 3.6 (van Doorn [18]). *$G_n(h_1, h_2)$ is connected if and only if $gcd(h_1, h_2, n) = 1$.*

Other immediate consequences of Theorem 3.4 are

Corollary 3.7 *An extended double loop network $G(n; a, e; b, f)$ is connected whenever $e - f$ is prime to n and $G(n'; a, e; b, f)$ is n'-uniform.*

Corollary 3.8 *Suppose n is square free. An extended double loop network $G(n; a, e; a, f)$ is connected if and only if $a \equiv 1 \pmod d$ and $gcd(e, f, n) = 1$, where $d = gcd(e - f, n)$.*

Let $G = G(n; a, e; b, f)$ be an extended double loop network with $gcd(n, a) = gcd(n, b) = 1$. Then G is closely related to a Cayley graph as follows. Denote by

$$A = \begin{pmatrix} a & e \\ 0 & 1 \end{pmatrix}, \quad B = \begin{pmatrix} b & f \\ 0 & 1 \end{pmatrix},$$

and by T the subgroup of $GL_2(\mathbb{Z}/\kappa\mathbb{Z})$ generated by A and B. Assume that the network G is connected. Identify the node i with the column vector $\binom{i}{1}$. Then the two outlinks at the node $\binom{i}{1}$ are $\left[\binom{i}{1}, A\binom{i}{1}\right]$ and $\left[\binom{i}{1}, B\binom{i}{1}\right]$. The group T acts on the nodes of G via matrix multiplication, that is, a matrix M in T sends the node $\binom{i}{1}$ to the node $M\binom{i}{1}$. It acts transitively since G is strongly connected. The stabilizer of the node $\binom{0}{1}$ is the subgroup H consisting of the diagonal matrices $\begin{pmatrix} d & 0 \\ 0 & 1 \end{pmatrix}$ in T. Hence the vertices of G may be identified with the right coset space T/H. As such, G can be viewed as a quotient graph of the Cayley graph $Cay(T; A, B)$ on T with generators A, B. If H is a normal subgroup of T, then G is also a Cayley graph. Denote by C the cyclic subgroup of T generated by $B^{-1}A$. It is the "arc-forcing subgroup" of the Cayley digraph $Cay(T; A, B)$ arising from considering outlinks as in §2. More precisely, for any node $x \in T$, the set $S(x)$ defined in §2 is the left coset Cx. Therefore the set $S(i)$ for a node i in G is a double coset CxH viewed as a union of right H cosets. Let D be the cyclic subgroup of T generated by AB^{-1}. It is the "arc-forcing subgroup" of $Cay(T; A, B)$ arising from considering inlinks instead of outlinks.

Lemma 3.9 *C is a normal subgroup of T if and only if $C = D$, i.e.,*

$$\langle B^{-1}A\rangle = \langle A^{-1}B\rangle = \langle AB^{-1}\rangle = \langle BA^{-1}\rangle .$$

Proof. Since $AB^{-1} = B(B^{-1}A)B^{-1}$ is conjugate to $B^{-1}A$, if C is normal, then $C = BCB^{-1} = D$. Conversely, suppose $C = D$. We want to show that C is a normal subgroup of T. It amounts to showing that $B^iCB^{-i} = C$ and $A^iCA^{-i} = C$ for all integers $i \geq 0$. Due to the symmetry in A and B, it suffices to prove the first statement. ¿From $B^iB^{-1}AB^{-i} = B^{i-1}(AB^{-1})B^{-(i-1)}$ it follows that $B^iCB^{-i} = B^{i-1}CB^{-(i-1)} = \cdots = C$.

If C is a normal subgroup of T, then $C \setminus T/H = T/CH$ and G is $\frac{|CH|}{|H|}$-uniform. Conversely, we shall see that the uniformity condition is weaker than the normality of C. Indeed, by the Lemma above, C is normal if and only if

$$B^{-1}A = \begin{pmatrix} b' & -fb' \\ 0 & 1 \end{pmatrix}\begin{pmatrix} a & e \\ 0 & 1 \end{pmatrix} = \begin{pmatrix} c & b'(e-f) \\ 0 & 1 \end{pmatrix}$$

is a power of

$$AB^{-1} = \begin{pmatrix} a & e \\ 0 & 1 \end{pmatrix}\begin{pmatrix} b' & -fb' \\ 0 & 1 \end{pmatrix} = \begin{pmatrix} c & e-cf \\ 0 & 1 \end{pmatrix} .$$

A direct computation shows that $B^{-1}A = (AB^{-1})^i$ if and only if

$$\begin{cases} c^{i-1} \equiv 1 \pmod{n} , \\ (c^{i-1} + \cdots + c + 1)(e - cf) \equiv b'(e - f) \pmod{n} . \end{cases} \qquad (*)$$

Suppose G is n/d-uniform. It follows from the algebraic criterion above that $gcd(e - cf, n) = gcd((e - f) + (1 - c)f, n) = d$ and the system $(*)$ is equivalent to

$$\begin{cases} \dfrac{c-1}{d}(c^{i-2} + \cdots + c + 1) \equiv 0 \left(\bmod \dfrac{n}{d}\right) , & \text{(a)} \\[2ex] (c^{i-2} + \cdots + c + 1)\dfrac{e - cf}{d} \equiv b'\dfrac{e - f}{d} - \dfrac{e - cf}{d} \left(\bmod \dfrac{n}{d}\right) , & \text{(b)} \end{cases}$$

By Theorem'3.5, the n/d numbers $1, c+1, c^2+c+1, \ldots, c^n/d-1+\cdots+c+1$ represent all the residues modulo n/d. Thus equation (b) has a unique solution mod n/d. Let i be a solution to (b). Then i satisfies (a) if and only if

$$\frac{c-1}{d}\left(b'\frac{e-f}{d} - \frac{e-cf}{d}\right) \equiv 0 \left(\bmod \frac{n}{d}\right) .$$

We have shown

Theorem 3.10 *Suppose that $G(n; a, e; b, f)$ is n/d-uniform, where $gcd(a, n) = gcd(b, n) = 1$ and $gcd(e - f, n) = d$. Then the cyclic group $C = \langle B^{-1}A\rangle$ is a normal subgroup of $T = \langle A, B\rangle$ if and only if*

$$\frac{c-1}{d}\left(\frac{e-f}{d} - \frac{e-cf}{d}b\right) \equiv 0 \left(\bmod \frac{n}{d}\right) , \qquad \text{(c)}$$

where $cb \equiv a (\bmod\ n)$.

In particular, when $a = b$, $G(n; a, e; a, f)$ is n/d-uniform and the group C is always a normal subgroup of T. We exhibit one example below to show that uniformity does not imply normality.

Example. Let p be an odd prime and $n = p^2$. Choose e, f such that $e - f = 1$. Then $d = 1$. Choose $c = p + 1$. Equation (c) in this case reads

$$p(1 - b + pfb) \equiv 0 \pmod{p^2},$$

which is equivalent to

$$1 \equiv b \pmod{p}.$$

If we choose $b \not\equiv 0, 1 \pmod{p}$ and $a \equiv bc \equiv b(1 + p) \pmod{p^2}$, then $G(p^2; a, e; b, f)$ is p^2-uniform, while $C = \langle B^{-1}A\rangle$ is not a normal subgroup of T.

Suppose C is a normal subgroup of T. Then the Cayley graph $Cay(T; A, B)$ is $|C|$-uniform and every path in the graph with length $|T|/|C|$ is basic. Thus, by Theorem 2.1, the Cayley graph is hamiltonian if and only if there is a word W in A, B of length $|T|/|C|$ which generates the group C. This is Rankin's criterion. The corresponding extended double loop network $G(n; a, e; b, f)$ is hamiltonian if and only if $G(d; a, e)$ is hamiltonian and there is a word W in A, B of length $|T|/|CH|$ such that W^m lies in H for $m = |CH|/|H|$, but not any $1 \leq m < |CH|/|H|$. In particular, if H is a subgroup of C, then the hamiltonian property of the extended double loop network follows from that of the Cayley graph.

We now turn to the general situation. Let $G = G(n; a, e; b, f)$ be an extended double loop network with $d = gcd(e - f, n)$ such that $G(n'; a, e; b, f)$ is n'/d-uniform. Suppose that G is connected, we want to investigate when it is hamiltonian. In view of Theorem 3.3, it suffices to consider the case $gcd(n, a) = gcd(n, b) = 1$, which we shall assume. By Theorem 3.4, every path in G of length d is basic. Therefore G is hamiltonian if and only if there is a path of length d with period n/d. As seen before, a path P in G of length d corresponds to a word $M = \begin{pmatrix} \alpha & \beta \\ 0 & 1 \end{pmatrix}$ in A, B of length d. Note that d divides β since P is basic. Further, P has period n/d if and only if the smallest positive integer m such that M^m lies in H, the stabilizer of $\binom{0}{1}$ in $T = \langle A, B\rangle$, is $m = n/d$. The uniformity assumption on G implies that for any node x in G, the set $S(x)$ consists of the nodes in G which are congruent to x modulo d. The matrix M acts on $S(x)$ for any x. It has period n/d if and only if this action is transitive. The latter can be easily checked algebraically as follows. The action of M on $S(0) = \{jd : 0 \leq j < n/d\}$ defines a directed graph G' on $S(0)$ with the out-neighbor from jd being $\alpha jd + \beta$ since $M\binom{jd}{1} = \binom{\alpha jd+\beta}{1}$. Clearly, M is transitive on $S(0)$ if and only if G' is hamiltonian. After dividing by the common factor d on nodes of G', the graph G' is nothing but the graph $G(n/d; \alpha, \beta/d)$. Therefore G is hamiltonian if and only if there is a word $M = \begin{pmatrix} \alpha & \beta \\ 0 & 1 \end{pmatrix}$ in A, B of length d such that $G(n/d; \alpha, \beta/d)$ is hamiltonian. Using the algebraic criterion for the hamiltonicity of $G(m; k, h)$, we summarize the above discussion in

Theorem 3.11 *Let $G = G(n; a, e; b, f)$ be an extended double loop network with $d = gcd(e - f, n)$. Suppose that $G(n'; a, e; b, f)$ is n'/d-uniform. Then G is hamiltonian if and only if*

(1) $gcd(e, f, n) = 1$, $gcd(a - 1, d) =: d'$ and d have the same prime factors, and 4 divides d' whenever 4 divides d;

(2) in case $gcd(n, a) = gcd(n, b) = 1$, i.e., $n' = n$, there is a word $M = \left(\begin{smallmatrix} \alpha & \beta \\ 0 & 1 \end{smallmatrix}\right)$ in $A = \left(\begin{smallmatrix} a & e \\ 0 & 1 \end{smallmatrix}\right)$ and $B = \left(\begin{smallmatrix} b & f \\ 0 & 1 \end{smallmatrix}\right)$ of length d such that $gcd(\beta, n) = d$, $gcd(\alpha - 1, n/d) =: d''$ and n/d have the same prime factors and 4 divides d'' whenever 4 divides n/d.

Further, the number of ways of getting M satisfying (2) is the number of hamiltonian circuits in G in the case $gcd(n, a) = gcd(n, b) = 1$.

When $a = b = 1$, the matrix $M = \left(\begin{smallmatrix} \alpha & \beta \\ 0 & 1 \end{smallmatrix}\right)$ has the form $\left[\begin{smallmatrix} 1 & ke+(d-k)f \\ 0 & 1 \end{smallmatrix}\right]$ for some $0 \le k \le d$, and the above theorem simplifies as

Corollary 3.12 *$G_n(h_1, h_2)$ is hamiltonian if and only if $gcd(h_1, h_2, n) = 1$ and there is an integer $0 \le k \le d$, where $d = gcd(h_1 - h_2, n)$, such that $gcd(kh_1 + (d-k)h_2, n) = d$.*

Note that $G_n(h_1, h_2)$ with $gcd(h_1 - h_2, n) > 1$ is not always hamiltonian, for instance, $G_30(5, 3)$ is hamiltonian, while $G_{60}(5, 3)$ is not.

Corollary 3.13 *Let $G = G(n; a, e; b, f)$ be an extended double loop network with $gcd(a, n) = gcd(b, n) = gcd(e - f, n) = 1$. Then G is hamiltonian if and only if at least one of $G(n; a, e)$, $G(n; b, f)$ is hamiltonian.*

Corollary 3.14 *Let $G = G(n; a, e; b, f)$ be an extended double loop network with $gcd(a, n) = gcd(b, n) = 2$ and $gcd(e - f, n) = 1$. Then G is hamiltonian.*

Both generalized deBruijn network and Imase-Itoh network have $gcd(e-f, n) = 1$. The results on these networks given in [1, 2] are special cases of Corollaries 3.13 and 3.14.

Remark. When the extended double loop network $G = G(n; a, e; b, f)$ is a Cayley graph, the criterion for G being hamiltonian obtained above is the same as that for a Cayley graph as studied in [13, 16, 17]. This is the case when $a = b = 1$ in particular. On the other hand, there are extended double loop networks which are not Cayley graphs, and we can still determine if they are hamiltonian using the theorem above. We exhibit one example below.

Example. One checks easily that $G = G(18; 11, 5; 11, 7)$ is an extended double loop network with $d = 2$ and $gcd(11, 18) = 1$. As $G(2; 11, 5)$ is hamiltonian, the network is connected. Let

$$A = \begin{pmatrix} 11 & 5 \\ 0 & 1 \end{pmatrix} \quad \text{and} \quad B = \begin{pmatrix} 11 & 7 \\ 0 & 0 \end{pmatrix} .$$

One finds $A^2 \equiv \left(\begin{smallmatrix}13 & 6\\ 0 & 1\end{smallmatrix}\right)$, $B^2 \equiv \left(\begin{smallmatrix}13 & 12\\ 0 & 1\end{smallmatrix}\right)$, $AB \equiv \left(\begin{smallmatrix}13 & 10\\ 0 & 1\end{smallmatrix}\right)$ and $BA \equiv \left(\begin{smallmatrix}13 & 8\\ 0 & 1\end{smallmatrix}\right)$ (mod 18). Since $gcd(13-1, 18/2) = 3$ and $9 = 18/2$ have the same prime factors and $gcd(6, 18) = 6$, $gcd(12, 18) = 6$, $gcd(10, 18) = 2$, $gcd(8, 18) = 2$, by Theorem 3.5, the network is hamiltonian with 2 hamiltonian circuits: one uses $[i, 11i + 7]$ at i even and $[i, 11i + 5]$ at i odd, and the other uses $[i, 11i + 7]$ at i odd and $[i, 11 + 5]$ at i even.

On the other hand, a Cayley graph is node symmetric. We shall show that G is not a Cayley graph by proving that it has only 6 cycles of length 2. Because then there are nodes with cycles of length 2 and there are nodes without cycles of length 2, and hence G is not node symmetric. Indeed, to find nodes at which there is a cycle of length 2 it suffices to solve

$$A^2\begin{pmatrix} i \\ 1 \end{pmatrix} \equiv \begin{pmatrix} i \\ 1 \end{pmatrix}, \quad B^2\begin{pmatrix} i \\ 1 \end{pmatrix} \equiv \begin{pmatrix} i \\ 1 \end{pmatrix}, \quad AB\begin{pmatrix} i \\ 1 \end{pmatrix} \equiv \begin{pmatrix} i \\ 1 \end{pmatrix} \quad \text{and} \quad BA\begin{pmatrix} i \\ 1 \end{pmatrix} \equiv \begin{pmatrix} i \\ 1 \end{pmatrix} \pmod{18}.$$

One verifies easily that each of the first 2 equations has 6 solutions, while the last two equations have none. This shows that G has only 6 cycles of length 2. By the way, this also explains why $G(18; 11, 5)$ and $G(18; 11, 7)$ are not hamiltonian.

In the example above, the Cayley graph $Cay(T; A, B)$ is also hamiltonian. It is unknown to us if the hamiltonian property of G is related to that of the associated Cayley graph when A, B are invertible.

4 Semi-Torus Networks

Let r, c be integers ≥ 2 and let $T(r, c)$ denote an $r \times c$ rectangular grid lying on the surface of a torus with $n = rc$ nodes labeled by the grid points (i, j) with i modulo r and j modulo c. Suppose that the row links are unidirectional, while the links on each column are unidirectional, but the directions may vary from column to column. We call $T(r, c)$ a semi-torus network. When the columns are also unidirectional, it is the familiar torus network. The directions on columns can be described by a ± 1 sequence $\{\epsilon_0, \epsilon_1, \ldots, \epsilon_{c-1}\}$ so that columns with the same direction have the same sign, and the opposite directions with opposite signs. The functions $g_1(i, j) = (i, j + 1)$ and $g_2(i, j) = (i + \epsilon_j, j)$ describe the two types of links on $T(r, c)$. Extend the definition of ϵ_k to integers k such that $\epsilon_k = \epsilon_j$, where j is the remainder of k divided by c. For any node (a, b) of $T(r, c)$, $g(a, b) = g_1^{-1} \circ g_2(a, b) = (a + \epsilon_b, b - 1)$, and consequently,

$$S_1(a, b) = \{(a + \epsilon_b + \epsilon_{b-1} + \cdots + \epsilon_{b-i+1}, b - i) : i = 0, 1, \ldots, m - 1\},$$

where m is the smallest positive integer k such that $b \equiv b - k \pmod{c}$ and $a + \epsilon_b + \epsilon_{b-1} + \cdots + \epsilon_{b-k+1} \equiv a \pmod{r}$. The first equation implies $m = cq$ for some integer q. Substituting it into the second equation yields

$$\epsilon_b + \epsilon_{b-1} + \cdots + \epsilon_{b-cq+1} = q(\epsilon_0 + \epsilon_1 + \cdots + \epsilon_{c-1}) = qs_c \equiv 0 \pmod{r},$$

where $s_c = \epsilon_0 + \epsilon_1 + \cdots + \epsilon_{c-1}$. Let $d = gcd(r, s_c)$. Then the second equation means that q is a multiple of r/d. Thus $m = rc/d = n/d$. We have shown

Lemma 4.1 *$T(r,c)$ is n/d-uniform.*

Corollary 4.2 *If $r, c \geq 2$ and $d = gcd(r, s_c) = 1$, then $T(r,c)$ is not hamiltonian.*

Theorem 4.3 *Let $s_c = \epsilon_0 + \epsilon_1 + \cdots + \epsilon_{c-1}$ and let $d = gcd(r, s_c)$. If $d = 2$, then every path in $T(r,c)$ of length 2 is basic. When $d \geq 3$, every path in $T(r,c)$ of length d is basic if and only if $\epsilon_0 = \epsilon_1 = \cdots = \epsilon_{c-1}$, that is, all columns are unidirectional.*

Proof. Let p be a path in $T(r,c)$ starting at the node (a,b). Its ending node has the form $(a + i_0\epsilon_b + i_1\epsilon_{b+1} + \cdots + i_k\epsilon_{b+k}, b+k)$ for some nonnegative integers $i_0, i_1, \ldots, i_k$, k with $i_0 + i_1 + \cdots + i_k + k = d$, and the intermediate nodes on p are represented by $(a + i_0\epsilon_b + \cdots + i_{t-1}\epsilon_{b+t-1} + i'_t\epsilon_{b+t},\ b+t)$, where $0 \leq i'_t \leq i_t$, $0 \leq t \leq k$, and $i_0 + \cdots + i_{t-1} + i'_t + t$ runs from 1 to $d-1$. Let $(a + i_0\epsilon_b + \cdots + i_{t-1}\epsilon_{b+t-1} + i'_t\epsilon_{b+t}, b+t)$ and $(a + i_0\epsilon_b + \cdots + i_{u-1}\epsilon_{b+u-1} + i''_u\epsilon_{b+u}, b+u)$ be two nodes on p with $d-1 \geq u + i''_u + i_{u-1} + \cdots + i_0 - (t + i'_t + i_{t-1} + \cdots + i_0) \geq 1$. Then either $u > t$ or else $u = t$ and $i''_u > i'_t$. We study the solvability of the following system of two congruence equations:

$$\begin{cases} b + t \equiv b + u - j \pmod{c}\ , \\ a + i_0\epsilon_b + \cdots + i_{t-1}\epsilon_{b+t-1} + i'_t\epsilon_{b+t} \equiv a + i_0\epsilon_b + \cdots + i_{u-1}\epsilon_{b+u-1} \\ \qquad + i''_u\epsilon_{b+u} + \epsilon_{b+u} + \epsilon_{b+u-1} + \cdots + \epsilon_{b+u-j+1} \pmod{r}\ . \end{cases}$$

¿From the first equation we get $t = u - j + qc$ for some integer q. Assume first that $u = t$, that is, the two nodes are on the same column. Then $j = qc$ and the second equation becomes $(i''_u - i'_u)\epsilon_{b+u} + \epsilon_{b+u} + \epsilon_{b+u-1} + \cdots + \epsilon_{b+u-qc+1} = (i''_u - i'_u)\epsilon_{b+u} + qs_c \equiv 0 \pmod{r}$. Since $d - 1 \geq i''_u - i_u > 0$, this equation is not solvable for q so the two nodes on the same column are not in the same class. Next we discuss the case $u > t$. In this case $j = u - t + qc$ and the second equation becomes

$$\begin{aligned} &(i_t - i'_t)\epsilon_{b+t} + i_{t+1}\epsilon_{b+t+1} + \cdots + i_{u-1}\epsilon_{b+u-1} + i''_u\epsilon_{b+u} + \epsilon_{b+u} + \epsilon_{b+u-1} + \cdots + \epsilon_{b+t-qc+1} \\ &= (i_t - i'_t)\epsilon_{b+t} + i_{t+1}\epsilon_{b+t+1} + \cdots + i_{u-1}\epsilon_{b+u-1} + i''_u\epsilon_{b+u} + qs_c + \epsilon_{b+u-qc} \\ &\qquad + \epsilon_{b+u-qc-1} + \cdots + \epsilon_{b+t-qc+1} \\ &= (i_t - i'_t)\epsilon_{b+t} + (i_{t+1}+1)\epsilon_{b+t+1} + \cdots + (i_{u-1}+1)\epsilon_{b+u-1} + (i''_u + 1)\epsilon_{b+u} + qs_c \\ &= M + qs_c \equiv 0 \pmod{r}\ , \end{aligned}$$

where

$$M = (i_t - i'_t)\epsilon_{b+t} + (i_{t+1}+1)\epsilon_{b+t+1} + \cdots + (i_{u-1}+1)\epsilon_{b+u-1} + (i''_u + 1)\epsilon_{b+u}\ .$$

As

$$|M| \le i_t - i'_t + (i_{t+1}+1) + \cdots + (i_{u-1}+1) + (i''_u + 1)$$

$$= (i_0 + \ldots + i_{u-1} + i''_u + u) - (i_0 + \cdots + i_{t-1} + i'_t + t) \le d-1 ,$$

the system has a solution, that is, the two nodes are in the same class, if and only if $M = 0$. When $\epsilon_0 = \cdots = \epsilon_{c-1}$, we have $M \neq 0$ and hence the path p is basic. Conversely, assume all paths of length $d \geq 3$ are basic, we want to show that $\epsilon_0 = \cdots = \epsilon_{c-1}$. Suppose otherwise, say, $\epsilon_{b+1} = -\epsilon_b$ for some $0 \leq b \leq c-2$. Consider a path p of length d starting at $(0,b)$, going to (ϵ_b, b), then to $(\epsilon_b, b+1)$ and onward to other nodes, so that $i_0 = 1$. Then $(0,b)$ and $(\epsilon_b, b+1)$ are in the same class since $M = (i_0 - 0)\,\epsilon_{b+0} + (0+1)\epsilon_{b+1} = 0$. Furthermore, $(\epsilon_b, b+1)$ is not the ending node of p since p has length ≥ 3. This shows that p is not basic.

Finally, we are left with the case $d = 2$. We have shown that two consecutive nodes (a,b) and $(a+\epsilon_b, b)$ on the same column are not in the same class. Next consider any two consecutive nodes on a row, say, (a,b) and $(a,b+1)$. Then $u = 1$, $t = 0$ and $M = \epsilon_{b+1}$ has absolute value 1. Again (a,b) and $(a,b+1)$ are not in the same class. This shows that in case $d = 2$ any path of length d is basic.

Concerning the existence of a hamiltonian circuit in a semi-torus network, we have the following result:

Theorem 4.4 *Suppose $r, c \geq 2$. Let $s_c = \epsilon_0 + \cdots + \epsilon_{c-1}$ and $d = gcd(s_c, r)$.*

(1) If $d = 1$, then $T(r,c)$ is not hamiltonian.

(2) If $d = 2$, then $T(r,c)$ is hamiltonian.

(3) If $d \geq 3$ and $T(r,c)$ is unidirectional in both rows and columns, then it is hamiltonian if and only if there exists an integer $k, 1 \leq k \leq d-1$, such that $gcd(k,r) = 1$ and $gcd(d-k, c) = 1$.

Proof. The first statement is Corollary 4.2. When $d = 2$, we know that any path in $T(r,c)$ of length 2 is basic by Theorem 4.3. Consider the path p from the node $(0,0)$ through the node $(0,1)$ to the node $(\epsilon_1, 1)$. It determines a factor F of $T(r,c)$. We shall show that p has period $n/2 = rc/d$ in F. As we have seen in the proof of Theorem 7 above, in $T(r,c)$ adjacent nodes are not in the same class. So the nodes lying on the path in F containing p belonging to the same class as $(0,0)$ have the form $(\epsilon_1 + \epsilon_2 + \cdots + \epsilon_i, i)$, and the period of p is the smallest positive solution to the system

$$\begin{cases} i \equiv 0 & \pmod{c} , \\ \epsilon_1 + \cdots + \epsilon_i \equiv 0 & \pmod{r} . \end{cases}$$

which is rc/d by a simple computation.

Finally, when rows and columns are unidirectional, $T(r,c)$ is the usual torus network. As any path of length k is basic by Theorem 4.3, a similar argument proves the third statement. However, the network is also the Cartesian product of two circuits of lengths r and c. The necessary and sufficient condition for such a network to be hamiltonian as described in (3) has been implied in the work of Rankin [13] and proved by Trotter and Erdös [17]. So no detail is given here.

While we do not have a general criterion for the case $d \geq 3$ and columns not unidirectional, the following construction gives a hamiltonian circuit for $T(6,9)$ with $d = 3$:

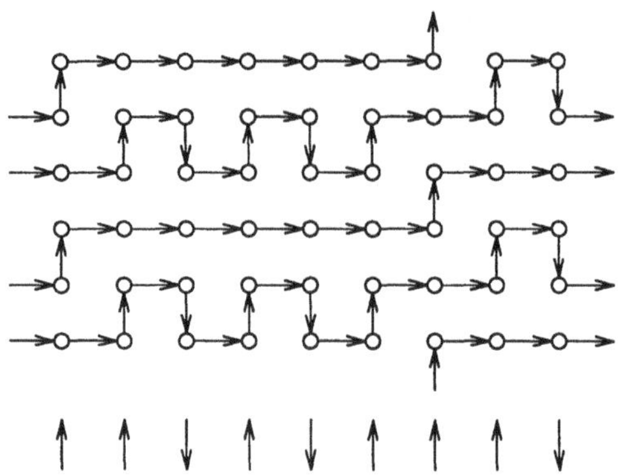

Figure 1: A Hamiltonian Circuit of $T(6,9)$

5 Semi-Manhattan Networks

A mesh with r rows and c columns is called a semi-Manhattan network $M(r,c)$ if all (except possibly one) rows alternate in direction but the columns have arbitrary directions. We are interested in the existence of a hamiltonian circuit in $M(r,c)$.

Observe from Corollary 4.2 that when r is odd, $r \geq 3$, and columns are unidirectional, there is no hamiltonian circuit in $M(r,c)$. The following theorem says that this is the only exception.

Theorem 5.1 *$M(r,c)$ is hamiltonian except when $r \geq 3$ is odd and the columns are unidirectional.*

Proof. Theorem 5.1 is easily checked when $r \leq 2$ or $c \leq 2$. So assume $r \geq 3$ and $c \geq 3$ and we shall prove the theorem by construction. In all the following constructions, the orientation of the hamiltonian circuit (clockwise or counterclockwise) depends on the direction of the last column and an arc in a figure represents a single edge.

Case 1. There are three adjacent columns alternating in direction. Without loss of generality, assume that these are the first two and the last columns. For even r we use the vertical comb in Fig. 2(a) and for odd r we use the vertical comb in Fig. 2(b).

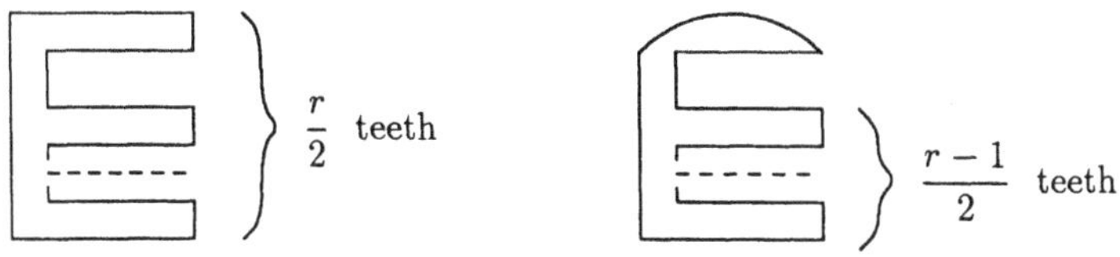

(a) (b)

Figure 2: Combs as Hamiltonian Circuits

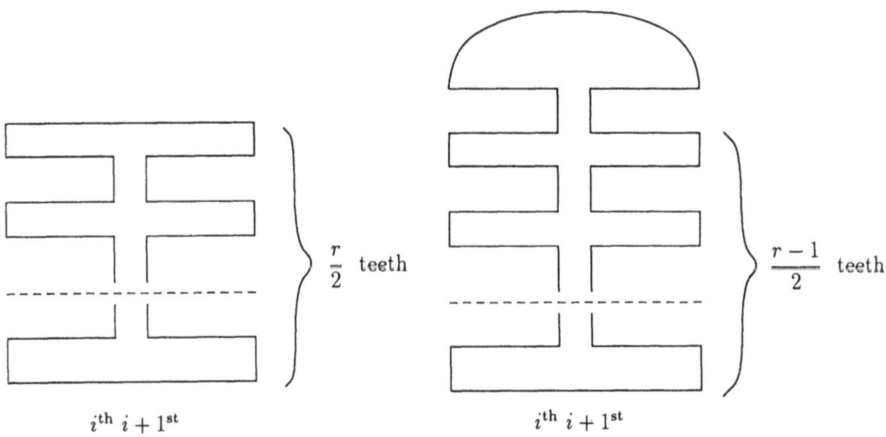

(a) (b)

Figure 3: 2-Sided Combs as Hamilton Circuits

Case 2. There are two pairs of adjacent columns, say, the first, the last, the ith and the $i+1^{\text{st}}$ such that the first and the ith have one direction, while the $i+1^{\text{st}}$ and the last have the other. For even r we use the 2-sided comb in Fig. 3(a) and for odd r the 2-sided comb in Fig. 3(b).

Case 3. r is even and there are two adjacent columns with the same direction, say, the first and the last. We use a different comb as shown in Fig. 4.

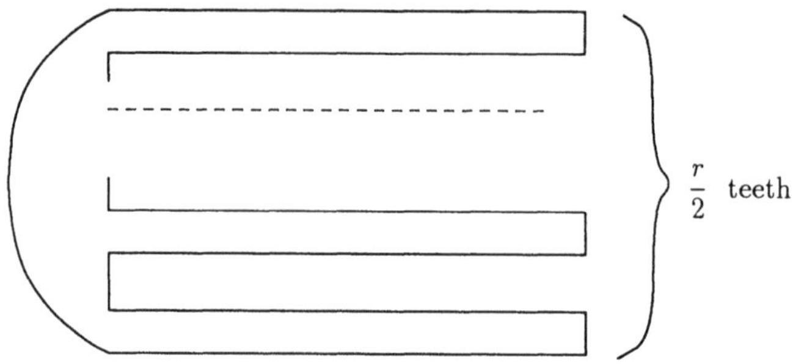

Figure 4: A Different Comb as a Hamiltonian Circuit

In general, when the columns are not unidirectional, partition the columns into consecutive runs of columns in the same direction. If there exists a run of length 1, then it falls in Case 1; if every run is of length at least 2, then it falls in Case 2. When all the columns are unidirectional, we have seen that $M(r,c)$ is not hamiltonian for odd r, for even r this is Case 3. The proof of Theorem 5.1 is now completed.

Corollary 5.2 *The Manhattan street network is hamiltonian.*

This is included in Case 1 above.

References

[1] D. Z. Du and F. K. Hwang, Generalized deBruijn digraphs, *Networks*, **18** (1988) 27–38.

[2] D. Z. Du, D. F. Hsu, F. K. Hwang and X. M. Zhang, The hamiltonian property of generalized deBruijn digraphs, *J. Combin. Theory, series B*, **52** (1991) 1–8.

[3] L. J. Guibas, H. T. Kung, and C. D. Thompson, Direct VLSI implementation of combinatorial algorithms, *Cal. Tech. Conf. VLSI*, Pasadena, Ca., Jan. 1979.

[4] F. K. Hwang, The hamiltonian property of linear functions, *OR Letters*, **6** (1987) 293–296.

[5] F. K. Hwang and W.-C. W. Li, 2-connectivities of extended double loop networks, submitted.

[6] M. Imase and M. Itoh, Design to minimize diameter on building-block network, *IEEE Trans. Comput.*, **C-30** (1981) 439–442.

[7] M. Imase and M. Itoh, A design for directed graphs with minimum diameters, *IEEE Trans. Comput.*, **C-32** (1983) 782–784.

[8] W. H. Kautz, K. N. Levitt and A. Waksman, Cellular interconnection arrays, *IEEE Trans. Comput.*, **C-17** (1968) 443–451.

[9] W. Liu, T. H. Hildebrandt and R. Cavin, III, Hamiltonian cycles in the shuffle-exchange network, *IEEE Trans. Comput.*, **C-38** (1989) 745–750.

[10] A. J. Martin, The TORUS: An exercise in constructing a processing surface, *Proc. VLSI Conf.* 1981.

[11] N. F. Maxemchuk, Regular mesh topologies in local and metropolitan area networks, *AT&T Tech. J.*, **65** (1985) 1659–1687.

[12] D. K. Pradhan, Fault-tolerant multiprocessor and VLSI-based systems communication architecture in *Fault Tolerant Computing Theory and Techniques*, Vol. II (ed. D. K. Pradhan), (Prentice-Hall, Englewood Cliff, NJ, 1986).

[13] R. A. Rankin, A campanological problem in group theory, *Proc. Camb. Phil. Soc.*, **44** (1948) 17–25.

[14] S. M. Reddy, D. K. Pradhan and J. G. Kuhl, Direct graphs with minimum diameter and maximal connectivity, School of Eng., Oakland Univ. Tech. Rep., July 1980.

[15] C. S. Raghavendra, M. Gerla and A. Avizienis, Reliable loop topologies for large local computer networks, *IEEE Trans. Comput.*, **C-34** (1985) 46–54.

[16] O. Serra, M. Mora and M. A. Fiol, On c-circulant digraphs, in *Proc. Combinatorics '88*, Ravello (Italy), F. Rocca and F. Mazzocca, eds., (Mediterranean Press, 1991) pp. 371–387.

[17] W. T. Trotter, Jr. and P. Erdős, When the Cartesian product of directed cycles is hamiltonian, *J. Graph Theory*, **2** (1978) 137–142.

[18] E. A. van Doorn, Connectivity of circulant digraphs, *J. Graph Theory*, **10** (1986) 9–14.

[19] D. Witte and J. A. Gallian, A survey: hamiltonian cycles in Cayley graphs, *Disc. Math.*, **51** (1984) 293–304.

Network Optimization Problems, pp. 111-123
Eds. D.-Z. Du and P.M. Pardalos

Equivalent Formulations for the Steiner Problem in Graphs

Bassam N. Khoury
Panos M. Pardalos
Donald W. Hearn
Department of Industrial and Systems Engineering, University of Florida, Gainesville, FL 32611 USA

Abstract

In the following paper, we present a collection of known and new formulations for the Steiner problem in undirected and directed graphs.
Keywords: Steiner problem, graphs, directed graphs, integer formulations, mixed integer formulations, continuous formulations.

1 Introduction

Providing an appropriate formulation for a problem is a crucial step toward solving it. In addition, different formulations can be used to obtain insight of the properties and the complexity of a problem. In this paper, we present a collection of known and new mathematical formulations for the Steiner problem in graphs (SPG), a problem that was first posed by Hakimi [11] and Levin [18] independently in 1971. Since its conception, the SPG has become of importance due to its numerous applications in areas like manufacturing systems, communication and transportation network design and phylogeny. For good sources of information on the SPG, the reader is referred to [13, 12, 30, 29, 19]. This paper is organized in the following way. In section 2 we define the SPG and a related problem called the Steiner problem in directed graphs (SPDG), and we establish the fact that any mathematical model for the SPDG can be used to model the SPG. Based on this fact, all the mathematical models given in the following sections, either for the SPG or for the SPDG, are considered potentially for the SPG. In section 3 mixed integer formulation models are presented. Integer formulation models are given in section 4. In section 5, a continuous formulation, which is the first of its kind, is presented.

2 The SPG and the SPDG

In this section, the SPG and the SPDG are defined and the relationship between the two problems is highlighted. The SPG definition given below has an interesting matroid interpretation [16, 9]. Given an undirected graph $G = (\mathcal{N}, \mathcal{A}, \mathcal{C})$, where

$\mathcal{N} = \{1, \ldots, n\}$ is a set of nodes, and

$\mathcal{A}$ is a set of undirected arcs (i, j) with each arc incident to two nodes, and

$\mathcal{C}$ is a set of nonnegative costs c_{ij} associated with undirected arcs (i, j),
then the SPG is defined as follows:

Instance: A graph $G = (\mathcal{N}, \mathcal{A}, \mathcal{C})$, a node subset $\mathcal{R} \subseteq \mathcal{N}$.
Question: Find the minimum cost tree, on G, that would connect all the nodes in $\mathcal{R}$.

The SPDG is the directed graph version of the Steiner problem. Apparently, [23] is the first reference to consider the SPDG. Consider $G^d = (\mathcal{N}, \mathcal{A}^d, \mathcal{C}^d)$ where

$\mathcal{N} = \{1, \ldots, n\}$ is a set of nodes, and

$\mathcal{A}^d$ is a set of directed arcs (i, j) incident to nodes in $\mathcal{N}$, and

$\mathcal{C}^d = \{c^d_{ij}\}$ is a set of costs associated with directed arcs (i, j).

Furthermore, let $\mathcal{R}$ (a set of regular nodes) be a subset of $\mathcal{N}$, and let r (a root node) be an arbitrary node in $\mathcal{R}$. Also, define a directed Steiner tree T rooted at node r, with respect to $\mathcal{R}$ on G, to be a directed tree where all nodes in $\mathcal{R}\backslash\{r\}$ can be reached from r via directed paths in T, and all leaf nodes are in $\mathcal{R}$. Next, we define the SPDG.

Instance: A graph $G^d = (\mathcal{N}, \mathcal{A}^d, \mathcal{C}^d)$, a node subset $\mathcal{R}$ and a root node r in $\mathcal{R}$.
Question: Find the minimum cost directed Steiner tree, on G^d with respect to $\mathcal{R}$, that has node r as root node.

In the light of the SPG and SPDG definitions given above, it is not hard for one to see that the SPDG is a generalization of the SPG. In fact, the SPG can be transformed to a special case of the SPDG where the arc cost structure is symmetric. This transformation can be done as follows. Given $G = (\mathcal{N}, \mathcal{A}, \mathcal{C})$, create the corresponding directed graph $G^d = (\mathcal{N}, \mathcal{A}^d, \mathcal{C}^d)$, where every undirected arc $(i, j) \in \mathcal{A}$ corresponds to two oppositely directed arcs (i, j) and $(j, i) \in \mathcal{A}^d$ associated with the same cost $(c^d_{ij} = c^d_{ji} = c_{ij})$. Potentially, any mathematical formulation for the SPDG can be used to formulate the SPG. Having highlighted this relationship, we can proceed to the next sections where different mathematical formulations are given for the SPDG and the SPG. Whenever used in the following sections, $\mathcal{O}(i)$ and $\mathcal{I}(i)$ denote the subsets of arcs that are going out of node i and into node i respectively.

3 Mixed Integer Formulations

One of the classes of mathematical models used to formulate the SPG is mixed integer programming, where some of the variables used in the model are integer, and others are continuous. Two known mixed integer formulations based on network flow requirements are given here.

3.1 Network Formulation 1

Given a directed graph $G^d = (\mathcal{N}, \mathcal{A}^d, \mathcal{C}^d)$ and a subset of nodes $\mathcal{R}$, consider a network flow problem in which the root node $r \in \mathcal{R}$ offers $|\mathcal{R}| - 1$ units of flow and each node in $\mathcal{R}\backslash\{r\}$ demands one unit. Let y_{ij} be the amount of flow on arc (i,j), and x_{ij} be a binary variable corresponding to arc (i,j). Then, consider the following formulation.

$$
\begin{array}{llrcll}
(\mathcal{F}_1) & \text{minimize} & \sum_{(i,j)\in\mathcal{A}^d} c_{ij}x_{ij} & & & \\
& \text{subject to:} & \sum_{(i,j)\in\mathcal{O}(i)} y_{ij} - \sum_{(j,i)\in\mathcal{I}(i)} y_{ji} & = & -1 & (\forall i \in \mathcal{R}\backslash\{r\}) \\
& & \sum_{(i,j)\in\mathcal{O}(i)} y_{ij} - \sum_{(j,i)\in\mathcal{I}(i)} y_{ji} & = & |\mathcal{R}| - 1 & (\text{for } i = r) \\
& & \sum_{(i,j)\in\mathcal{O}(i)} y_{ij} - \sum_{(j,i)\in\mathcal{I}(i)} y_{ji} & = & 0 & (\forall i \in \mathcal{N}\backslash\mathcal{R}) \\
& & 0 \le y_{ij} \le (|\mathcal{R}| - 1)x_{ij} & & & (\forall (i,j) \in \mathcal{A}^d) \\
& & x_{ij} \in \{0,1\} & & & (\forall (i,j) \in \mathcal{A}^d).
\end{array}
$$

The constraints involving variables y_{ij} only are simply flow conservation constraints. The coupling constraints enforce the fact that an arc can carry a positive flow only if it is used in the tree, meaning that its corresponding binary variable x_{ij} is 1.

There are formulations of the same type used in location problems and in the traveling salesman problem [8, 25]. This formulation for the SPDG is given in [2] by Arpin et al. where relaxation methods are applied to it.

3.2 Network Formulation 2

Given a directed graph $G^d = (\mathcal{N}, \mathcal{A}^d, \mathcal{C}^d)$ and a subset of nodes $\mathcal{R}$, consider a multicommodity flow problem in which the root node $r \in \mathcal{R}$ offers one unit of every commodity k ($k = 1 \cdots |\mathcal{R}| - 1$), where commodities correspond to nodes in $\mathcal{R}\backslash\{r\}$. Let y^k_{ij} be the amount of commodity k (the amount of flow between nodes r and k) on arc (i,j), and let x_{ij} be a binary variable that indicates whether or not an arc (i,j) is included in the solution. Then, the SPDG second formulation is given below.

$$
\begin{aligned}
(\mathcal{F}_2) \qquad & \text{minimize} && \sum_{(i,j)\in\mathcal{A}^d} c_{ij}x_{ij} && \\
& \text{subject to:} && \sum_{(i,j)\in\mathcal{O}(i)} y_{ij}^k - \sum_{(j,i)\in\mathcal{I}(i)} y_{ji}^k = 1 && (\text{for } i = r, \forall k \in \mathcal{R}\backslash\{r\}) \\
& && \sum_{(i,j)\in\mathcal{O}(i)} y_{ij}^k - \sum_{(j,i)\in\mathcal{I}(i)} y_{ji}^k = -1 && (\text{for } i = k, \forall k \in \mathcal{R}\backslash\{r\}) \\
& && \sum_{(i,j)\in\mathcal{O}(i)} y_{ij}^k - \sum_{(j,i)\in\mathcal{I}(i)} y_{ji}^k = 0 && (\text{for } i \notin \{r,k\}, \forall k \in \mathcal{R}\backslash\{r\}) \\
& && 0 \le y_{ij}^k \le x_{ij} && (\forall (i,j) \in \mathcal{A}^d) \\
& && x_{ij} \in \{0,1\} && (\forall (i,j) \in \mathcal{A}^d).
\end{aligned}
$$

The coupling constraints indicate that flow of any commodity is allowed on arc (i,j) only if that arc is included in the solution. In a way, we can view x_{ij} as the flow capacity for commodity k on arc (i,j). The first three constraints, involving y_{ij}^k only, dictate that one unit of commodity k be routed between nodes r and k.

This formulation is given in [31] by Wong where a dual ascent approach is applied to it. Note that formulation $\mathcal{F}_2$ can be reduced in size by aggregating the coupling constraints [31]. However, a number of studies of related problems such as plant location and fixed charge network design [7, 21, 27, 28] have concluded that the disaggregated formulation, although it is larger in size, is much more useful for developing efficient algorithms.

4 Integer Formulations

The SPG can be approached as an instance in the realm of integer programming. In integer programming, formulating a problem is crucially important to solving it [24]. In this section, we present several integer formulations for the SPG and the SPDG. Some of these formulations have been approached with relaxation methods used to develop lower bounds for a branch and bound scheme in some previous studies.

4.1 An Elementary Path Formulation

Given an undirected graph $G = (\mathcal{N}, \mathcal{A}, \mathcal{C})$, a subset of regular nodes $\mathcal{R}$ and an arbitrary node $r \in \mathcal{R}$, associate with every arc (i,j) two binary variables x_{ij} and y_{ijk}; x_{ij} is one if arc (i,j) is used in the solution tree, and zero otherwise; y_{ijk} is one if arc (i,j) is on the unique elementary path in the solution tree from node r to node k $(k \in \mathcal{R}\backslash\{r\})$, and zero otherwise. The SPG is formulated as follows.

$$
\begin{array}{lll}
(\mathcal{F}_3) & \text{minimize} & \sum_{(i,j)\in\mathcal{A}} c_{ij}x_{ij} \\
& \text{subject to:} & \text{there exists an elementary path } [y_{ijk}] \text{ from } r \text{ to } k\ , \forall k \in \mathcal{R}\backslash\{r\} \\
& & x_{ij} \geq y_{ijk} \qquad (\forall k \in \mathcal{R}\backslash\{r\}, \forall (i,j) \in \mathcal{A}) \\
& & x_{ij} \in \{0,1\} \qquad (\forall (i,j) \in \mathcal{A}) \\
& & y_{ijk} \in \{0,1\} \qquad (\forall k \in \mathcal{R}\backslash\{r\}, \forall (i,j) \in \mathcal{A}).
\end{array}
$$

The first constraint ensures that the solution is connected. The second constraint dictates that an arc is on some path only if that arc is used in the tree. This formulation is given in [4] by Beasley where a Lagrangean relaxation technique is used to develop lower bounds for a branch and bound algorithm. In this approach, redundant "box constraints" are added to the formulation to strengthen lower bounds.

4.2 A Minimum Spanning Tree Formulation

This formulation is given in [5] by Beasley where a branch and bound algorithm is devised using lower bounds obtained from Lagrangean relaxation techniques. Given an undirected graph $G = (\mathcal{N}, \mathcal{A}, \mathcal{C})$, a subset of nodes $\mathcal{R}$ and an arbitrary node $r \in \mathcal{R}$, the formulation is motivated by the following. The SPG is to choose a subset of $\mathcal{A}$ that would connect together all nodes in $\mathcal{R}$ with the minimum cost. It is well known that edges in this subset of $\mathcal{A}$ will form a minimum cost spanning tree, the Steiner tree, on some subset of vertices $\mathcal{R} \cup \mathcal{S}$, where $\mathcal{S} \subseteq \mathcal{N}\backslash\mathcal{R}$ is called the set of Steiner vertices. In order to exploit this valuable information, let us add an artificial node 0 to the graph. Also, for each node $i \in \mathcal{N}\backslash\mathcal{R}$, add an edge $(0,i)$ of cost zero. For node r, add an edge $(0,r)$ of cost zero. Clearly, the SPG reduces to finding the minimum spanning tree (MST) on the augmented graph $G^0 = (\mathcal{N}^0, \mathcal{A}^0, \mathcal{C}^0)$ with the additional restriction that in this MST, any node $i \in \mathcal{N}\backslash\mathcal{R}$ connected by the arc $(0,i)$ to vertex 0 must have degree 1 (see Figure 1). In more rigorous mathematical terms, this can expressed by the following formulation.

$$
\begin{array}{llrcl}
(\mathcal{F}_4) & \text{minimize} & \sum_{(i,j)\in\mathcal{A}^0} c_{ij}x_{ij} & & \\
& \text{subject to:} & [x_{ij}] & \equiv & \text{spanning tree on } G^0 = (\mathcal{N}^0, \mathcal{A}^0, \mathcal{C}^0) \\
& & x_{0i} + x_{pq} & \leq & 1 \qquad (\text{for } p \text{ or } q = i, \forall i \in \mathcal{N}^0\backslash\mathcal{R}) \\
& & x_{ij} \in \{0,1\} & & \qquad (\forall (i,j) \in \mathcal{A}^0).
\end{array}
$$

4.3 A Set Covering Formulation

Given a directed graph $G^d = (\mathcal{N}, \mathcal{A}^d, \mathcal{C}^d)$ and a subset of nodes $\mathcal{R}$, consider the partition of $\mathcal{N} = \mathcal{X} \cup \overline{\mathcal{X}}$ such that $\mathcal{X} \cap \overline{\mathcal{X}} = \emptyset$, $\mathcal{R} \cap \mathcal{X} \neq \emptyset$ and $\mathcal{R} \cap \overline{\mathcal{X}} \neq \emptyset$. Let

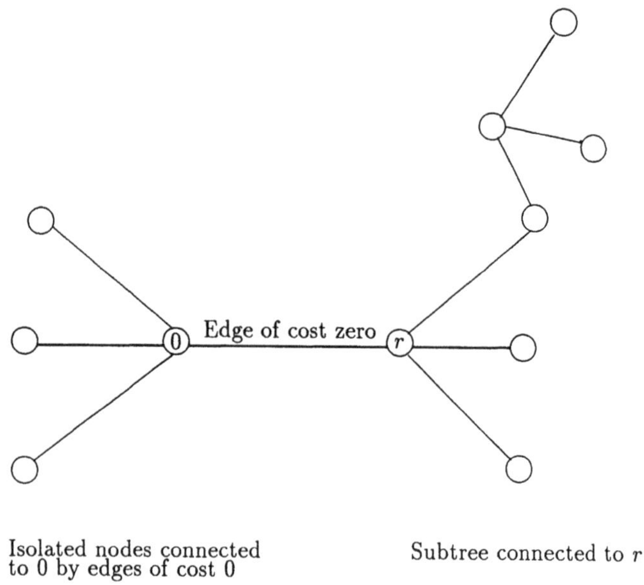

Figure 1: An optimal solution for the restricted MST problem.

$\mathcal{P} = (\mathcal{X}, \overline{\mathcal{X}})$ denote the cut-set of edges between $\mathcal{X}$ and $\overline{\mathcal{X}}$. Obviously, there are exponentially many sets. Suppose that we enumerate such sets to be $\mathcal{P}_1, \cdots, \mathcal{P}_\pi$. Define constants a_{pij} ($p \in \{1 \cdots \pi\}$, $(i,j) \in \mathcal{A}^d$) to be 1 if arc (i,j) is in cut set $\mathcal{P}_p$, and zero otherwise. Now, the SPDG can be formulated as the following set-covering problem:

$$(\mathcal{F}_5) \quad \text{minimize} \quad \sum_{(i,j)\in\mathcal{A}^d} c_{ij}x_{ij}$$

$$\text{subject to:} \quad \sum_{(i,j)\in\mathcal{A}^d} a_{pij}x_{ij} \geq 1 \qquad (\forall p \in \{1, \cdots, \pi\}$$

$$x_{ij} \in \{0,1\} \qquad (\forall (i,j) \in \mathcal{A}^d).$$

This formulation is given in [1] by Aneja where a set covering algorithm for the SPG is proposed. In addition, Chopra et al. [6] have used this formulation to find lower bounds, useful for a branch and bound algorithm, by solving the relaxed linear problem. The relaxed version has potentially an exponential number of constraints; however, one need not consider all of them, and this can be done using a row generation technique [1].

4.4 A Rooted Tree Formulation

In the following, a rooted-tree formulation, which was proposed recently by Khoury et al. [14] in the context of a test problem generator for the SPG based on the Karush-Kuhn-Tucker optimality conditions, is presented. Given a directed graph $G^d = (\mathcal{N}, \mathcal{A}^d, \mathcal{C}^d)$, a subset of nodes $\mathcal{R}$ and a root node $r \in \mathcal{R}$, the SPDG can be formulated as follows.

$$(\mathcal{F}_6) \quad \text{minimize} \quad \sum_{(i,j)\in\mathcal{A}^d} c_{ij}x_{ij}$$

$$\text{subject to:} \quad \sum_{(i,j)\in\mathcal{I}(j)} x_{ij} = 1 \qquad (\forall j \in \mathcal{R}\backslash\{r\})$$

$$\sum_{(k,i)\in\mathcal{I}(i)\backslash\{(j,i)\}} x_{ki} \geq x_{ij} \qquad (\forall (i,j) \in \mathcal{A}^d\backslash\mathcal{O}(r))$$

$$x_{ij} \in \{0,1\} \qquad (\forall (i,j) \in \mathcal{A}^d)$$

$$[x_{ij}] \text{ contains no cycles.}$$

Whether or not an arc (i,j) is used in the tree is translated to assigning the value 1 or 0 to its corresponding binary variable x_{ij}. An aggregated form of the above formulation, where aggregation is performed on the second set of constraints, is given by Nastansky et al. [23]. The first type of constraint implies that at least one arc is going into every regular node i, $i \in \mathcal{R}\backslash\{r\}$. The second type of constraint dictates that for a directed arc (i,j) to be used in a tree, at least one tree arc has to be going into node i, $i \in \mathcal{N}\backslash\{r\}$. This constraint and the last constraint together imply the structure of a rooted tree. Note that there are different ways to represent the last constraint mathematically. This type of constraint is encountered in formulating the traveling salesman person where cycles corresponding to subtours are to be eliminated [17].

Although this formulation does not involve flow variables, it has the feature of rooted trees in common with the network flow formulations given earlier. This combination of the integer and rooted-tree features gives the formulation a unique ingredient. Also, note that this formulation does not involve potentially an exponential number of constraints like $\mathcal{F}_5$.

4.5 A Node-Arc Integer Formulation with Valid Cuts

Here, we present a new formulation motivated by the fact that we can see an undirected Steiner tree on an undirected graph as a directed tree, on the corresponding directed graph, rooted at any regular node (we have $|\mathcal{R}|$ such directed trees). Given an undirected graph $G = (\mathcal{N}, \mathcal{A}, \mathcal{C})$, its corresponding symmetric directed graph $G^d = (\mathcal{N}, \mathcal{A}^d, \mathcal{C}^d)$ and a subset of regular nodes $\mathcal{R}$, associate with every undirected arc (i,j) in G a binary variable x_{ij}, and associate with every directed arc (i,j) in G^d binary variables y_{ij}^k; x_{ij} is one if undirected arc (i,j) is used in the solution tree, and zero otherwise; y_{ij}^k is one if directed arc (i,j) is in a tree rooted at $k \in \mathcal{R}$, and zero otherwise. Also, with every node $i \in \mathcal{N}\backslash\mathcal{R}$, associate a binary variable δ_i that is one if node i is in the solution tree, and zero otherwise. The SPG is formulated as follows.

$$
\begin{array}{llrcll}
(\mathcal{F}_7) & \text{minimize} & \displaystyle\sum_{(i,j)\in\mathcal{A}} c_{ij}x_{ij} & & & \\
& \text{subject to:} & \displaystyle\sum_{(i,j)\in\mathcal{A}} x_{ij} & \geq & 1 & (\forall i \in \mathcal{R}) \\
& & \displaystyle\sum_{(i,j)\in\mathcal{A}} x_{ij} & \geq & 2\delta_i & (\forall i \in \mathcal{N}\backslash\mathcal{R}) \\
& & \delta_i & \geq & x_{ij} & (\forall i \in \mathcal{N}\backslash\mathcal{R}, \forall (i,j) \in \mathcal{A}) \\
& & y_{ij}^k + y_{ji}^k & = & x_{ij} & (\forall k \in \mathcal{R}, \forall (i,j) \in \mathcal{A}^d) \\
& & \displaystyle\sum_{(i,j)\in\mathcal{I}(j)} y_{ij}^k & = & 1 & (\forall k \in \mathcal{R}, \forall j \in \mathcal{R}\backslash\{k\}) \\
& & \displaystyle\sum_{(i,j)\in\mathcal{I}(j)} y_{ij}^k & = & \delta_j & (\forall k \in \mathcal{R}, \forall j \in \mathcal{N}\backslash\mathcal{R}) \\
& & y_{ik}^k & = & 0 & (\forall k \in \mathcal{R}, \forall (i,k) \in \mathcal{I}(k)) \\
& [y_{ij}^k] \text{ contains no cycles} & & & & (\forall k \in \mathcal{R}).
\end{array}
$$

The first three constraints ensure that in the undirected tree, every regular node is connected to the tree, and that every Steiner node has at least a degree of two. The fourth set of constraints enforce the fact that every rooted tree solution is derived from the unrooted tree by appropriately assigning directions to arcs in the undirected tree and vice versa. The fifth, sixth, seventh and eighth sets of constraints ensure that the directed interconnecting network corresponding to node $k \in \mathcal{R}$ is a directed tree rooted at k connecting its root node to all other regular nodes in $\mathcal{R}\backslash\{k\}$, a property which is present in formulation $\mathcal{F}_6$. Although this formulation contains a relatively large number of constraints, most of these constraints are valid cuts. As a result, we expect this formulation to develop good lower bounds for a branching tree.

5 Continuous Formulations

Every mathematical formulation that is not continuous because of integer zero-one variables x_i can be made continuous by relaxing the integrality constraints, replacing them with zero-one "box constraints" and adding to the objective function a large penalty term corresponding to violated integer constraints. This is usually done by adding a large M term, if we are dealing with a minimization problem, of the form $\underline{x}^t M(\underline{1} - \underline{x})$. Naturally, the question now is how large should M be? How difficult is it to deal with a very large M factor in the minimization process? In general, these questions are very difficult to answer, see [26]. Here, we present a practical continuous formulation, which was proposed by Khoury [15] in the context of continuous approaches to the SPG. However, in this formulation, the source of concavity is still attached to the objective function. Indeed, a series of continuous ϵ-concave optimization problems is presented, and a proof that the limit of the series when ϵ goes to zero is the fixed charge formulation of the SPDG (see [10]) is given. Also, a rough error bound on approximating the solution of the SPDG by the solution of a sequence in the series is presented below.

5.1 An ϵ-Concave Optimization Approach

Assume we are given a directed graph $G = (\mathcal{N}, \mathcal{A}^d, \mathcal{C}^d)$. Let $\mathcal{R}$ be a subset of nodes on the graph, set of regular nodes. Furthermore, let node r, the root node, be an arbitrary node in $\mathcal{R}$. In addition, let y_{ij} be a flow variable that is associated with arc (i,j). Moreover, let ϵ be a strictly positive real number. Now, define $\mathcal{F}_8(\epsilon)$, a concave minimization problem, as follows.

$$
\begin{aligned}
(\mathcal{F}_{8,\epsilon}) \qquad & \text{minimize } f_8(\epsilon, Y) = \sum_{(i,j)\in\mathcal{A}^d} c_{ij}\frac{y_{ij}}{y_{ij}+\epsilon} \\
& \text{subject to:} \quad AY = B \\
& \qquad\qquad\qquad\; Y \geq 0.
\end{aligned}
$$

where A is an $|\mathcal{N}| \times |\mathcal{A}^d|$ node-arc incidence matrix, and Y is an $|\mathcal{A}^d| \times 1$ vector whose entries are the y_{ij}s, and B is an $|\mathcal{N}| \times 1$ vector whose entries b_i correspond to the set of nodes; furthermore, $b_i = 0$ if $i \in \mathcal{N}\backslash\mathcal{R}$, and $b_i = -1$ if $i \in \mathcal{R}\backslash\{r\}$, and $b_r = |\mathcal{R}| - 1$. This is basically a minimum cost flow problem where the objective function is a separable concave function. Note that due to the strict concavity of $f_\epsilon(Y)$, the unimodularity of A and the integrality of B, the optimal solution to $\mathcal{F}_{8,\epsilon}$ is always integer.

Now, if we let $x_{ij} = \dfrac{y_{ij}}{y_{ij}+\epsilon}$, the limit of $\mathcal{F}_{8,\epsilon}$ when ϵ goes to zero becomes

$$
\begin{aligned}
(\mathcal{F}_8) \quad & \text{minimize } f_8(Y) = \sum_{(i,j)\in\mathcal{A}^d} c_{ij}x_{ij} \\
& \text{subject to: } \quad AY = B \\
& \qquad\qquad\qquad x_{ij} = 0(1) \text{ for } y_{ij} = 0(>0)\ (\forall(i,j)\in\mathcal{A}^d) \\
& \qquad\qquad\qquad Y \geq 0.
\end{aligned}
$$

Observe that $\mathcal{F}_8$, which is a fixed formulation of the SPG, is an equivalent tight formulation of $\mathcal{F}_1$.

5.2 ϵ-Error Bound

Here, we give an ϵ-error bound on the approximation of the solution of the SPDG by the solution of one sequence $\mathcal{F}_{8,\epsilon}$ of the series of concave optimization problems given above.

Theorem: The error on approximating f_8^* by $f_8^*(\epsilon_0)$ is less than $\dfrac{\epsilon_0}{1+\epsilon_0}\sum_{(i,j)\in\mathcal{A}^d} c_{ij}$.

Proof

Since $f_8(\epsilon, Y)$ is a convex and decreasing function of ϵ, we have

$$
\begin{aligned}
0 < f_8^* - f_8^*(\epsilon_0) &< \lim_{\epsilon\to 0} f_8(\epsilon, Y^*(\epsilon_0)) - f_8^*(\epsilon_0) \\
&\leq (1 - \frac{1}{1+\epsilon_0}) \sum_{(i,j)\in\mathcal{A}^d} c_{ij} \\
&\leq \frac{\epsilon_0}{1+\epsilon_0} \sum_{(i,j)\in\mathcal{A}^d} c_{ij}. \square
\end{aligned}
$$

6 Concluding Remarks

In this paper, we have presented eight different formulations for the SPDG and the SPG. Those formulations fall into three categories, namely mixed integer, integer and continuous categories. In the mixed integer category, two network formulations are presented, where one of them is based on flow requirements and the other on commodity requirements. Under the second category, we have presented three known formulations, the elementary path, the set covering and the minimum spanning tree formulations. Also, we have proposed two new formulations. One of the proposed formulations, the rooted-tree integer formulation, combines the properties of involving only integer variables and having the structure of rooted tree, a property present in the network formulations mentioned above. The second proposed integer formulation is more complicated in nature; however, it contains more valid cuts, and it is expected to give good lower bounds. In addition, we have presented, in the category of continuous formulations, a new approach where the SPDG is approximated by a sequence of a series of concave optimization problems.

Every one of these formulations is valid. However, some formulations are better than others due to a better distribution of bad and good properties between the constraint set and the objective function. It is well known that manipulating an optimization model can be done by either manipulating the constraint set or manipulating the objective function (see also [22]). Sometimes a formulation handles good optimality properties through its objective function. In this case, it is a good idea to express those properties in the feasible domain by adding more constraints to the problem to further tighten the feasible region. An alternative situation is when the set of constraints has bad properties while the objective function has nice properties. It would be a good idea in this case to take some of the bad properties out of the constraint set and express them in the objective function. The moral of the story here is that with better shared information between the objective function and the constraint set, mathematical optimization models would be better off. Criteria to identify better models and systematic ways to formulate them are, however, open questions.

References

[1] Aneja, Y.P. (1980), "An Integer Linear Programming Approach to the Steiner Problem in Graphs," *Networks, Vol. 10*, 167-178.

[2] Arpin, D., Maculan, N. and Nguyen, S. (1983), " Le Problème de Steiner sur un Graphe Orienté: Formulations et Relaxations," Publication 315, Centre de Recherche sur les Transports, Université de Montréal.

[3] Balakrishnan, A. (1982), "Formulations and Algorithms for the Steiner Network Problem," Unpublished manuscript, Sloan School Management, MIT.

[4] Beasley, J.E. (1984), "An Algorithm for the Steiner Problem in Graphs," *Networks, Vol. 14*, 147-159.

[5] Beasley, J.E. (1989), "An SST-Based Algorithm for the Steiner Problem on Graphs," *Networks, Vol. 19*, 1-16.

[6] Chopra, S., Gorres E.R. and Rao, M.R. (1992), " Solving the Steiner Tree Problem on Graphs Using Branch and Cut," *ORSA Journal on Computing, Vol. 4, No. 3*, 320-335.

[7] Cornuejolis, G., Fisher, M.L. and Nemhauser, G.L. (1977), "Location of Bank Accounts to Optimize Float: An Analytic Study of Exact and Approximate Algorithms," *Management Science, Vol 23*, 789-810.

[8] Gavish, R. and Graves, S.C. (1982), " Scheduling and routing in transportation systems: Formulations and new relaxations," Graduate School of Management, University of Rochester, N.Y.

[9] Gondran, M. and Minoux, M. (1984), *Graphs and Algorithms*, John Wiley & Sons.

[10] Guisewite, G.M. and Pardalos, P.M. (1991), "Algorithms for the Single-Source Uncapacitated Minimum Concave-Cost Network Flow Problem," *Journal of Global Optimization, Vol. 1*, 245-265.

[11] Hakimi, S.L. (1971), "Steiner's Problem in Graphs and its Implications," *Networks, Vol. 1*, 113-133.

[12] Hwang, F.K. and Richards, D.S. (1992), "Steiner Tree Problems," *Networks, Vol. 2*, 55-89.

[13] Hwang, F.K., Richards, D.S. and Winter, P. (1992), *The Steiner Tree Problem*, Elsevier, Amsterdam.

[14] Khoury, B.N., Pardalos, P.M. and Du, D.-Z (1993, to appear), "A Test Problem Generator for the Steiner Problem in Graphs," *ACM TOMS*.

[15] Khoury, B.N. (1993), "The Steiner Problem in Graphs," Ph.D. Dissertation (under preparation), Department of Industrial and Systems Engineering, University of Florida, Gainesville, Florida.

[16] Lawler, E.L. (1976), *Combinatorial Optimization Networks and Matroids*, Holt, Rinehart and Winston.

[17] Lengauer, T. (1990), *Combinatorial Algorithms for Integrated Circuit Layout*, John Wiley & Sons, Inc. .

[18] Levin, A.Ju. (1971), "Algorithms for the Shortest Connection of a Group of Graph vertices," *Soviet Math. Doklady, Vol. 12*, 1477-1481.

[19] J. MacGregor Smith and P. Winter (Eds.), (1991), *Topological Network Design, Vol. 31, No. 1-4 of Annals of Operations Research.*

[20] Maculan, N. (1987), "The Steiner Problem in Graphs," *Ann. Discrete Math., Vol. 31*, 185-222.

[21] Magnanti, T.L. and Wong, R.T. (1981), "Accelerating Benders Decomposition: Algorithmic Enhancement and Model Selection Criteria," *Operations Research, Vol. 29*, 464-484.

[22] Martin R.K. (1991), "Using Separation Algorithms to Generate Mixed Integer Model Reformulations," *Operations Research Letters, Vol. 10*, 119-128.

[23] Nastansky, L., Selkow, S.M. and Stewart, N.F. (1974), "Cost-Minimal Trees in Directed Acyclic Graphs," *Zeitschrift für Operations Research, Vol. 18*, 59-67.

[24] Nemhauser, G.L. and Wolsey L.A. (1988), *Integer and Combinatorial Optimization*, John Wiley & Sons, Inc.

[25] Nobert, Y. (1982), "Construction d'Algorithmes Optimaux pour des Extensions du Problém de Voyageur de Commerce," PhD Dissertation, Publication 297, Centre de Recherche sur les Transport, Université de Montréal.

[26] Pardalos, P.M. and Rosen, J.B. (1987), *Constrained Global Optimization, Algorithms and Applications*, Springer-Verlag, Lecture Notes in Computer Science 268.

[27] Rardin, R.L. (1982), "Tight Relaxations of Fixed Charge Network Flow Problems," Technical Report J-82-3, School of Industrial and Systems Engineering, Georgia Institute of Technology, Atlanta, GA.

[28] Rardin, R.L., Parker R.G. and Lim, W.K. (1982), "Some Polynomially Solvable Multi-Commodity Fixed Charge Network Flow Problems," Technical Report J-82-2, School of Industrial and Systems Engineering, Georgia Institute of Technology, Atlanta, GA.

[29] Voß, S. (1990), *Steiner-Probleme in Graphèn* [in German], Hain, Frankfurt.

[30] Winter, P. (1987), "Steiner Problem in Networks: A Survey," *Networks, Vol. 17*, 129-167.

[31] Wong, R.T. (1984), "A dual Ascent Approach to Steiner Tree Problems on a Directed Graph," *Mathcmatical Programming Vol. 28*, 271-287.

Network Optimization Problems, pp. 125-145
Eds. D.-Z. Du and P.M. Pardalos

Minimum Concave-Cost Network Flow Problems with a Single Nonlinear Arc Cost

Bettina Klinz
Institut für Mathematik B, TU Graz, Kopernikusgasse 24, A-8010 Graz, Austria

Hoang Tuy
Institute of Mathematics, Vien Toan Hoc, P.O.B. 631 Bo Ho, 10.000 Hanoi, Vietnam

Abstract

We apply the *parametric method* recently developed for rank two quasiconcave minimization problems to minimum *concave-cost* network flow problems in which all arcs except one have linear cost. The main step of the resulting algorithm consists of solving a specially structured *parametric linear* network flow problem. We derive upper bounds on the number of breakpoints in the optimal value function of this parametric problem and obtain the following results: For general networks with integer capacities and demands, we get a pseudo-polynomial bound on both the number of operations and of objective function evaluations required by our algorithm. On single source, uncapacitated networks with n nodes, m arcs and $|T|$ sinks, our method can be shown to run in $O(S(n,m))$ operations and at most $|T|+1$ evaluations of the single nonlinear arc cost function, where $S(n,m)$ is the time needed to solve a single source shortest path problem on the given network. These bounds depend on n and m only and are thus *strongly polynomial.*

1 Introduction

Quasiconcave rank-two minimization. In view of the difficulty inherent to global optimization, it is natural to look for specially structured global optimization problems that can be solved practically by properly adapted methods. In [23] an important class of these problems has been singled out whose nonconvexity is located in a subspace of low dimension and hence can be handled very efficiently. This class includes problems which have the following general formulation:

$$(P) \qquad \text{minimize} \quad f(x) \quad \text{subject to} \quad x \in D$$

where D is a polyhedron in $\mathbb{R}^d$ while $f(x)$ is a quasiconcave function defined on an open convex set $\Omega \supset D$ and satisfying the following "*rank two condition*":

> There exist two linearly independent vectors c^1 and c^2 such that if $\langle c^i, w\rangle \geq 0$ $(i = 1,2)$ then $f(x + w) \geq f(x)$ for all $x \in \Omega$. (1)

Here $\langle y, z\rangle$ denotes the scalar product of the two vectors y and z. For solving problems of type (P), Tuy and Tam [22] proposed two algorithms. The first one, using a polyhedral annexation technique, is most suitable when the computation of the values of $f(x)$ is time consuming (which is the case, in particular, if $f(x)$ is defined in some implicit way). The second one, which practically reduces the problem to solving a parametric linear program, has proven to be very efficient when the values of $f(x)$ are easy to compute (which occurs e.g. if $f(x)$ is the product of two affine functions, as considered in [5, 7, 14, 18, 21, 22, 23]).

In the present paper we shall discuss a special global optimization problem for which the above approach seems to be particularly useful. Specifically, we will study the *minimum concave-cost network flow problem* $(MCCFP)$ in the case when *only one arc* has a *nonlinear cost*. It is well known that the general $(MCCFP)$ is $\mathcal{NP}$-hard. Nevertheless, as will be shown below, the special version of this problem that we shall be considering falls into the above class of rank two quasiconcave minimization problems and therefore can be efficiently solved by the *parametric method* as developed by Tuy et al. in [22, 23].

Organization of the paper. In the next section we will first present the theoretical foundations of the parametric approach to quasiconcave minimization problems satisfying the rank two condition (1). Then in Section 3 we show that the minimum concave-cost network flow problem with a single concave arc cost $(MCCFP_1)$ belongs to this class and how it can be solved through a two-phase algorithm. In the first phase we solve a specially structured *parametric-cost linear* network flow problem (TP_α), i.e. we determine an optimal flow for each optimality interval (=interval between two adjacent breakpoints in the optimal value function of (TP_α)). Due to the results obtained in Section 2, in the second phase it suffices to minimize the concave cost function over the set of flows determined in the first step. To analyze this algorithm, we derive a pseudo-polynomial upper bound on the number of breakpoints of (TP_α) for the case of integer-valued capacities and demands. Based on this result, the number of operations and of evaluations of the single nonlinear arc cost function performed by our algorithm can be shown to be pseudo-polynomial as well.

In Section 4 we discuss the special case of a single source, uncapacitated (SSU) network as considered by Guisewite and Pardalos [8] and prove that for SSU networks, $|T|$, the number of sinks in the network, provides an upper bound on the number of breakpoints of (TP_α). Finally, in Section 5 we describe an implementation of the parametric algorithm for SSU networks with n nodes and m arcs which requires $O(S(n,m))$ operations and at most $|T|+1$ evaluations of the single nonlinear arc cost function, where $S(n,m)$ denotes the time needed to solve a single source shortest path problem on the given network. In contrast to the algorithm of Guisewite and Pardalos [8], these time bounds depend on n and m only and not on the size of the total demand, and are thus *strongly polynomial.* We close the paper with a short summary of our main results and some concluding remarks in Section 6.

2 A Parametric Method for Rank Two Quasiconcave Minimization

We first prove a theorem which serves as the foundation of the parametric approach to quasiconcave minimization problems satisfying the rank two condition (1). In [22], this proposition was indirectly derived from a polyhedral annexation procedure for solving (P). The proof below is more direct and may provide insight into the structure of the problem.

Consider the parametric linear program (Q_α), $\alpha \in [0,1]$, where

$$(Q_\alpha) \qquad \text{minimize} \quad \langle c^1 + \alpha(c^2 - c^1), x\rangle \qquad \text{s.t.} \qquad x \in D.$$

Assuming $D \neq \emptyset$, let $\phi(\alpha)$ denote the *optimal cost curve* of (Q_α). From the theory of parametric linear programming (cf. e.g. Murty [15]) it is well known that $\phi(\alpha)$ is a continuous, concave and piecewise linear function in the parameter α.

Let $\alpha_0 = 0 < \alpha_1 < \ldots < \alpha_{N-1} < \alpha_B = 1$ be the sequence of *breakpoints* of $\phi(\alpha)$. For each interval $\Delta_k = [\alpha_{k-1}, \alpha_k]$ $(k = 1, \ldots, B)$ (=optimality interval) let us choose any vertex $x^{(k)}$ of D which is an optimal solution of (Q_α) for all $\alpha \in \Delta_k$. If (Q_α) has unbounded optimal value for $\alpha \in \Delta_k$, then $x^{(k)}$ is taken to be a vertex "at infinity", i.e. an unbounded edge of D over which the objective function of (Q_α) is unbounded below. It follows from the properties of $\phi(\alpha)$ mentioned above that this may happen only for $k \in \{1, B\}$. When $x^{(k)}$ is an unbounded edge of D, we agree to denote by $f(x^{(k)})$ the infimum of $f(x)$ over this edge.

Let E be a convex set in $\mathbb{R}^d$. Then we denote by E^* the *polar* of E, i.e. the set of all v such that $\langle v, x\rangle \leq 1$ for all $x \in E$.

Theorem 2.1 *We have*

$$\min\{f(x) : x \in D\} = \min\{f(x^{(k)}) : k = 1, \ldots, B\}.$$

Proof. Let $k^* \in \arg\min\{f(x^{(k)}) : k = 1, \ldots, B\}$ and $\gamma = f(x^{(k^*)})$. Define

$$C = \{x \in \Omega : f(x) \geq \gamma\}.$$

We have to show that $D \subseteq C$. Since this is trivial if $\gamma = -\infty$, we may assume $\gamma > -\infty$. Arguing by contradiction, suppose there exists a point $z \in D \setminus C$. Take any point $x^0 \in D \cap C$ and let y be the intersection of the line segment $[x^0; z]$ with the boundary of the convex set C. Then there exists a vector $v \neq 0$ such that

$$\langle v, x - y \rangle \leq 0 \qquad \text{for all } x \in C, \tag{2}$$

$$\langle v, z - y \rangle > 0. \tag{3}$$

(So $\langle v, x - y \rangle = 0$ is a supporting hyperplane for C at point y.) By multiplying v by a positive factor, if necessary, we may always assume $\langle v, y - x^0 \rangle \leq 1$. Then from (2) we have $\langle v, x - x^0 \rangle = \langle v, x - y \rangle + \langle v, y - x^0 \rangle \leq 1$ for all $x \in C$, i.e. $\langle v, \tilde{x} \rangle \leq 1$ for all $\tilde{x} \in C - x^0$ and hence,

$$v \in (C - x^0)^*. \tag{4}$$

But in view of the rank two condition (1), if $K := \{w : \langle c^i, w \rangle \geq 0,\ i = 1, 2\}$ then $f(x^0 + w) \geq f(x^0) \geq \gamma$ for all $w \in K$, i.e. $x^0 + K \subseteq C$. Consequently, $K \subset C - x^0$, which in turn implies

$$(C - x^0)^* \subseteq K^*. \tag{5}$$

On the other hand, it is known from convex analysis that the polar of the cone $K = \{w : \langle c^i, w \rangle \geq 0,\ i = 1, 2\}$ is the cone generated by the vectors $-c^1$ and $-c^2$. From (4) and (5) we deduce $v = -t_1 c^1 - t_2 c^2$, with $t_1, t_2 \geq 0$, $t_1 + t_2 > 0$ (note that $v \neq 0$). Setting $\alpha = t_2/(t_1 + t_2)$, so that $\alpha \in [0, 1]$, we have

$$v = -(t_1 + t_2)[c^1 + \alpha(c^2 - c^1)]. \tag{6}$$

Let $\alpha \in \Delta_k$. Since $x^{(k)}$ solves (Q_α) for $\alpha \in \Delta_k$, it follows from the above expression (6) that $x^{(k)}$ solves also the linear program $\max\{\langle v, x \rangle : x \in D\}$. In view of the fact $z \in D$, the inequality (3) implies $\max\{\langle v, x \rangle : x \in D\} \geq \langle v, z \rangle > \langle v, y \rangle$. Hence, in case $x^{(k)}$ is a vertex, $\langle v, x^{(k)} - y \rangle > 0$ follows, and consequently, by virtue of (2), $x^{(k)} \notin C$, i.e. $f(x^{(k)}) < \gamma$. If $x^{(k)}$ is an unbounded edge of D, then by taking a sufficiently remote point $\bar{x}^{(k)}$ on this edge, we have $\langle v, \bar{x}^{(k)} - y \rangle > 0$, hence $\bar{x}^{(k)} \notin C$, i.e. $f(x^{(k)}) \leq f(\bar{x}^{(k)}) < \gamma$. We arrived at a contradiction to the definition of γ in both cases and thus $D \subseteq C$, i.e. $x^{(k^*)}$ solves (P). □

Theorem 2.1 gives rise to the following procedure for solving (P):

Algorithm 1 (parametric method):

1. Compute the successive breakpoints $\alpha_0, \alpha_1, \ldots, \alpha_B$ of the optimal cost curve of the parametric linear program (Q_α) and the associated sequence $x^{(1)}, \ldots, x^{(B)}$ such that $x^{(k)}$ is a basic optimal solution of (Q_α) for all $\alpha \in \Delta_k = [\alpha_{k-1}, \alpha_k]$ or an unbounded edge of D over which (Q_α) is unbounded below for all $\alpha \in \Delta_k$, respectively.

2. Find $k^* \in \arg\min\{f(x^{(k)}) : k = 1, \dots, B\}$, where for an unbounded edge z $f(z)$ is defined as the infimum of $f(x)$ over this edge. Then $x^{(k^*)}$ solves (P).

Thus solving a quasiconcave minimization problem with rank two property requires almost the same amount of work as solving a parametric linear program. In the next section we will examine a case when this procedure seems to be particularly efficient.

Remark. At this point it should be mentioned that the rank two condition (1) is only a special case of the following more general "*rank κ condition*" given by Tuy in [24]: There exists a cone K whose polar K^* has dimension κ such that

$$\tilde{x} + K \subseteq \{x \in \Omega : f(x) \geq f(\tilde{x})\} \qquad \text{for all } \tilde{x} \in D. \tag{7}$$

In our case $\kappa = 2$ and $K = \{w : \langle c^i, w\rangle \geq 0,\ i = 1, 2\}$. In [23, 24] Tuy showed that both the parametric approach described above and the polyhedral annexation method mentioned in the introduction, work also for problems of type (P) satisfying the more general rank κ condition (7). (For details and even further generalizations see [23].) □

At first sight it might be surprising that the rank two condition allows the reduction of the original d-dimensional problem (P) to a problem of dimension 2. However, we may gain some intuition by observing that from condition (1) it follows that

$$\langle c^i, w\rangle = 0 \ (i = 1, 2) \qquad \Longrightarrow \qquad f(x + w) = f(x) \qquad \text{for all } x \in \Omega, \tag{8}$$

i.e. the objective function $f(x)$ is constant on a subspace of $\mathbb{R}^d$ whose codimension is 2. Note that (8) can also be interpreted as rank two condition in the sense of (7), though it is obviously weaker than (1). (Set simply $K = \{w : \langle c^i, w\rangle = 0,\ i = 1, 2\}$.)

Remark. A careful inspection of the proof of Theorem 2.1 shows that only a small change to Algorithm 1 is necessary for solving problems of type (P) which satisfy only the weaker rank two condition (8) — one simply has to compute all breakpoints of the parametric linear program (Q_α) for $\alpha \in \mathbb{R}$ instead of for $\alpha \in [0, 1]$ as above. □

To conclude this section, let us remark that an alternative parametric procedure for solving problem (P) can be developed by considering the following parametric *right-hand-side* linear program.

$$(R_\beta) \qquad \text{minimize } \langle c^2, x\rangle \qquad \text{s.t.} \qquad x \in D \text{ and } \langle c^1, x\rangle = \beta$$

for $\beta \in \mathbb{R}$. It is interesting to note that there holds a duality relationship between the parametric problems (Q_α) and (R_β).

3 A Special Minimum Concave-Cost Network Flow Problem

3.1 Basic Notation and Problem Statement

Minimum concave cost flow problems

Let $G = (N, A)$ be a connected directed graph with node set N and arc set A, where $|N| = n$ and $|A| = m$. Associated with each arc $(i, i') \in A$ are a nonnegative integer *capacity* $u_{ii'}$ ($u_{ii'} = \infty$ is possible) and a *concave cost function* $h_{ii'} : \mathbb{R}_0^+ \to \mathbb{R}$. Furthermore, an integer number b_i, the *demand*, is attached to each node $i \in N$, where we assume $\sum_{i \in N} b_i = 0$. Let $S := \{i \in N : b_i < 0\}$ and $T := \{i \in N : b_i > 0\}$ denote the set of *sources* and the set of *sinks*, respectively.

The *minimum concave-cost network flow problem* $(MCCFP)$ can now be formulated as follows:

$$
\begin{array}{llll}
(MCCFP) & \min & \sum_{(i,i') \in A} h_{ii'}(x_{ii'}) & \\
& \text{s.t.} & \sum_{(i',i) \in A} x_{i'i} - \sum_{(i,i') \in A} x_{ii'} = b_i & \text{for all } i \in N \\
& & 0 \le x_{ii'} \le u_{ii'} & \text{for all } (i, i') \in A.
\end{array}
$$

By introducing A_G, the node-arc incidence matrix of the graph G, and using vector notation for the flow variables $x_{ii'}$, the capacities $u_{ii'}$ and the demands b_i, we obtain the following more compact formulation for the constraints of $(MCCFP)$:

$$A_G x = b \tag{9}$$
$$0 \le x \le u. \tag{10}$$

Complexity of minimum concave cost flow problems

It is well known that $(MCCFP)$ is $\mathcal{NP}$-hard in general. Unfortunately, also many special cases of this problem, such as the fixed-charge network flow problem where the costs $h_{ii'}(\cdot)$ are restricted to be fixed-charge costs and the single source uncapacitated (SSU) network case in which $|S| = 1$ and $u_{ii'} = \infty$ for all arcs $(i, i') \in A$ remain $\mathcal{NP}$-hard. (For further details on complexity issues in connection with $(MCCFP)$ the interested reader is referred to the survey article of Guisewite and Pardalos [9] and the references given therein.)

The inherent difficulty of solving the general $(MCCFP)$ provides the motivation for looking for efficiently solvable special cases. Optimization problems with general nonlinear functions in the objective, like $(MCCFP)$, however, pose certain difficulty as far as their complexity is concerned. The main problem that arises is that there is no complexity model available for the description of general functions (though some attempts to define such a model can be found in the book of Nemirowsky and Yudin [16]). For nonanalytical functions or functions without explicit regular representation, the length of the input may even become infinite. Likewise, the length of the output may become infinite, if the computation of the exact optimal solution is required. For $(MCCFP)$ with integer capacities and demands, however, the latter case cannot occur and there is thus no need to deal with the concept of ε-accurate optimal solutions.

One way out of the problems with encoding the objective function, is to assume that there exists an oracle providing us with the required function values (cf. Hochbaum and Shantikumar [12]), since then no representation is necessary. In this complexity model an algorithm is called *(strongly) polynomial* if both the *number of operations* (additions, multiplications, comparisons etc.) and the *number of objective function evaluations* it performs are (strongly) polynomial in the input length. Of course, if the involved functions can be evaluated in (strongly) polynomial time, the algorithm becomes (strongly) polynomial in the more traditional sense.

As the above definition of polynomiality and strong polynomiality, respectively, seems to be most appropriate for our purpose, we will use it throughout the paper. The same convention is apparently adopted in many papers on polynomially solvable nonlinear optimization problems, in particular on concave-cost flow problems (see e.g. [4, 8, 27]), though mostly only implicitly. (For a more detailed discussion of this topic see Hochbaum and Shantikumar [12]).

Up to now not many special cases of $(MCFFP)$ are known to be solvable in polynomial time (in the sense explained above). Almost all rely on a reduction to some kind of shortest path problem. In this regard, Zangwill [27] gave a polynomial time algorithm for SSU networks with only one sink and Erickson, Monma and Veinott [4] derived from their general send-and-split method polynomial time algorithms for $(MCCFP)$ on k-planar graphs (k fixed), and for several special production and inventory planning problems.

Minimum concave-cost flows with a single nonlinear arc cost

In the following we will consider another interesting special case, namely minimum concave cost flow problems where *all cost functions* $h_{ii'}(\cdot)$, *except a single one, are linear functions.* Recently Guisewite and Pardalos [8] developed a polynomial time algorithm for this problem on SSU networks. In this section we will give a pseudo-polynomial time algorithm for general networks and in the subsequent sections we shall show that for SSU networks our algorithm is in fact strongly polynomial and improves upon the algorithm obtained by Guisewite and Pardalos.

For the ease of exposition, assume from now on that the arcs in A are labeled $\{1, 2, \ldots, m\}$ and that the single arc with nonlinear cost is the first arc in this order.

Furthermore, let $f_1(\cdot)$ be a concave monotone nondecreasing function and denote by $c_j \in \mathbb{R}$ the unit cost of arc j, $j = 2, \ldots, m$. Then the problem under consideration can be written as follows:

$$(MCCFP_1) \qquad \min \; f(x) = f_1(x_1) + \sum_{j=2}^{m} c_j x_j \qquad (11)$$

$$\text{s.t.} \quad (9) \text{ and } (10).$$

Remark. Note that in the bounded solution case the assumption that f_1 is nondecreasing can be made without loss of generality because in that case the flow on an arc is bounded above by the sum of the flows leaving the source. We may e.g. proceed as follows: If necessary, we split arc 1 into two arcs and assign the cost functions $f_1 + g$ and $-g$, respectively, to the newly created arcs, where the linear function g is chosen such that $f_1 + g$ becomes nondecreasing.
An alternative approach to problems with cost functions f_1 which do not satisfy the required monotonicity assumption would be to use the weaker rank two condition (8) instead of condition (1) used in the current paper. In many practical applications, however, f_1 will be nondecreasing from the beginning, anyway. □

3.2 A Parametric Algorithm for Minimum Concave Cost Flow Problems with a Single Nonlinear Arc Cost

It is not hard to see that for $(MCCFP_1)$ the rank two condition (1) holds with

$$c^1 = (1, 0, \ldots, 0) \qquad \text{and} \qquad c^2 = (0, c_2, \ldots, c_m).$$

Indeed if $\langle c^1, w\rangle = w_1 \geq 0$ and $\langle c^2, w\rangle = \sum_{j=2}^{m} c_j w_j \geq 0$ then $f(x + w) \geq f(x)$ for any $x \in \mathbb{R}^m_+$. Observe that for the weaker rank two condition (8) to hold, it is not necessary to require f_1 to be monotone. For efficiency reasons (smaller parameter interval in the associated parametric program), we prefer to use the stronger condition (1) nevertheless.

Consequently, $(MCCFP_1)$ is a special quasiconcave rank-two minimization problem and hence we may apply the parametric method developed in the previous section (cf. Algorithm 1). Recall that Algorithm 1 consists of the following two steps:

1. Solve the parametric linear program (Q_α) which in the case of $(MCCFP_1)$ becomes

$$(TP_\alpha) \qquad \min \; (1-\alpha)x_1 + \alpha \sum_{j=2}^{m} c_j x_j$$

$$\text{s.t.} \quad (9) \text{ and } (10).$$

Note that (TP_α) is a parametric-cost linear network flow problem with arc costs $d_j(\alpha)$ where

$$d_j(\alpha) = \begin{cases} 1 - \alpha & \text{for } j = 1 \\ c_j\alpha & \text{for } j = 2, \ldots, m. \end{cases} \tag{12}$$

Let $\phi(\alpha)$ denote the optimal value function of problem (TP_α) and, as previously, let $\alpha_0 = 0 < \alpha_1 < \ldots < \alpha_{B-1} < \alpha_B = 1$ and $x^{(1)}, \ldots, x^{(B)}$ be the associated sequences of breakpoints and corresponding optimal solutions, respectively. Depending on whether or not there exists a finite optimal solution of (TP_α) for $\alpha \in [\alpha_{k-1}, \alpha_k]$, the optimal solution $x^{(k)}$ is either a vertex or an unbounded edge of the set of feasible flows given by (9) and (10), respectively.

2. Determine $k^* \in \arg\min\{f(x^{(k)}) : k = 1, \ldots, B\}$. Therein $f(x^{(k)})$ is given by $f_1(x_1^{(k)}) + \sum_{j=2}^m c_j x_j^{(k)}$ if $x^{(k)}$ is a vertex and by the minimum of $f_1(x_1) + \sum_{j=2}^m c_j x_j$ over the edge $x^{(k)}$ if $x^{(k)}$ is an unbounded edge (cf. (11)).

   Then $x^{(k^*)}$ is an optimal solution of $(MCCFP_1)$.

Analysis of the parametric algorithm

Step 1: Using an algorithm often attributed to Eisner and Severance [3] the breakpoints α_k and the associated optimal flows $x^{(k)}$ $(k = 1, \ldots, B)$ can be computed in $O(BM(n, m))$ time, where $M(n, m)$ is the time needed to solve a classical minimum cost flow problem on a network with n nodes and m arcs. The currently best strongly polynomial minimum cost flow algorithm with time complexity $M(n, m) = O(m \log n\,(n \log n + m))$ is due to Orlin [17]. (This algorithm essentially consists of a sequence of $O(m \log n)$ shortest path calculations). Hence, Step 1 can be implemented in $O(\,Bm \log n(n \log n + m)\,)$ time, i.e. polynomial in the number of breakpoints of the optimal value function $\phi(\alpha)$. (Other algorithms to solve the parametric-cost problem (TP_α) can e.g. be found in Gusfield [10, 11].)

Step 2: We need to evaluate $f(x^{(k)})$ for $k = 1, \ldots, B$. For $x^{(k)}$ being a vertex, this task simply boils down to one evaluation of the single nonlinear arc cost function f_1. The situation is, however, slightly more complicated in the case of an unbounded edge where we have to decide whether or not the original nonlinear objective function is unbounded below on this edge. Whereas till now it was sufficient to assume that the function values $f_1(t)$ are computable (given to us by an oracle), we now need to require that also the following limit is available: $\lim_{t\to\infty} f_1{}'_+(t)$ where $f_1{}'_+(t)$ denotes the right-hand-side derivative of f_1 at t. Then the infimum of $f(x)$ along an unbounded edge can be computed as follows:

Lemma 3.1 *Let z be an unbounded edge, say $\Gamma = \{q + \lambda y : \lambda \geq 0\}$, of the feasible set and let $\xi = \lim\limits_{t\to\infty} f_1{}'_+(t)$. Then $f(z) = \inf\{f(x) : x \in \Gamma\}$ is given by*

$$f(z) = \begin{cases} -\infty & \textit{if } \xi y_1 + \sum_{j=2}^{m} c_j y_j < 0, \\ f(q) & \textit{otherwise.} \end{cases} \tag{13}$$

Proof. The concave function $f(x)$ either attains its minimum over Γ at the origin q of Γ, or is unbounded below over Γ; the latter case occurs if and only if we have $\inf_{\lambda \geq 0} \lambda^{-1} f(\lambda y) < 0$ (see e.g. Rockafellar [19], Theorem 8.6.). But since $f_1(\lambda y_1) \leq f_1(ty_1) + (\lambda - t) y_1 f_1{}'_+(ty_1)$ for all $t \geq 0$ with equality for $t = \lambda$, we have

$$\begin{aligned} \inf_{\lambda \geq 0} \frac{f(\lambda y)}{\lambda} &= \sum_{j=2}^{m} c_j y_j + \inf_{\lambda \geq 0} \inf_{t \geq 0} \frac{f_1(ty_1) + (\lambda - t) y_1 f_1{}'_+(ty_1)}{\lambda} = \\ &= \sum_{j=2}^{m} c_j y_j + \inf_{t \geq 0} y_1 f_1{}'_+(ty_1) = \sum_{j=2}^{m} c_j y_j + y_1 \xi. \end{aligned}$$

(Note that due to the concavity of f_1 the right-hand-side derivative $f_1{}'_+(t)$ is monotone nonincreasing.) □

Remark. Let us have another look at the condition $\xi y_1 + \sum_{j=2}^{m} c_j y_j < 0$ in (13) from a more combinatorial point of view. For that purpose, observe that due to the fact that $\Gamma = \{q + \lambda y : \lambda \geq 0\}$ is an edge of the flow polyhedron, the arc set $\{j \in A : y_j > 0\}$ induces a cycle C. Hence (13) simply checks whether or not C has negative weight w.r.t. the arc weights c_j for arcs $j = 2, \ldots, m$ and $\xi = \lim_{t \to \infty} f_1{}'_+(t)$ for arc 1, respectively. Note the close relationship of (13) to conditions guaranteeing the existence of a finite optimal solution for the general minimum concave-cost flow problem (cf. e.g. Erickson et al. [4]). □

Summarizing we have obtained the following result:

Theorem 3.2 *Let a network with n nodes and m arcs be given and assume that $\phi(\alpha)$, the optimal value function of the parametric-cost network flow problem (TP_α), has $B - 1$ breakpoints aside from 0 and 1. Furthermore, assume that the limit $\lim_{t \to \infty} f_1{}'_+(t)$ is available. Then the parametric algorithm computes an optimal solution of $(MCCFP_1)$ on this network within $O(Bm \log n(n \log n + m))$ operations and at most B evaluations of the concave cost function f_1.*

Proof. First, note that in Algorithm 1, the parametric algorithm, Step 1 dominates Step 2 with respect to time complexity. As explained above, Step 1 can be implemented to run in $O(Bm \log n(n \log n + m))$ time. The upper bound of B for the number of objective function evaluations is an immediate consequence of the above discussion. □

Remark. As already remarked in Section 2, due to the concavity of $\phi(\alpha)$, an unbounded edge $x^{(k)}$ can only occur for $k = 1$ or $k = B$. In the special case of (TP_α)

dealt with here, however, only $k = B$ is possible, for if there exists a feasible flow x at all, i.e. a vector x satisfying the constraints (9) and (10), then it follows that $\phi(0) > -\infty$. (In fact, we have $\phi(0) = \min\{x_1 : x \text{ is a feasible flow}\} \geq 0$.) Obviously, for nonnegative arc costs and nonnegative concave cost function f_1, we know beforehand that $\phi(\alpha) > -\infty$ for all $\alpha \in [0, 1]$. Hence, in this case, it is not necessary to have the limit $\lim_{t\to\infty} f'_{1+}(t)$ available. □

Upper bounds on the number of breakpoints

Theorem 3.2 provides the motivation for investigating how large the number of breakpoints of $\phi(\alpha)$ might become in the worst case. Unfortunately, a class of parametric-cost linear network flow problems of type (TP_α) for which the number of breakpoints is exponential in the number n of nodes of the underlying network can be derived in the following way:

We start from the famous pathological network class of Zadeh, see [26]. For every integer $r \geq 2$ this class contains a network $\mathcal{N}_r$ with $n = 2r+2$ nodes and $m = r^2+r+2$ arcs. While the network topology and the capacities remain unchanged, the costs have to be modified in order to obtain a problem of type (TP_α). For simplicity, assume that the arc from the source 1 to the sink n in $\mathcal{N}_r$ is arc 1 and let c_j^Z denote the unit cost of arc j in Zadeh's original construction. Then the parametric costs $d_j(\alpha)$ are constructed as follows: We set $d_1(\alpha) = 1 - \alpha$ and $d_j(\alpha) = c_j^Z \alpha$ for $j = 2, \ldots, m$. Furthermore, we require a flow of value $v = 2^r + 2^{r-2} - 2$ by defining the right hand side as follows: $b_1 = -v$, $b_i = 0$ for $i \neq 1, n$ and $b_n = v$. Using similar arguments as in Carstensen [2] and Ruhe [20], it can now easily be shown that in this problem, $\phi(\alpha)$ has $2^r - 2$ breakpoints in the interval $[0, 1]$, aside from the endpoints 0 and 1.

One might ask now whether there exist special cases for which there is a polynomial upper bound on the number of breakpoints. Since $\mathcal{N}_r$ has only one source, there is no hope to get such a bound for single source capacitated networks. By using a classical transformation for getting rid of the capacities (see e.g. Ahuja, Magnanti and Orlin [1]), the same negative result follows for multi source uncapacitated networks. However, as we will see in the next section, there is a nice polynomial bound for the single source uncapacitated case.

Since in the bad network class described above the size of the demands depends exponentially on the problem size n, the question whether there also exist bad networks with "small" demands arises. The answer is "no" and is provided by the pseudo-polynomial bound given in Theorem 3.3 below. The key observation will be that due to the special structure of the costs in (TP_α), the number of breakpoints is bounded above by the number of different values for the flow on arc 1.

Let V denote the set of basic feasible solutions of (TP_α) and define

$$X_1 := \{x_1 : x \in V\},$$

i.e. X_1 is the set of values the flow on arc 1 might attain in a basic feasible solution. Recall that T denotes the set of sinks and c_j and b_i denote the capacity of arc j and the demand of sink i, respectively. We now can show

Theorem 3.3

(i) The optimal cost curve of (TP_α) has at most $|X_1| - 1$ breakpoints aside from 0 and 1.

(ii) Let $R := \min\{u_1, \sum_{i \in T} b_i\}$. Then we have

$$|X_1| \leq R + 1.$$

Proof. As remarked above we have $\phi(\alpha) > -\infty$ for all $\alpha \in [0, \alpha_{B-1}]$ provided there exists a feasible flow at all. From the linearity of the objective of (TP_α) it follows now that for $\alpha \in [0, \alpha_{B-1}]$

$$\phi(\alpha) = \min\{\phi_x(\alpha) : x \in V\}, \tag{14}$$

where $\phi_x(\alpha) = (1-\alpha)x_1 + \alpha \sum_{j=2}^m c_j x_j$. For short we write $\phi_x(\alpha) = x_1 + q_x \alpha$, with $q_x = -x_1 + \sum_{j=2}^m c_j x_j$ (in the sequel x_1 and q_x will be referred to as the intercept and slope, respectively, of the affine function $\phi_x(\alpha)$).
For each $\ell \in X_1$ let V_ℓ be the collection of all $x \in V$ such that $x_1 = \ell$ and let $q_\ell = \min\{q_x : x \in V_\ell\}$ and $\phi_\ell(\alpha) = \ell + q_\ell \alpha$. Since the functions $\phi_x(\alpha)$, $x \in V_\ell$, have equal intercepts, it is clear that $\phi_\ell(\alpha) = \min\{\phi_x(\alpha) : x \in V_\ell\}$ for all $\alpha \in [0,1]$, hence by (14),

$$\phi(\alpha) = \min\{\ell + q_\ell \alpha : \ell \in X_1\}. \tag{15}$$

Now each interval $\Delta_r = [\alpha_{r-1}, \alpha_r]$, $(r = 1, \ldots, B-1)$, corresponds to a value $\ell_r \in X_1$ such that

$$\phi(\alpha) = \ell_r + q_{\ell_r}\alpha \qquad \text{for all } \alpha \in \Delta_r.$$

Furthermore, since $\phi(\alpha)$ is concave and continuous, its slope on Δ_r, i.e. q_{ℓ_r}, must decrease with r, while its intercept ℓ_r must increase. That is, two different intervals correspond to two distinct elements from X_1. As the number of breakpoints aside from 0 and 1 is one less than the number of intervals, (i) follows immediately.

To give an upper bound on $|X_1|$ note that clearly

$$X_1 \subseteq \{\ell : \ell \text{ integer}, 0 \leq \ell \leq R = \min\{u_1, \sum_{i \in T} b_i\}\}. \tag{16}$$

Property (ii) is now a direct consequence. □

Combining Theorems 3.2 and 3.3 we obtain

Corollary 3.4 *Let $R := \min\{u_1, \sum_{i\in T} b_i\}$. In order to compute an optimal solution for $(MCCFP_1)$, the parametric algorithm requires $O(\,Rm\log n\,(n\log n + m))$ operations and at most $R+1$ evaluations of the concave function f_1, again provided that $\lim_{t\to\infty} f'_{1+}(t)$ is available for problems in which the existence of an unbounded optimal solution cannot be ruled out beforehand.*

The bounds in Corollary 3.4 imply that $(MCCFP_1)$ can be solved in pseudo-polynomial time. This is in sharp contrast to a result by Guisewite and Pardalos [9] for the general $(MCCFP)$ in which they show that this problem cannot be solved by a pseudo-polynomial algorithm being polynomial in the sum of the arc capacities unless $\mathcal{P} = \mathcal{NP}$.

4 A Strongly Polynomial Time Algorithm for SSU Networks

In [8] Guisewite and Pardalos gave a polynomial time algorithm for $(MCCFP_1)$ in the case of a single source, uncapacitated (SSU) network. In the sequel we will show that the parametric approach presented above yields another polynomial time algorithm for this special case.

Let node 1 be the only source, i.e. $b_1 < 0$ and $b_i \geq 0$ $(i = 2,\ldots,n)$. Since in the SSU case there are no capacity restrictions we set $u_j = \infty$ for $j = 1,\ldots,m$. Theorem 3.3 can now be strengthened as follows:

Theorem 4.1 *For a SSU network with $|T|$ sinks, $\phi(\alpha)$, the optimal cost curve of (TP_α), has at most $|T|$ breakpoints aside from the two endpoints of the interval $[0,1]$.*

Proof. Recall that for fixed $\alpha \in [0,1]$ the problem (TP_α) is a classical minimum cost flow problem. For the proof it is now essential to observe that the problem of finding a flow of minimum cost on a SSU network can be reduced to a single source shortest path problem. (This can for example easily be seen by applying the classical successive shortest path algorithm, see e.g. Ahuja et al. [1].)

More specifically, for each sink $i \in T$ consider the subproblem $(TP_{\alpha,i})$ obtained from (TP_α) by replacing b_1 by -1, b_i by 1 and b_ℓ $(\ell \neq 1, i)$ by 0. This problem is in fact to find the shortest path $P_i(\alpha)$ from node 1 to node i with respect to the arc weights $d_j(\alpha)$ defined in (12).

Fix $\tilde{\alpha} \in [0,1]$ and let $\phi_i(\alpha)$ denote the optimal value function of problem $(TP_{\alpha,i})$. (In other words, $\phi_i(\tilde{\alpha})$ is the weight of $P_i(\tilde{\alpha})$, the shortest path from 1 to i, in case it exists and $-\infty$ otherwise.) If $y^{(i)}$ is a basic optimal solution of $(TP_{\alpha,i})$ for $\alpha = \tilde{\alpha}$, then, as can easily be checked, $x = \sum_{i\in T} b_i y^{(i)}$ is an optimal solution of (TP_α) for

$\alpha = \tilde{\alpha}$. (x is the flow one obtains by sending b_i units of flow along the path $P_i(\tilde{\alpha})$ for every sink $i \in T$. Hence,

$$\phi(\tilde{\alpha}) = \sum_{i \in T} b_i \phi_i(\tilde{\alpha}) \tag{17}$$

whenever $\phi_i(\tilde{\alpha}) > -\infty$ for all $i \in T$, i.e. whenever there exists a shortest path $P_i(\tilde{\alpha})$ from node 1 to node i for all $i \in T$.

But since the subproblem $(TP_{\alpha,i})$ involves just one sink, with demand 1, for any basic feasible solution x to this subproblem one must have $x_1 \in \{0, 1\}$. ($x_1 = 1$ corresponds to the case where the arc 1 is included in the path $P_i(\tilde{\alpha})$ while for $x_1 = 0$ this arc is not contained in $P_i(\tilde{\alpha})$.)

Relation (16) and Theorem 3.3 now imply that the function $\phi_i(\alpha)$ has at most one breakpoint aside from the two endpoints of the interval $[0, 1]$. formula (17) then shows that $\phi(\alpha)$ has at most $|T|$ breakpoints aside from 0 and 1. □

The following corollary is now an immediate consequence of Theorems 3.2 and 4.1.

Corollary 4.2 *For SSU networks with n nodes and m arcs and $\lim_{t \to \infty} f_1{}'_+(t)$ available, the parametric algorithm computes an optimal solution of problem $(MCCFP_1)$ in $O(\,|T| m \log n \,(n \log n + m)\,)$ operations and at most $|T|+1$ evaluations of the concave cost function f_1.*

As the above bounds depend on n and m only (note $|T| \leq n-1$), $(MCCFP_1)$ can be solved in strongly polynomial time. What we have not exploited so far is that in the SSU network case the parametric network flow problem (TP_α) has a very special structure and hence can be solved more efficiently than the general problem. The key idea appears already in the proof of Theorem 4.1 where it is shown how a solution to (TP_α) can be obtained by shortest path computations (cf. formula (17)). The next section will deal with an efficient implementation of this idea.

5 An Improved Algorithm for SSU Networks

5.1 An Efficient Algorithm for a Special Parametric Shortest Path Problem

In the sequel it will be shown that the parametric shortest path problem that occurred in the last section can be solved asymptotically in the same time already needed for solving a shortest path problem with non-parametric weights.

For $\alpha \in [0, 1]$ consider the weighted digraph $G_\alpha = (N, A, d(\alpha))$ with parametric weights

$$d_j(\alpha) = \begin{cases} 1 - \alpha & \text{for } j = 1 \\ c_j \alpha & \text{for } j = 2, \ldots, m. \end{cases}$$

The problem (SP_α) is then to determine for each sink $i \in T$ the shortest path from the source node 1 to sink i with respect to the weights $d_j(\alpha)$ or to find a negative cycle in G_α, where in the first case the shortest paths are to be chosen such that they form a tree, the so-called *shortest path tree.*

For the ease of exposition we introduce the class $\mathcal{P}^\ell$ of all paths containing arc 1 exactly ℓ times. The following observations follow now easily from the special structure of the weights $d_j(\alpha)$:

1. If G_α contains no negative cycle, then for each sink i, the shortest path from 1 to i either belongs to class $\mathcal{P}^1$ or to class $\mathcal{P}^0$, i.e. it either uses arc 1 exactly once or does not use this arc at all.

2. There exists a maximum value of α, say α^*, such that there are no cycles of negative weight in the graph G_α. (If G_α contains no negative cycle for all $\alpha \in [0,1]$, we set $\alpha^* = 1$.)

3. G_0 contains no negative cycle as $d_j(0) = 0$ for $j = 2, \ldots, m$ and $d_1(0) = 1$.

A closely related parametric shortest path problem, $(SP1_\lambda)$ for short, was treated by Karp and Orlin [13] and subsequently by Young, Tarjan and Orlin [25]. Let $A' \subseteq A$ and $e_j \in \mathbb{R}$ for $j \in A$ and consider the following weights with $\lambda \in \mathbb{R}$:

$$w_j(\lambda) = \begin{cases} e_j - \lambda & \text{for } j \in A' \\ e_j & \text{for } j \in A \setminus A'. \end{cases}$$

Actually, our problem (SP_α) can easily be transformed such that it fits into the framework of $(SP1_\lambda)$. (In this special case, the set A' contains only one arc, namely arc 1.)

Let us therefore first give a brief overview of what is known for problem $(SP1_\lambda)$. In [13] Karp and Orlin showed that the optimal value function of this parametric shortest path problem has at most $n(n-1)/2 = O(n^2)$ breakpoints. Furthermore, they developed two algorithms, the first, a direct approach, running in $O(n^3)$ time and the second, a specialized parametric simplex algorithm, running in $O(mn \log n)$ time. Recently, Young et al. [25] improved the time complexity of the latter approach to $O(mn + n^2 \log n)$ by using Fibonacci-heaps (see Fredman and Tarjan [6]). These time bounds rely however on the worst case assumption of $O(n^2)$ breakpoints whereas (SP_α) has only $O(n)$ breakpoints (cf. Theorem 4.1).

In the following we will deal with the simplifications and improvements that are possible for problem (SP_α). Therein we assume some familiarity with the above algorithms.

The parametric simplex method

The parametric simplex algorithm proceeds tree by tree. First we compute an initial shortest path tree T_0 for an appropriately chosen starting value $\alpha = \alpha_0$. In our case the choice $\alpha_0 = 0$ suggests itself. Assume now that for $\alpha = \alpha_k$ an optimal tree T_k is available. Then our task is to determine the next breakpoint α_{k+1}, where T_k ceases to be a shortest path tree, along with a shortest path tree T_{k+1} optimal at some $\alpha > \alpha_{k+1}$.

It is definitely beyond the scope of this paper to describe in full detail how the transition from tree T_k to its successor T_{k+1} is made in [25]. The main idea, however, is rather simple: It is clear that if T_k ceases to be a shortest path tree, there is a node i which acquires a new shortest path. Obviously in our case as long as $\alpha \leq \alpha^*$, this can happen only if the new shortest path from 1 to i contains arc 1 while this arc was not contained in the old path from T_k. Hence there are at most $n-1$ path changes since afterwards all shortest paths contain arc 1.

Conceptually the parametric simplex algorithm proceeds as follows: For each arc j currently not in T_k, it computes a critical value $\hat{\alpha}$ such that the subgraph we obtain by inserting j into the tree and deleting the unique arc of T_k which is directed into the head of j, has smaller weight than T_k for $\alpha > \hat{\alpha}$. Then an arc j^* with minimum critical value is selected to enter the tree. If the resulting subgraph contains a cycle, we have found α^* and the corresponding cycle whose weight is negative for $\alpha > \alpha^*$. Otherwise, we obtain the next breakpoint α_{k+1} along with the next tree T_{k+1}.

To analyse this algorithm, note first that the initial shortest path problem for $\alpha_0 = 0$ can be solved in $O(m)$ time since only one arc has a weight different from zero. (In the general problem the same task needs $O(nm)$ time since we must be able to detect a negative cycle.) Second, as in our case there at most $n-1$ path changes as opposed to $O(n^2)$ such changes for $(SP1_\lambda)$, we can get rid of a factor of n in the time taken by the algorithm after initialization. Based on the analysis in [25] we thus have obtained an $O(n \log n + m)$ algorithm for solving the parametric shortest path problem (SP_α). This is the same time as already needed for solving a single shortest path problem with non-parametric weights.

The direct method

Let $D_i^\ell(\alpha)$ denote the length of the shortest path w.r.t the weights $d_j(\alpha)$ that starts in node 1, terminates in node i and uses the special arc 1 exactly ℓ times. Further let $\delta(i, i', \alpha)$ designate the length of the shortest path from i to i' not containing arc 1 and assume that arc $1 = (i_1, i_2)$. The following lemma can now be easily established directly (it follows also from the results of Karp and Orlin [13]).

Lemma 5.1

(i) Let $\alpha \leq \alpha^*$. Then we have

$$D_i^\ell(\alpha) = \delta(1, i_1, \alpha) + \ell \cdot (1 - \alpha) + (\ell - 1) \cdot \delta(i_2, i_1, \alpha) + \delta(i_2, i, \alpha) \tag{18}$$

for all $i \in N$ and $1 \leq \ell \leq n$.

(ii) If the graph $G = (N, A \setminus \{1\})$ contains a negative cycle then $\alpha^* = 0$. Otherwise let W be such that $\delta(i_2, i_1, \alpha) = W\alpha$. Then

$$\alpha^* = \begin{cases} 1/(1-W) & \text{for } W < 1 \\ 1 & \text{for } W \geq 1\,. \end{cases} \tag{19}$$

Proof. *(i):* Let us start with $\ell = 1$. Observe that a shortest path from 1 to i using arc (i_1, i_2) exactly once is obviously composed of a shortest path P_1 from node 1 to node i_1, the arc (i_1, i_2) and a shortest path P_2 from node i_2 to node i. (Of course neither P_1 nor P_2 contain arc 1.) As the weight of arc 1 is $d_1(\alpha) = 1 - \alpha$, formula (18) is correct for $\ell = 1$. The correctness for $\ell > 1$ can be established by induction. Just note that a shortest path from 1 to i in class $\mathcal{P}^\ell$ can be split into a shortest path from 1 to i_1 belonging to the class $\mathcal{P}^{\ell-1}$, the arc $1 = (i_1, i_2)$ and a shortest path from i_2 to i in class $\mathcal{P}^0$.

(ii): In case that $G = (N, A \setminus \{1\})$ contains a negative cycle, then this cycle will be negative for all $\alpha \in (0, 1]$. Hence $\alpha^* = 0$. From (18) it follows that

$$D_i^{\ell_2} - D_i^{\ell_1} = (\ell_2 - \ell_1) \cdot ((1 - \alpha) + \delta(i_2, i_1, \alpha))$$

for $1 \leq \ell_1 < \ell_2 \leq n$. The expression $((1 - \alpha) + \delta(i_2, i_1, \alpha))$ equals however the weight of the cycle C which consists of arc 1 and the shortest path from i_2 to i_1 within the class $\mathcal{P}^0$. It follows from Theorem 2 of [13] that α^* is equal to the value of α for which the weight $W(C, \alpha)$ of C becomes 0. By the definition of W the weight of C is however equal to $W(C, \alpha) = (1 - \alpha) + W \cdot \alpha$. For $W < 1$, $W(C, \alpha)$ becomes negative for $\alpha > 1/(1 - W)$, whereas for $W \geq 1$, $W(C) \geq 0$ for all $\alpha \in [0, 1]$. □

Recall that for $\alpha \leq \alpha^*$ the length of the shortest path from node 1 to node i is given by $\min\{D_i^0(\alpha), D_i^1(\alpha)\}$. Hence it is an immediate consequence of Lemma 5.1 that the direct method solves (SP_α) essentially by solving two single source shortest path problems, one with source 1 (to compute $\delta(1, i, \alpha)$ for $i \in N$) and the other with source i_2 (to determine $\delta(i_2, i_1, \alpha)$). In case that no negative weights are involved in these shortest path computations (i.e. for $c_j \geq 0$ for all arcs $j = 2, \ldots, m$), we get an $O(n \log n + m)$ algorithm, thus the same time bound as we already obtained for the parametric simplex approach. In the general case we need $O(nm)$ steps to solve (SP_α). One advantage of the direct approach over the parametric simplex method is however that it might be easier to implement.

5.2 An Improved SSU Implementation

From the results above it follows that Corollary 4.2 can be strengthened as follows:

Corollary 5.2 *For SSU networks with n nodes and m arcs and* $\lim_{t \to \infty} f'_{1+}(t)$ *available, the parametric algorithm can be implemented to solve problem* $(MCCFP_1)$ *in* $O(n \log n + m)$ *operations and at most* $|T| + 1$ *evaluations of* f_1.

Comparison with the algorithm of Guisewite and Pardalos

To obtain additional insight into the problem structure, let us briefly compare our parametric algorithm for the *SSU* case and the algorithm of Guisewite and Pardalos [8], henceforth called *GP* algorithm for short. Though obtained by fundamentally different approaches, our algorithm and the GP algorithm share several interesting properties. Both are two-phase approaches where in the first phase a set of up to $|T|+1$ candidate flows is determined and in the second phase the objective function is minimized over this candidate set only. The essential difference lies merely in the method used for constructing the candidate set.

Define $T' \subseteq T$ to contain all sinks i for which the breakpoint of the optimal value function $\phi_i(\alpha) = \min\{D_i^0(\alpha), D_i^1(\alpha)\}$ (cf. also Section 4) is contained in the interval $[0,1]$. (These are the sinks for which a path change occurs in the course of the parametric simplex algorithm.) What our algorithm and the GP algorithm have in common is that for both there exists an ordering $t_{k_1}, t_{k_2}, \ldots, t_{k_q}$ of the set T', $q = |T'|$, such that the set of candidate flows used in the second phase can be described as follows:

(i) For the first candidate flow we send b_i units of flow from the source 1 to each sink i along the shortest path in the class $\mathcal{P}^0$ whenever this class contains at least one path from 1 to i. For sinks i with no such path, the shortest path from 1 to i in the class $\mathcal{P}^1$ is used instead.

(ii) The r-th candidate flow is obtained from the $(r-1)$-st flow by rerouting the flow to be sent to sink t_{k_r} along the shortest path from 1 to t_{k_r} belonging to class $\mathcal{P}^1$. Hence more and more flow is sent on arc 1.

In our algorithm the ordering of T' is chosen such that the breakpoints of the functions $\phi_i(\alpha), i \in T'$, are in increasing order. As the weights in the parametric shortest path problem (SP_α) do not depend on the single nonlinear concave cost function f_1, this ordering is independent of f_1.

In the GP algorithm the ordering of the set T' is obtained by a binary search approach which in contrast to the method described in this paper depends on the cost function f_1. (Actually, Guisewite and Pardalos construct an ordering of the whole set T and not only of T', but it can easily be shown that this is not necessary since for sinks not in T', arc 1 will never become beneficial in the sense of Guisewite and Pardalos.) Hence, in general the orderings obtained by the GP algorithm and by our algorithm will differ.

Though in [8] no tight analysis is given, it is easy to see that the GP algorithm can be implemented to run in $O(S(n,m) + |T|\log R + |T|\log|T|)$ operations and $O(|T|\log R)$ evaluations of f_1, where R is the total demand, thus $R = \sum_{i\in T} b_i$, and $S(n,m)$ denotes again the time needed to solve a single source shortest path problem. (Thus $S(n,m) = O(n\log n + m)$ or $S(n,m) = O(nm)$ depending on whether or not the given arc costs c_j are restricted to be nonnegative.) Note that both the number of operations and the number of evaluations of f_1 depend on R and are thus not strongly polynomial.

6 Summary and Concluding Remarks

In this paper we considered $(MCCFP_1)$, the minimum concave-cost flow problem when all arcs except one have linear cost. By specializing a method recently developed for rank two quasiconcave minimization problems, we derived a pseudo-polynomial time algorithm for this problem. For a network with n nodes and m arcs, our algorithm runs in $O(\,Rm\log n\,(n\log n+m)\,)$ time and performs at most R evaluations of the single nonlinear arcs cost function, where R is an upper bound on the flow on the special arc (i.e. $R=\min\{\,u_1,\sum_{i\in T}b_i\,\}$). In the case of a single source uncapacitated network with $|T|$ sinks we described a strongly polynomial time implementation requiring $O(n\log n+m)$ operations and $|T|+1$ evaluations of the single nonlinear arc cost function.

There are, however, still quite a few issues that deserve further attention. Let us briefly mention some of them.

(1) Since for general networks our algorithm to solve $(MCCFP_1)$ is only pseudo-polynomial, the question on the complexity status of this problem arises — is it NP-hard or can a polynomial time algorithm be developed?

(2) In $(MCCFP_1)$ we allowed only one arc to have a nonlinear cost. A natural generalization would be to consider $k\geq 2$ arcs with nonlinear cost for fixed k. In two forthcoming papers it will be shown how the techniques developed in this paper can be generalized to solve the more general problem efficiently. It turns out that for fixed k the SSU network case can again be solved in strongly polynomial time.

(3) It appears to be a challenging research question to investigate to what other types of combinatorial optimization problems the dimension reduction technique based on the rank two and rank κ properties respectively could be applied successfully.

(4) Are there any other interesting special cases of $(MCCFP_1)$ apart from the SSU network case for which the number of breakpoints of the optimal value function $\phi(\alpha)$ is bounded polynomially in the number of nodes of the given network?

Acknowledgement: This work was initiated during Hoang Tuy's stay at the Technical University of Graz in the Winter term 1991. Bettina Klinz acknowledges partial financial support by the Christian Doppler Laboratorium "Diskrete Optimierung" and by the Fonds zur Förderung der wissenschaftlichen Forschung, Project P8971-PHY.

References

[1] R.K. Ahuja, T. Magnanti and J.B. Orlin, Network Flows, in: G.L. Nemhauser et al. (eds.), *Handbooks in OR & MS*, Vol. 1, (North Holland, Amsterdam, 1989) pp. 211–369.

[2] P.J. Carstensen, Complexity of some parametric integer and network programming problems, *Mathematical Programming* **5** (1983) 64–75.

[3] M.J. Eisner and D.G. Severance, Mathematical techniques for efficient record segmentation in large shared databases, *Journal of the Association for Computing Machinery* **23** (1976) 619–635.

[4] R.E. Erickson, C.L. Monma and A.F. Veinott, Send-and-split method for minimum-concave-cost network flows, *Mathematics of Operations Research* **12** (1987) 634-664.

[5] F. Forgo, The solution of a special quadratic problem, (in Hungarian), *Szigma* (1975) 53–59.

[6] M.L. Fredman and R.E. Tarjan, Fibonacci heaps and their uses in improved network optimization algorithms, *Journal of the Association for Computing Machinery* **34** (1987) 596–615.

[7] R. Gabasov and F.M. Kirillova, *Linear Programming Methods*, Part 3 (special problems) (in Russian) (Minsk, 1980).

[8] G.M. Guisewite and P.M. Pardalos, A polynomial time solvable concave network flow problem, Technical Report CS-90–48, Department of Computer Science, Pennsylvania State University, PA (1990), submitted to *Networks.*

[9] P.M. Guisewite and P.M. Pardalos, Minimum concave-cost network flow problems: applications, complexity and algorithms, *Annals of Operations Research* **25** (1990) 75–100.

[10] D. Gusfield, Sensitivity analysis for combinatorial optimization, UCN/ERL M80/22, Research Memorandum, Electronics Research Laboratory, Berkeley, CA (1980).

[11] D. Gusfield, Parametric combinatorial computing and a problem of program module distribution, *Journal of the Association for Computing Machinery* **30** (1983) 551–563.

[12] D.S. Hochbaum and J.G. Shantikumar, Convex separable optimization is not much harder than linear optimization, *Journal of the Association for Computing Machinery* **37** (1990) 843–862.

[13] R.M. Karp, R.M. and J.B. Orlin, Parametric shortest path algorithms with an application to cyclic staffing, *Discrete Applied Mathematics* **3** (1981) 37–45.

[14] H. Konno and T. Kuno, Linear Multiplicative Programming, *Mathematical Programming* **56** (1992) 51–64.

[15] K.G. Murty, *Linear Programming* (John Wiley & Sons, New York, 1983).

[16] A.S. Nemirowsky and D.D. Yudin, *Problem Complexity and Method Efficiency in Optimization* (John Wiley & Sons, New York, 1983).

[17] J.B. Orlin, A faster strongly polynomial minimum cost flow algorithm, *Proc. 20-th Annual Symp. Theory of Computing* (1988) 377–387.

[18] P.M. Pardalos, Polynomial time algorithms for some classes of constrained nonconvex quadratic problems, *Optimization* **21** (1990) 843–853.

[19] R.T. Rockafellar, *Convex Analysis* (Princeton University Press, Princeton, 1970).

[20] G. Ruhe, *Algorithmic Aspects of Flows in Networks*, Mathematics and Its Applications, Volume 69 (Kluwer Academic Publishers, Doortrecht, 1991).

[21] P.T. Thach, R.E. Burkard and W. Oettli, Mathematical programs with a two-dimensional reverse convex constraint, *Journal of Global Optimization* **1** (1991) 145–154.

[22] H. Tuy and B.T. Tam, An efficient solution method for rank two quasiconcave minimization problems, *Optimization* **24** (1992) 43–56.

[23] H. Tuy, The complementary convex structure in global optimization, *Global Optimization* **2**, 21–40.

[24] H. Tuy, Polyhedral annexation, dualization and dimension reduction technique in global optimization, *Journal of Global Optimization* **1** (1991), 229–244.

[25] N.E. Young, R.E. Tarjan and J.B. Orlin, Faster parametric shortest path minimum-balance algorithms, *Networks* **21** (1991) 205–221.

[26] N. Zadeh, A bad network problem for the simplex method and other minimum cost flow algorithms, *Mathematical Programming* **5** (1973) 255-266.

[27] W. Zangwill, Minimum concave cost flows in certain networks, *Management Science* **14** (1968) 429–450.

Network Optimization Problems, pp. 147-167
Eds. D.-Z. Du and P.M. Pardalos

A Method for Solving Network Flow Problems with General Nonlinear Arc Costs

Bruce W. Lamar
Department of Management, University of Canterbury, Christchurch, New Zealand

Abstract

General nonlinear network flow problems concern the minimization of costs over networks involving arcs with arbitrary cost functions. Finding the optimal solution to such problems is challenging because the arc cost functions are, in general, neither convex nor concave. This paper presents a procedure for converting any network with general nonlinear arc cost functions into an equivalent network containing only concave arc cost functions. This conversion procedure permits established solution methods for concave minimum cost network flow problems to be applied to networks with arbitrary arc cost functions.

1 Introduction

Single commodity network flow problems can be classified according to the complexity of their solution procedure. The salient feature of this taxonomy is the form of the arc cost function. Accordingly, we consider four classes of minimum cost network flow problems (MCNFP's) based on the arc costs: (*i*) linear, (*ii*) convex, (*iii*) concave, and (*iv*) general nonlinear.

Linear MCNFP's have constant marginal arc costs. Because linear MCNFP's are a special form of linear program, this class of problems can be solved in polynomial time. Moreover, basis updating can be performed using simple list processing techniques [13]. Thus, linear MCNFP's can be solved very efficiently. Algorithms and applications for linear MCNFP's are discussed in [1, 4, 15, 20].

Convex MCNFP's have nondecreasing marginal arc costs. This class of problems involves the minimization of a convex function over a convex feasible region. Thus, a local optimum is also a global optimum. Iterative procedures, wherein each subproblem is a linear MCNFP, are used to solve convex MCNFP's. This means that, although convex MCNFP's are harder to solve than linear MCNFP's, large scale convex MCNFP's (on the order of 1000's of arcs) are still tractable. For convex MCNFP algorithms and applications see [12, 21].

Concave MCNFP's have nonincreasing marginal arc costs. Concave MCNFP's are considerably harder to solve than either linear MCNFP's or convex MCNFP's. This increased complexity arises from the fact that—although an optimal solution (if one exists) always occurs at an extreme point of the feasible region—identification of the optimal point requires, in the worst case, a complete enumeration of all the extreme points in the feasible region. In fact, the concave MCNFP is known to be NP-hard [5]. So, except for special cases, exact algorithms for concave MCNFP's run in exponential time. For moderate sized problems (involving 100's of arcs), exact procedures—using branch and bound or dynamic programming techniques—are available. Large scale problems, however, require heuristic methods. Recent work on concave MCNFP's is contained in [7, 8, 9, 10, 11, 14, 16, 19]. See Guisewite and Pardalos [6] for a comprehensive survey of algorithms and applications for concave MCNFP's.

Lastly, general nonlinear MCNFP's allow for arbitrary arc cost functions including functions that are neither concave nor convex [1]. Because of the arbitrary nature of the arc cost functions, general nonlinear MCNFP's are the most difficult of the four classes of MCNFP's to solve optimally. Yet, many practical problems involving the transport of people, commodities, or information have a general nonlinear arc cost structure. For example, consider the transport of passengers on vehicles with limited capacity. The costs are typically dominated by fixed charges (such as driver cost, insurance, fuel charges) that are proportional to the number of vehicles (not the number of passengers). Thus, if arc flow is in units of passengers, the arc cost function is in the form of a "staircase" as shown in Figure 1. For a second example, consider the distribution system of a wholesaler. The costs in the distribution network may typically include setup costs and quantity discounts. Larger volume flows on an arc will have lower marginal costs. But there will be discontinuities at the volume "breakpoints" making the arc cost functions neither concave nor convex. Instead, this cost structure will result in a "sawtooth" arc cost function of the form shown in Figure 2. See [3] for a discussion of quantity discounting and other forms of discontinuity in cost functions.

The focus of this paper is on the last—and most challenging—class of MCNFP's. The paper is organized as follows. Section 2 formulates the general nonlinear MCNFP as a nonlinear integer programming problem and describes an equivalent formulation of the problem as a concave MCNFP. The equivalence between the two problems is based on converting each arc with an arbitrary cost function in the original problem

[1]Note that a general nonlinear MCNFP is different than a "generalized MCNFP". A "generalized MCNFP" refers to a network with flow losses or gains and either linear or convex arc cost functions. "Generalized MCNFP's" are not discussed in the paper.

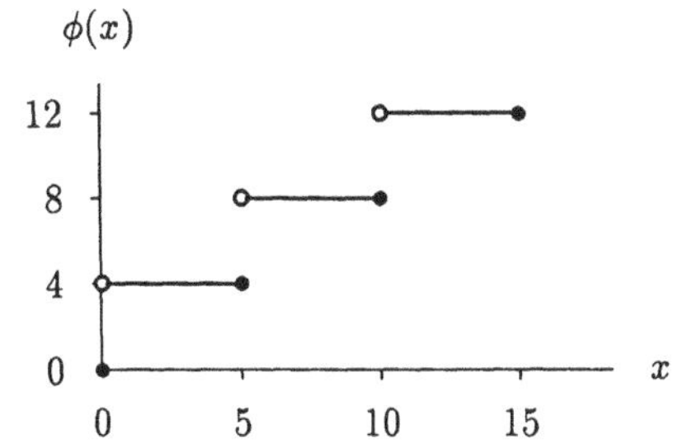

Figure 1: "Staircase" Arc Cost Function.

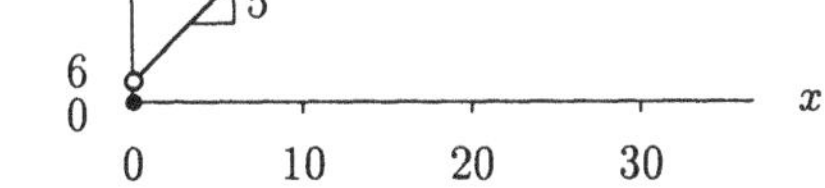

Figure 2: "Sawtooth" Arc Cost Function.

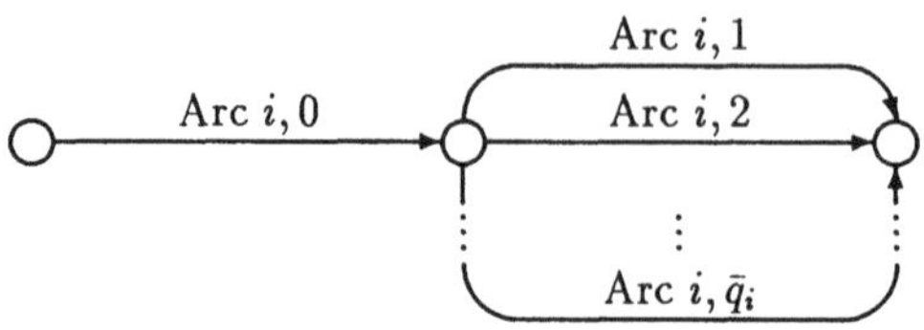

Figure 3: Arcs in Expanded Network.

into an arc with a concave cost function in series with a set of parallel arcs, each with a linear (and therefore concave) arc cost function. This conversion procedure is described in Section 3. Section 4 illustrates this procedure with two numerical examples. Finally, Section 5 summarizes the paper.

2 Problem Formulation

In this section we formulate two problems, denoted P and Q. Problem P is the general nonlinear MCNFP. Problem Q is equivalent to Problem P. However, Problem Q is a concave MCNFP. To make the two problems equivalent, each arc in Problem P is replaced with an arc in series with a set of parallel arcs, as shown in Figure 3. Thus, although Problem Q involves a larger network, it can be solved using established solution techniques for concave MCNFP's.

2.1 General Nonlinear MCNFP Formulation

To formulate Problem P, let

- A denote the (directed) arc set with generic element i;
- N denote the node set with generic element n;
- L_n denote the set of arcs leaving node n;
- E_n denote the set of arcs entering node n;
- x_i denote the flow carried on arc i;
- u_i denote the flow capacity for arc i;
- b_n denote the net supply at node n (if $b_n > 0$ then n is supply node; if $b_n < 0$ then n is a demand node);

- $\phi_i(x_i)$ denote the general nonlinear cost function for arc i defined over the domain of x_i (i.e., $[0, u_i]$).

The decision variables in the problem are the set of arc flows $\{x_i\}$. Using this notation, Problem P (the general nonlinear MCNFP) is formulated as follows:

Problem P:

$$z^* = \text{globalmin} \sum_{i \in A} \phi_i(x_i) \tag{1}$$

subject to:

$$\sum_{i \in L_n} x_i - \sum_{i \in E_n} x_i = b_n \text{ for all } n \in N \tag{2}$$

$$0 \leq x_i \leq u_i \text{ for all } i \in A \tag{3}$$

$$x_i \text{ integer for all } i \in A \tag{4}$$

The objective function (1) minimizes costs. Constraints (2) represent the conservation of flow at each node. Constraints (3) restrict the flow decision variables $\{x_i\}$ to feasible values on each arc. Because of the arbitrary form of the objective function, the optimal value of the decision variables will, in general, not be integer-valued unless explicitly constrained as in (4).

2.2 Equivalent Concave MCNFP Formulation

We refer to the network for Problem P as the "original network". To construct Problem Q, we replace each arc i in arc set A in the original network with an arc in series with a set of parallel arcs (again, refer to Figure 3). We refer to the network for Problem Q as the "expanded network". For each arc i in the original network, we refer to the series arc in the expanded network as arc $i, 0$. Also, for each arc i in the original network, let q be the index of the set of parallel arcs in the expanded network and let $\bar{q}_i$ be the number of parallel arcs. The parallel arcs are denoted as arcs $i, 1$ through $i, \bar{q}_i$. Let x_i be the flow on arc $i, 0$ and let $x_{i,q}$ be the flow on arc i, q for $q = 1, \ldots, \bar{q}_i$. In addition, let u_i be the capacity of arc $i, 0$; and let $u_{i,q}$ be the capacity of arc i, q for $q = 1, \ldots, \bar{q}_i$. We define $r_i(x_i)$ to be the cost function for arc $i, 0$; and we define $s_{i,q}(x_{i,q})$ to be the cost function for arc i, q for $q = 1, \ldots, \bar{q}_i$. By construction, $r_i(x_i)$ will be a piecewise-linear-concave function and each $s_{i,q}(x_{i,q})$ will be a linear (and thus concave) function for $q = 1, \ldots, \bar{q}_i$.

The decision variables in Problem Q are the arc flows in the expanded network; that is, $\{x_i\}$ and $\{x_{i,q}\}$. We now formulate Problem Q (the concave MCNFP equivalent of Problem P) as follows:

Problem Q:

$$z^* = \text{globalmin} \sum_{i \in A} r_i(x_i) + \sum_{i \in A} \sum_{q=1}^{\bar{q}_i} s_{i,q}(x_{i,q}) \tag{5}$$

subject to:

$$\sum_{i \in L_n} x_i - \sum_{i \in E_n} x_i = b_n \text{ for all } n \in N \tag{6}$$

$$x_i = \sum_{q=1}^{\bar{q}_i} x_{i,q} \text{ for all } i \in A \tag{7}$$

$$0 \leq x_i \leq u_i \text{ for all } i \in A \tag{8}$$

$$0 \leq x_{i,q} \leq u_{i,q} \text{ for all } i \in A, q = 1, \ldots, \bar{q}_i \tag{9}$$

The objective function (5) minimizes costs in the expanded network. Constraints (6) and (7) ensure that flow is conserved at all nodes in the expanded network. Constraints (8) and (9) allow only feasible flows on the arcs in the expanded network. Note that if the net supplies $\{b_n\}$ and the capacities $\{u_i\}$ and $\{u_{i,q}\}$ are integer valued, then the integer constraints (4) are not required in Problem Q. This is because every extreme point of the feasible region will be integer valued; and the minimal value of a concave function over a convex region will always occur at an extreme point [2]. In other words, Problem Q is a concave minimization problem whereas Problem P is a general nonlinear integer programming problem. Because of the equivalence between the two problems, the optimal solution $\{x_i^*\}$ to Problem Q will also be optimal in Problem P; and the optimal objective function value, z^*, will be the same for the two problems. However, because Problem Q is a concave MCNFP, it can be solved by established solution techniques for this class of problems (see [6]).

Note that some concave MCNFP algorithms (e.g., [22]) assume that the arc cost functions are continuously differentiable. In such cases, the piecewise-linear-concave function $r_i(x_i)$ can be approximated by a continuously differentiable function, denoted $\tilde{r}_i(x_i)$. We let Problem $\tilde{Q}$ denote the approximation of Problem Q in which the summation of $r_i(x_i)$ is objective function (5) is replaced by the summation of $\tilde{r}_i(x_i)$.

The next section describes a general procedure for determining the cost functions $\{r_i(x_i)\}$ and $\{s_{i,q}(x_{i,q})\}$ and the capacities $\{u_{i,q}\}$ in the expanded network of Problem Q (and the approximate functions $\tilde{r}_i(x_i)$ in Problem $\tilde{Q}$).

3 Conversion Procedure

In this section, we demonstrate how an arc in the original network with a general nonlinear cost function can be converted into a set of arcs in the expanded network consisting of an arc with a concave cost structure in series with a set of parallel arcs, each with a linear arc cost function. Figure 4 provides an overview of the five steps involved in the conversion procedure. We point out, however, that these steps are for conceptual purposes only. That is, they are intended solely to explain the rationale behind the conversion procedure. The actual mechanics of converting an arc in the original network with a general nonlinear cost structure into the form described above are summarized in Subsection 3.6.

In this section we consider the conversion procedure for a generic arc i contained in the original arc set A. Because it is understood that in this section we are always referring to a given arc i, the subscript i is omitted on all coefficients and variables defined and used in this section. Thus, we let x be the flow on arc i, u be the flow capacity for arc i, and $\phi(x)$ be the general nonlinear cost function for arc i. We allow $\phi(x)$ to be any possible functional form. However, we restrict out attention to flows x on arc i that are integer-valued (as required by constraints (4)).

3.1 Step 1

In the first step of the conversion procedure, we observe that, without loss of generality, we can replace $\phi(x)$ with a piecewise-linear-continuous function, denoted $f(x)$, such that $f(x) \equiv \phi(x)$ for x integer.

Note that, like $\phi(x)$, the domain of $f(x)$ for arc i is $[0, u]$. We now divide this domain into contiguous intervals and we divide each interval into contiguous segments. Let j be the index for the intervals and let k be the index for the segments. Also let $\bar{j}$ denote the number of intervals for arc i and let $\bar{k}_j$ denote the number of segments in the j-th interval for arc i. In addition, let $m_{j,1}$ and $m_{j+1,1}$ denote, respectively, the left and right endpoints for the j-th interval for arc i; and define the set I_j as

$$I_j = \{x : m_{j,1} \leq x \leq m_{j+1,1}\} \tag{10}$$

With this notation, we define the intervals for arc i such that the number of segments $\bar{j}$ is as small as possible given that the following two conditions are satisfied:

- If j is odd, then $f(x)$ is a piecewise-linear-*concave* function for $x \in I_j$;
- If j is even, then $f(x)$ is a piecewise-linear-*convex* function for $x \in I_j$.

For the j-th interval of arc i, let $m_{j,k}$ and $m_{j,k+1}$ denote, respectively, the left and right endpoints of the k-th segment. Note that, because the intervals and segments are contiguous, the right endpoint of the last segment of the j-th interval always equals the left endpoint of the first segment of the $(j+1)$-st interval. That is, $m_{j,\bar{k}_j+1} \equiv m_{j+1,1}$. Moreover, the left endpoint of first segment of the first interval of arc i is always zero and the right endpoint of the last segment of the last interval of arc i is always equal to the capacity u. That is, $m_{1,1} \equiv 0$ and $m_{\bar{j},\bar{k}_{\bar{j}}+1} \equiv m_{\bar{j}+1,1} \equiv u$.

To specify the segments for the j-th interval, we define the set $I_{j,k}$ as

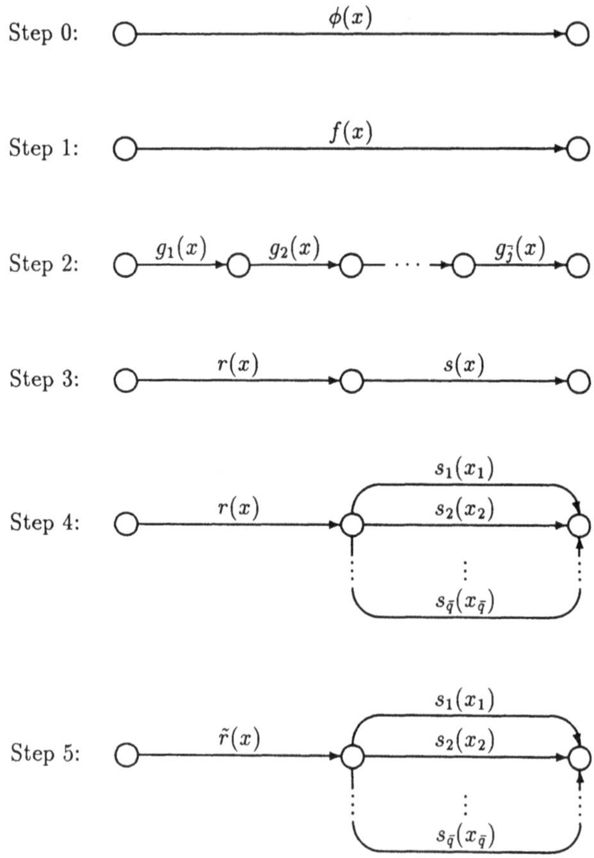

Figure 4: Steps in Conversion Procedure.

$$I_{j,k} = \{x : m_{j,k} \leq x \leq m_{j,k+1}\} \tag{11}$$

The segments of interval j for arc i are defined such that $\bar{k}_j$, the number of segments for the j-th interval, is as small as possible given that $f(x)$ is linear for each $x \in I_{j,k}$.

3.2 Step 2

Having defined the intervals and segments for arc i, we can now describe the second step in the conversion procedure (again, refer to Figure 4). In the second step, we replace the single arc i with a series of $\bar{j}$ arcs, one associated with each interval for arc i. We refer to the j-th arc in this series as arc i, j and we let $g_j(x)$ denote the cost function for arc i, j. To specify $g_j(x)$, let $\Delta_{j,k}$ denote the slope of the (linear) function $f(x)$ for $x \in I_{j,k}$ and let $\hat{\Delta}_j$ denote the slope of $f(x)$ in the $\bar{k}_j$-th (i.e., last) segment of interval j. We define $\hat{\Delta}_0 \equiv 0$. Note that $g_j(x)$ must be defined for x in the domain $[0, u]$, not just $[m_{j,1}, m_{j+1,1}]$. Thus, for each arc i, j we define two additional segments—the *zero*-th segment and the $(\bar{k}_j + 1)$-st segment. For each arc i, j, we also define the sets $I_{j,0}$ and $I_{j,\bar{k}_j+1}$ as

$$I_{j,0} = \{x : 0 \leq x \leq m_{j,1}\} \tag{12}$$

$$I_{j,\bar{k}_j+1} = \{x : m_{j+1,1} \leq x \leq u\} \tag{13}$$

We now specify the cost function $g_j(x)$ for arc i, j recursively as

$$g_j(x) = 0 \text{ for all } x \in I_{j,0} \tag{14}$$

$$g_j(x) = g_j(m_{j,k}) + (\Delta_{j,k} - \hat{\Delta}_{j-1}) \cdot (x - m_{j,k}) \text{ for all } x \in I_{j,k}, k = 1, 2, \ldots, \bar{k}_j \tag{15}$$

$$g_j(x) = g_j(m_{j+1,1}) + (\hat{\Delta}_j - \hat{\Delta}_{j-1}) \cdot (x - m_{j+1,1}) \text{ for all } x \in I_{j,\bar{k}_j+1} \tag{16}$$

Note that, for j odd, $g_j(x)$ is a piecewise-linear-concave function for x in the domain $[0, u]$. Similarly, for j even, $g_j(x)$ is a piecewise-linear-convex function for x in the domain $[0, u]$.

The logic underlying the specification of the cost functions $g_j(x)$ is as follows. For flows x in the first interval, i.e., for $x \in I_1$, the function $g_1(x)$ is the same as the function $f(x)$ and the other arc cost functions $g_2(x), \ldots, g_{\bar{j}}(x)$ all have zero cost. For flows x in the second interval, i.e., for $x \in I_2$, the *sum* of the functions $g_1(x)$ plus $g_2(x)$ is the same as the function $f(x)$ and the remaining functions $g_3(x), \ldots, g_{\bar{j}}(x)$ all have zero cost. For flows $x \in I_3$, the sum of $g_1(x)$ plus $g_2(x)$ plus $g_3(x)$ equals $f(x)$ and the remaining functions $g_4(x), \ldots, g_{\bar{j}}(x)$ are all zero; and so on. Thus, for any x between zero and u, the cost of sending x units of flow through the series of arcs $i, 1$

through $i, \bar{j}$ is the same as the cost of sending the same amount of flow through the single arc i.

3.3 Step 3

In the third step of the conversion process, we simply note that the *order* of the arcs in the series $i,1$ through $i,\bar{j}$ is unimportant. This means that we may equivalently rearrange the order of the series of arcs such that the arcs i,j for j odd are to the left of the arcs i,j for j even. Furthermore, we can recombine the arcs i,j with j odd into a single arc. We refer to this recombined arc as arc $i,0$. Let $r(x)$ denote the cost function for arc $i,0$ where

$$r(x) = \sum_{j=1,3,\ldots}^{\bar{j}} g_j(x) \tag{17}$$

Observe that $r(x)$ is the sum of piecewise-linear-concave functions and so is itself a piecewise-linear-concave function. Moreover, the endpoints of the linear segments of $r(x)$ are given by $m_{j,k}$ for $k = 1,\ldots,\bar{k}_j$ and j odd. However, to simplify notation, let p be the index of the linear segments of the function $r(x)$, let $\bar{p}$ be the number of segments, and let v_p and v_{p+1} be the left and right endpoints of the p-th segment.

In a similar fashion, we can replace the arcs i,j for j even with a single arc. Let $s(x)$ denote the cost function for this recombined arc where

$$s(x) = \sum_{j=2,4,\ldots}^{\bar{j}} g_j(x) \tag{18}$$

Here, $s(x)$ is a piecewise-linear-convex function and the endpoints of the linear segments are given by $m_{j,k}$ for $k = 1,\ldots,\bar{k}_j$ and j even. Once again, we simplify notation by letting q be the index of the linear segments of $s(x)$, letting $\bar{q}$ be the number of segments, and letting w_q and w_{q+1} be the left and right endpoints of the q-th segment.

3.4 Step 4

In the fourth step of the conversion procedure, we use the established method of converting a piecewise-linear-convex arc cost function with $\bar{q}$ linear segments into a set of $\bar{q}$ parallel arcs, each with a linear cost function (see [12, p. 80]). We denote the q-th parallel arc as arc i,q, we let x_q denote the flow on arc i,q, and we let $s_q(x_q)$ be the (linear) cost function for arc i,q. The functional form of $s_q(x_q)$ is

$$s_q(x_q) = \delta_q \cdot x_q \tag{19}$$

where δ_q is the slope of the q-th linear segment of the piecewise-linear function $s(x)$. We let u_q denote the flow capacity for arc i,q. Here, u_q is given by

$$u_q = w_{q+1} - w_q \tag{20}$$

3.5 Step 5

Up to this point, the conversion procedure we have described is exact. That is, solving a network with each arc i of the form shown in Step 0 in Figure 4 (i.e., the original network of Problem P) is identical to solving a network in which each arc i is replaced with the set of arcs shown in Step 4 (i.e., the expanded network of Problem Q). As mentioned at the end of Section 2, however, it may be desirable to approximate $r(x)$ with a continuously differentiable concave function, denoted $\tilde{r}(x)$. Thus, the (optional) fifth step in the conversion procedure is to use an appropriate curve fitting technique to approximate $r(x)$ by $\tilde{r}(x)$. Then $\tilde{r}(x)$ is used as the cost function for arc $i, 0$ in the expanded network for Problem $\tilde{Q}$.

3.6 Computational Summary

We conclude this section by pointing out that the form of the functions $r(x)$ and $s_q(x_q)$ in the expanded network (i.e., Step 4) can be computed directly from the general nonlinear function $\phi(x)$ in the original network (i.e., Step 0). To describe these computations, let $\theta^L(x)$ and $\theta^R(x)$ denote, respectively, the slope of the piecewise-linear function $f(x)$ to the left and right of x for $x = 1, 2, \ldots, u-1$. These slopes are computed directly from the general nonlinear function $\phi(x)$ as

$$\theta^L(x) = \phi(x) - \phi(x-1) \tag{21}$$

$$\theta^R(x) = \phi(x+1) - \phi(x) \tag{22}$$

In addition, let $\theta(x)$ denote the difference between the right and left slopes. That is,

$$\theta(x) = \theta^R(x) - \theta^L(x) \tag{23}$$

To determine the intervals j and segments k, and the associated endpoints $m_{j,k}$, we initially set $j \leftarrow 1$, $k \leftarrow 1$, and $m_{1,1} \leftarrow 0$. Then, for x sequentially set to $1, 2, \ldots, u-1$ (as in a "DO-loop") we perform the following four tests:

- If j is odd and $\theta(x) < 0$, then set $m_{j,k+1} \leftarrow x$ and increment $k \leftarrow k+1$;
- If j is odd and $\theta(x) > 0$, then set $m_{j,k+1} \leftarrow x$, set $m_{j+1,1} \leftarrow x$, and set $\bar{k}_j \leftarrow k$; then increment $j \leftarrow j+1$ and reset $k \leftarrow 1$;
- If j is even and $\theta(x) > 0$, then set $m_{j,k+1} \leftarrow x$ and increment $k \leftarrow k+1$;
- If j is even and $\theta(x) < 0$, then set $m_{j,k+1} \leftarrow x$, set $m_{j+1,1} \leftarrow x$, and set $\bar{k}_j \leftarrow k$; then increment $j \leftarrow j+1$ and reset $k \leftarrow 1$;

Note that when $\theta(x) = 0$ the conditions of none of the four tests are satisfied. After performing these four tests for $x = 1, 2, \ldots, u-1$, we then set $\bar{j} \leftarrow j$, $\bar{k}_j \leftarrow k$, $m_{j,k+1} \leftarrow u$ and $m_{j+1,1} \leftarrow u$.

To compute the piecewise-linear-concave function $r(x)$, the numerical value of the slopes $\Delta_{j,k}$ and $\hat{\Delta}_j$ used in eqs. (14) through (16) are obtained from the slopes $\theta^L(x)$ and $\theta^R(x)$ given in eqs. (21) and (22). Specifically,

$$\Delta_{j,k} = \theta^R(m_{j,k}) \tag{24}$$

$$\hat{\Delta}_j = \theta^L(m_{j+1,1}) \tag{25}$$

Then the segments $k = 1, \ldots, \bar{k}_j$ for j odd are reindexed as $p = 1, \ldots, \bar{p}$ and the endpoints v_p are set equal to $m_{j,k}$ for the appropriate j and k.

In a similar way, to compute the linear functions $s_q(x_q)$, the segments $k = 1, \ldots, \bar{k}_j$ for j even are reindexed as $q = 1, \ldots, \bar{q}$, the slopes δ_q in eq. (19) are determined from $\Delta_{j,k}$ and $\hat{\Delta}_j$ using eqs. (24) and (25), the endpoints w_q are set equal to $m_{j,k}$ for the appropriate j and k, and the capacities u_q are computed using eq. (20).

The next section illustrates the conversion procedure with two numerical examples.

4 Numerical Examples

In this section, we apply the conversion procedure described in Section 3 to the "staircase" arc cost function shown in Figure 1 and the "sawtooth" arc cost function shown in Figure 2. To illustrate the technique, the conversion steps are shown in full for the "staircase" function, but just briefly summarized for the "sawtooth" function. As in Section 3, because we are referring to a given arc i in the original arc set A, we omit the subscript i in the notation in this section.

4.1 "Staircase" Function

For the function in Figure 1, we have

$$\phi(x) = \begin{cases} 0 & \text{for } x = 0 \\ 4 & \text{for } 0 < x \leq 5 \\ 8 & \text{for } 5 < x \leq 10 \\ 12 & \text{for } 10 < x \leq 15 \end{cases}$$

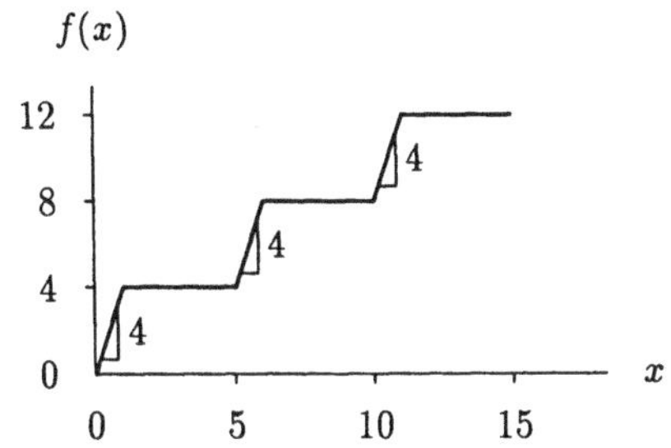

Figure 5: Piecewise-Linear Equivalent for "Staircase" Function.

We assume $\phi(x)$ is defined over the domain $[0,15]$; i.e, $u = 15$. The function $f(x)$, the piecewise-linear-continuous equivalent of $\phi(x)$, is given by

$$f(x) = \begin{cases} 4 \cdot x & \text{for } 0 \leq x \leq 1 \\ 4 & \text{for } 1 \leq x \leq 5 \\ 4 + 4 \cdot (x-5) & \text{for } 5 \leq x \leq 6 \\ 8 & \text{for } 6 \leq x \leq 10 \\ 8 + 4 \cdot (x-10) & \text{for } 10 \leq x \leq 11 \\ 12 & \text{for } 11 \leq x \leq 15 \end{cases}$$

and is shown in Figure 5. The domain $[0,15]$ is divided into five intervals; i.e., $\bar{j} = 5$. They are

$$I_1 = \{x : m_{1,1} \leq x \leq m_{2,1}\} = \{x : 0 \leq x \leq 5\}$$

$$I_2 = \{x : m_{2,1} \leq x \leq m_{3,1}\} = \{x : 5 \leq x \leq 6\}$$

$$I_3 = \{x : m_{3,1} \leq x \leq m_{4,1}\} = \{x : 6 \leq x \leq 10\}$$

$$I_4 = \{x : m_{4,1} \leq x \leq m_{5,1}\} = \{x : 10 \leq x \leq 11\}$$

$$I_5 = \{x : m_{5,1} \leq x \leq m_{6,1}\} = \{x : 11 \leq x \leq 15\}$$

The first interval contains two segments; i.e., $\bar{k}_1 = 2$. They are

$$I_{1,1} = \{x : m_{1,1} \leq x \leq m_{1,2}\} = \{x : 0 \leq x \leq 1\}$$

$$I_{1,2} = \{x : m_{1,2} \leq x \leq m_{1,3}\} = \{x : 1 \leq x \leq 5\}$$

Each of the other intervals contains exactly one segment. That is, $I_{2,1} = I_2$, $I_{3,1} = I_3$, $I_{4,1} = I_4$, and $I_{5,1} = I_5$. For each $j = 1, \ldots, \bar{j}$, we also define the sets $I_{j,0}$ and $I_{j,\bar{k}_j+1}$ as follows:

$$I_{1,0} = \{x : 0 \leq x \leq m_{1,1}\} = \{x : 0 \leq x \leq 0\}$$

$$I_{1,3} = \{x : m_{2,1} \leq x \leq u\} = \{x : 5 \leq x \leq 15\}$$

$$I_{2,0} = \{x : 0 \leq x \leq m_{2,1}\} = \{x : 0 \leq x \leq 5\}$$

$$I_{2,2} = \{x : m_{3,1} \leq x \leq u\} = \{x : 6 \leq x \leq 15\}$$

$$I_{3,0} = \{x : 0 \leq x \leq m_{3,1}\} = \{x : 0 \leq x \leq 6\}$$

$$I_{3,2} = \{x : m_{4,1} \leq x \leq u\} = \{x : 10 \leq x \leq 15\}$$

$$I_{4,0} = \{x : 0 \leq x \leq m_{4,1}\} = \{x : 0 \leq x \leq 10\}$$

$$I_{4,2} = \{x : m_{5,1} \leq x \leq u\} = \{x : 11 \leq x \leq 15\}$$

$$I_{5,0} = \{x : 0 \leq x \leq m_{5,1}\} = \{x : 0 \leq x \leq 11\}$$

$$I_{5,2} = \{x : m_{6,1} \leq x \leq u\} = \{x : 15 \leq x \leq 15\}$$

Applying eqs. (24) and (25), the slopes $\Delta_{j,k}$ and $\hat{\Delta}_j$ are as follows:

$$\begin{array}{rclcrcl}
\Delta_{1,1} & = & 4 & \quad & \hat{\Delta}_0 & \equiv & 0 \\
\Delta_{1,2} & = & 0 & & \hat{\Delta}_1 & = & 0 \\
\Delta_{2,1} & = & 4 & & \hat{\Delta}_2 & = & 4 \\
\Delta_{3,1} & = & 0 & & \hat{\Delta}_3 & = & 0 \\
\Delta_{4,1} & = & 4 & & \hat{\Delta}_4 & = & 4 \\
\Delta_{5,1} & = & 0 & & \hat{\Delta}_5 & = & 0
\end{array}$$

Using the slopes and intervals given above, the functions $g_j(x)$ defined in eqs. (14) through (16) are as follows:

$$g_1(x) = \begin{cases} 4 \cdot x & \text{for } 0 \leq x \leq 1 \\ 4 & \text{for } 1 \leq x \leq 15 \end{cases}$$

$$g_2(x) = \begin{cases} 0 & \text{for } 0 \leq x \leq 5 \\ 4 \cdot (x - 5) & \text{for } 5 \leq x \leq 15 \end{cases}$$

$$g_3(x) = \begin{cases} 0 & \text{for } 0 \leq x \leq 6 \\ -4 \cdot (x - 6) & \text{for } 6 \leq x \leq 15 \end{cases}$$

$$g_4(x) = \begin{cases} 0 & \text{for } 0 \leq x \leq 10 \\ 4 \cdot (x - 10) & \text{for } 10 \leq x \leq 15 \end{cases}$$

$$g_5(x) = \begin{cases} 0 & \text{for } 0 \leq x \leq 11 \\ -4 \cdot (x - 11) & \text{for } 11 \leq x \leq 15 \end{cases}$$

Recombining the functions $g_j(x)$ according to eqs. (17) and (18) gives

$$r(x) = g_1(x) + g_3(x) + g_5(x)$$

$$s(x) = g_2(x) + g_4(x)$$

Numerically, this yields

$$r(x) = \begin{cases} 4 \cdot x & \text{for } 0 \leq x \leq 1 \\ 4 & \text{for } 1 \leq x \leq 6 \\ 4 - 4 \cdot (x - 6) & \text{for } 6 \leq x \leq 11 \\ -16 - 8 \cdot (x - 11) & \text{for } 11 \leq x \leq 15 \end{cases}$$

$$s(x) = \begin{cases} 0 & \text{for } 0 \leq x \leq 5 \\ 4 \cdot (x - 5) & \text{for } 5 \leq x \leq 10 \\ 20 + 8 \cdot (x - 10) & \text{for } 10 \leq x \leq 15 \end{cases}$$

The functions $r(x)$ and $s(x)$ are shown in Figures 6 and 7, respectively.

Since function $s(x)$ contains three linear segments, we have $\bar{q} = 3$ and, using eqs. (19) and (20), we compute the linear functions $s_q(x_q)$ and capacities u_q as follows:

$$\begin{aligned} s_1(x_1) &= 0 & u_1 &= 5 \\ s_2(x_2) &= 4 \cdot x_2 & u_2 &= 5 \\ s_3(x_3) &= 8 \cdot x_3 & u_3 &= 5 \end{aligned}$$

Finally, if desired, the piecewise-linear-concave function $r(x)$ can be approximated by a continuously differentiable function $\tilde{r}(x)$. For instance, to approximate $r(x)$ by a quadratic, the following function can be used:

$$\tilde{r}(x) = 5 - 0.4 \cdot (x - 3.5)^2$$

This approximation is shown by the dotted line in Figure 6.

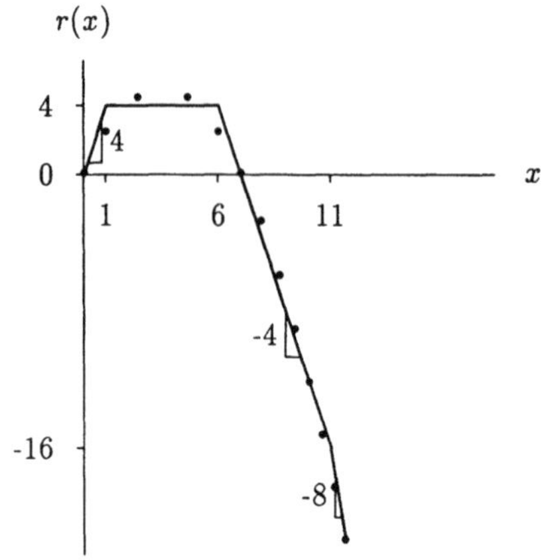

Figure 6: Concave Component for "Staircase" Function.

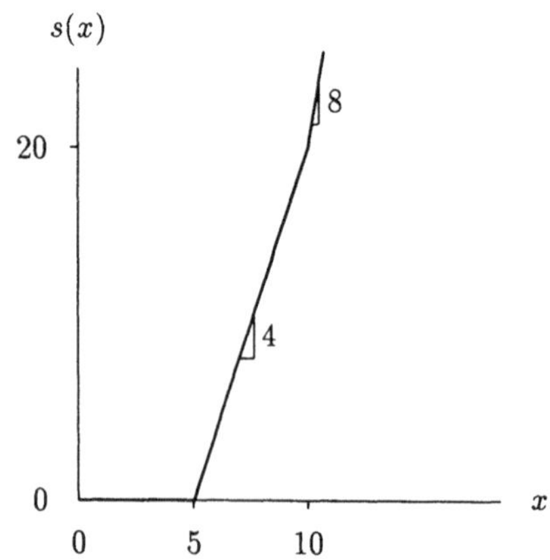

Figure 7: Convex Component for "Staircase" Function.

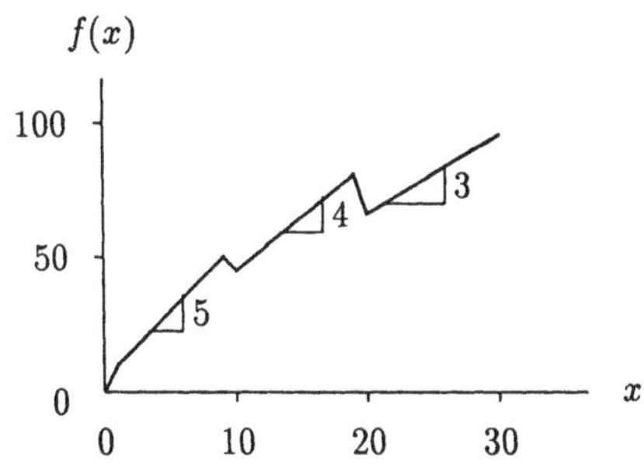

Figure 8: Piecewise-Linear Equivalent for "Sawtooth" Function.

4.2 "Sawtooth" Function

For the "sawtooth" function in Figure 2, $\phi(x)$ is given by

$$\phi(x) = \begin{cases} 0 & \text{for } x = 0 \\ 6 + 5 \cdot x & \text{for } 0 < x < 10 \\ 6 + 4 \cdot x & \text{for } 10 \le x < 20 \\ 6 + 3 \cdot x & \text{for } 20 \le x \le 30 \end{cases}$$

We assume that this function is defined over the domain $[0, 30]$; i.e, $u = 30$. The function $f(x)$, the piecewise-linear-continuous equivalent of $\phi(x)$, is given by

$$f(x) = \begin{cases} 11 \cdot x & \text{for } 0 \le x \le 1 \\ 11 + 5 \cdot (x - 1) & \text{for } 1 \le x \le 9 \\ 51 - 5 \cdot (x - 9) & \text{for } 9 \le x \le 10 \\ 46 + 4 \cdot (x - 10) & \text{for } 10 \le x \le 19 \\ 82 - 16 \cdot (x - 19) & \text{for } 19 \le x \le 20 \\ 66 + 3 \cdot (x - 20) & \text{for } 20 \le x \le 30 \end{cases}$$

as shown in Figure 8. The domain $[0, 30]$ is divided into four intervals; i.e., $\bar{j} = 4$. They are

$$I_1 = \{x : m_{1,1} \le x \le m_{2,1}\} = \{x : 0 \le x \le 10\}$$

$$I_2 = \{x : m_{2,1} \le x \le m_{3,1}\} = \{x : 10 \le x \le 19\}$$

$$I_3 = \{x : m_{3,1} \le x \le m_{4,1}\} = \{x : 19 \le x \le 20\}$$

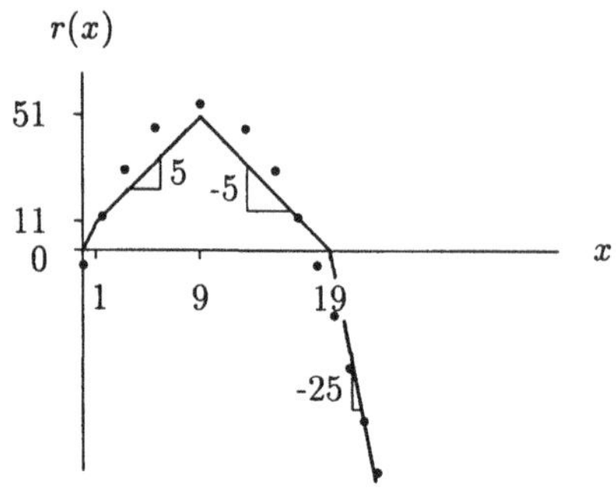

Figure 9: Concave Component for "Sawtooth" Function.

$$I_4 = \{x : m_{4,1} \le x \le m_{5,1}\} = \{x : 20 \le x \le 30\}$$

The first interval contains three segments; i.e., $\bar{k}_1 = 3$. Each of the other intervals contains exactly one segment. Using eqs. (14) through (16), the functions $g_j(x)$ are determined for $j = 1, \ldots, 4$. Then using eqs. (17) and (18), functions $r(x)$ and $s(x)$ are given by

$$r(x) = g_1(x) + g_3(x)$$

$$s(x) = g_2(x) + g_4(x)$$

This yields

$$r(x) = \begin{cases} 11 \cdot x & \text{for } 0 \le x \le 1 \\ 11 + 5 \cdot (x - 1) & \text{for } 1 \le x \le 9 \\ 51 - 5 \cdot (x - 9) & \text{for } 9 \le x \le 19 \\ 1 - 25 \cdot (x - 19) & \text{for } 19 \le x \le 30 \end{cases}$$

$$s(x) = \begin{cases} 0 & \text{for } 0 \le x \le 10 \\ 9 \cdot (x - 10) & \text{for } 10 \le x \le 20 \\ 90 + 28 \cdot (x - 20) & \text{for } 20 \le x \le 30 \end{cases}$$

The functions $r(x)$ and $s(x)$ are shown in Figures 9 and 10, respectively. Since function $s(x)$ contains three linear segments, we have $\bar{q} = 3$ and the linear functions $s_q(x_q)$ and capacities u_q are given as follows:

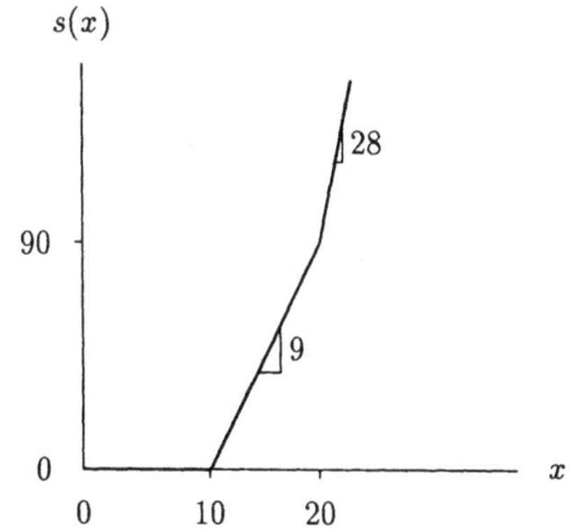

Figure 10: Convex Component for "Sawtooth" Function.

$$\begin{aligned} s_1(x_1) &= 0 & u_1 &= 10 \\ s_2(x_2) &= 9 \cdot x_2 & u_2 &= 10 \\ s_3(x_3) &= 28 \cdot x_3 & u_3 &= 10 \end{aligned}$$

Lastly, if desired, the piecewise-linear-concave function $r(x)$ can be approximated by a continuously differentiable function $\tilde{r}(x)$. For instance, to approximate $r(x)$ by a quadratic, the following function can be used:

$$\tilde{r}(x) = 55 - 0.75 \cdot (x - 9)^2$$

This approximation is shown by the dotted line in Figure 9.

The next section summarizes the paper.

5 Summary

This paper has described how a network containing arcs with general nonlinear cost functions can be converted into an expanded network involving only concave arc cost functions. The functional form of the concave arc cost functions can be calculated very efficiently from the original problem data. Although the expanded network will be larger than the original network, it can be solved using established solution techniques for concave minimum cost network flow problems. Special cases of this conversion procedure have been applied to less-than-truckload motor carrier networks and to cash flow management problems [17, 18]. This paper has extended the technique to *any* minimum cost network flow problem with arbitrary arc cost functions.

References

[1] D.P. Bertsekas (1991), *Linear Network Optimization*, MIT Press, Cambridge, MA.

[2] A. Charnes and W.W. Cooper (1961), *Management Models and Industrial Applications of Linear Programming*, John Wiley and Sons, New York, NY.

[3] R.J. Dolan (1987), "Quantity Discounts: Managerial Issues and Research Opportunities," *Marketing Science*, vol. 6, pp. 1-22.

[4] J.R. Evans and E. Minieka (1992), *Optimization Algorithms for Networks and Graphs*, Marcel Dekker, New York, NY.

[5] M.R. Garey and D.S. Johnson (1979), *Computers and Intractability: A Guide to the Theory of NP-Completeness*, Freeman and Co., San Francisco, CA.

[6] G.M. Guisewite and P.M. Pardalos (1990), "Minimum Concave-Cost Network Flow Problems: Applications, Complexity, and Algorithms," *Annals of Operations Research*, vol. 25, pp. 75-100.

[7] G.M. Guisewite and P.M. Pardalos (1991a), "Single-Source Uncapacitated Minimum Concave Cost Network Flow Problems," in H.E. Bradley (ed.), *Operational Research '90*, Pergamon Press, Oxford, England, pp. 703-713.

[8] G.M. Guisewite and P.M. Pardalos (1991b), "Algorithms for the Single-Source Uncapacitated Minimum Concave-Cost Network Flow Problem," *Journal of Global Optimization*, vol. 1, pp. 245-265.

[9] G.M. Guisewite and P.M. Pardalos (1991c), "Global Search Algorithms for Minimum Concave Cost Network Flow Problems," *Journal of Global Optimization*, vol. 1, pp. 309-330.

[10] G.M. Guisewite and P.M. Pardalos (1991d), "A Polynomial Time Solvable Concave Network Flow Problem," *Networks*, forthcoming.

[11] G.M. Guisewite and P.M. Pardalos (1992), "Performance of Local Search in Minimum Concave-Cost Network Flow Problems," in C.A. Floudas and P.M. Pardalos (eds.), *Recent Advances in Global Optimization*, Princeton University Press, Princeton, NJ, pp. 50-75.

[12] P.A. Jenson and J.W. Barnes (1980), *Network Flow Programming*, John Wiley and Sons, New York, NY.

[13] E.L. Johnson (1966), "Networks and Basic Solutions," *Operations Research*, vol. 14, pp. 619-623.

[14] D.B. Khang and O. Fujiwara (1991), "Approximate Solutions of Capacitated Fixed-Charge Minimum Cost Network Flow Problems," *Networks*, vol. 21, pp. 689-704.

[15] J.L. Kennington and R.V. Helgason (1980), *Algorithms for Network Programming*, John Wiley and Sons, New York, NY.

[16] B.W. Lamar (1992), "An Improved Branch and Bound Algorithm for Minimum Concave Cost Network Flow Problems," *Journal of Global Optimization*, forthcoming.

[17] B.W. Lamar and S. Jorjani (1990), "Incorporating Discounting Into Network-Based Cash Flow Management Models," working paper, Graduate School of Management, University of California, Irvine, CA.

[18] B.W. Lamar and Y. Sheffi (1988), "An Implicit Enumeration Method for LTL Network Design," *Transportation Research Record*, no. 1120, pp. 1-16.

[19] B.W. Lamar, Y. Sheffi, and W.B. Powell (1990), "A Capacity Improvement Lower Bound for Fixed Charge Network Design Problems," *Operations Research*, vol. 38, pp. 704-710.

[20] D.T. Phillips and A. Garcia-Diaz (1981), *Fundamentals of Network Analysis*, Prentice-Hall, Englewood Cliffs, NJ.

[21] Y. Sheffi (1985), *Urban Transportation Networks: Equilibrium Analysis with Mathematical Programming Methods*, Prentice-Hall, Englewood Cliffs, NJ.

[22] B. Yaged, Jr. (1971), "Minimum Cost Routing for Static Network Models," *Networks*, vol. 1, pp. 139-172.

Network Optimization Problems, pp. 169-175
Eds. D.-Z. Du and P.M. Pardalos

Application of Global Line Search in Optimization of Networks

Jonas Mockus
Department of Optimal Decision Theory, Institute of Mathematics and Informatics, Akademijos 4, Vilnius 2600, Lithuania

Abstract

In this paper a review of application of global line search to optimization of networks is given. Advantages and disadvantages of this approach are discussed. It is shown that global line search provides global minimum after finite number of steps in two cases of piecewise linear cost functions of arcs. The first case is where all cost functions are convex. The second case is where all costs are equal to zero at zero flow and equal to some constant at non-zero flow. In other cases the global line search approaches a global minimum with small average error.

The extension of the method to vector demands is given. The application of the method to the optimization of high-voltage net of power system is described.

1 Global Line Search

Suppose that the objective function $f(x), x = (x_j, j = 1, ..., J)$ may be approximately expressed as a sum of components depending on one variable x_j.

$$f(x) = \sum_{j=1}^{J} f_j(x_j) \tag{1}$$

Then the original J-dimensional optimization problem can be reduced to a sequence of one-dimensional optimization problems. If the decomposition (1) is exact, then we shall get the global optimum after J steps of optimization. If the sum (1) represents $f(x)$ approximately, then generally we shall get some approximation of a global optimum.

The result of step i we regard as an initial point for the $i+1$-sth step of optimization. The optimization stops, if no change happens during J steps. The difference from the classical version of line search method is that search is not local but global. Generally it helps to approach a global minimum closer. There are important cases when global line search gets global minimum and local line search does not. One of such cases is the following problem of network optimization.

2 Optimization of Networks

Suppose that the cost of network $f(x)$ can be expressed as a sum (1), where $f_j(x_i)$ is a cost of arc j and x_j is a flow of arc j. Sum of flows of arcs connected to each node i has to be equal to the nodes demand. It means that:

$$\sum_{j=1}^{J} a_{ij}x_j = a_i c_i, i = 1, ..., I \tag{2}$$

where

$$\sum_{i=1}^{i=I} a_i c_i = 0$$

Here J is a number of arcs, I is a number of nodes, c_i is demand of node i,

$$a_{ij} = \begin{cases} +1, & \text{if arc } j \text{ goes to node } i \\ -1, & \text{if arc } j \text{ goes from node } i \\ 0, & \text{if arc } j \text{ is not connected to } i. \end{cases}$$

and

$$a_i = \begin{cases} +1, & \text{if node } i \text{ is source} \\ -1, & \text{if node } i \text{ is sink} \end{cases}$$

The J-dimensional problem of network cost minimization (1) under $I-1$ of conservation-of-flow equations (2) can be reduced to $J-I+1$-dimensional unconstrained minimization problem:

$$f(x) = \sum_{k=1}^{K} f_k(x_k) + \sum_{l=K+1}^{J} f_l(\sum_{k=1}^{K} b_{lk}x_k + x_{l0}) \tag{3}$$

Here $K = J-I+1$ is a number of loops, x_{l0} is from (2) taking $x_k = 0, k = 1, ..., K$,

$$b_{lk} = \begin{cases} +1, & \text{if arc } l \text{ is directed as loop } k \\ -1, & \text{if arc } l \text{ is directed opposite to loop } k \\ 0, & \text{if arc } l \text{ don't belong to loop } k \end{cases}$$

Loop k is generated connecting a pair of nodes of some tree containing arcs $l, l = K+1, ..., J$ by some additional arc $k, k = 1, ..., K$. The flow x_k of arc k generating a loop k is called loop flow. Loop flows $x_k, k = 1, ..., K$ can be changed independently during optimization.

Expression (3) depends on the tree which generates loops $k, k = 1, ..., K$. Assume that the costs of arcs are piecewise linear functions. Then it is convenient to carry out the global line search along the bounds of linear areas. It can be done changing the tree after each step of optimization. An obvious rule is to remove some arc from the tree, if the arcs flow happens to be on a bound of linear parts of piecewise linear cost function. Here we do not consider degenerate problems.

It is shown, see Mockus (1967), that the algorithm provides global minimum in two cases:

1. Cost functions of all arcs are convex

2. Cost functions of all arcs are constant, with exception of zero flow point. At zero flow the value of cost function has to be zero. It means that cost functions are very "non-convex".

It is easy to see that in the linear case the algorithm is as simple and efficient as conventional algorithms of linear programming. The difference is that global line search algorithm works almost as well also in non-linear convex problems and in some special non-convex problems.

The generalization of the algorithm to S-dimensional demand is straightforward theoretically: we just replace the scalar demand c_i by vector demand $c_i = (c_{i1}, ..., c_{iS})$. To keep conservation-of-flow equations (2) we have to replace scalar flows x_j by corresponding vector flows $x_j, x_j = (x_{j1}, ..., x_{jS})$. For practical applications the straightforward generalization is not convenient by two reasons:

The first reason is the exponential growth of calculations when S is large. The second reason is the practical difficulties defining general S-dimensional piecewise linear functions $f_j(x_{j1}, ..., x_{jS})$.

In special S-dimensional demand cases global line search can be carried out more conveniently using some heuristics. The convergence proof can be extended directly only for straightforward generalization. However the experience shows that global line search is efficient in solving S-dimensional load problems of network design. One of such problems is optimization of the high-voltage net of large power system.

3 The Optimization of High-Voltage Net of Power System

The arcs may represent components of network such as power transmission lines and transformers. The nodes represent demands. Demands are sources (generators or higher voltage substations) or sinks (users or lower voltage substations). The demands of nodes $c_i = (c_{in}, n = 1, ..., N)$ in real life power systems are some vector-valued functions of time $c_{in} = c_{in(t)}, t \in [1, T]$. Usually those functions are approximated as some step functions of time, so

$$c_i = (c_{int}, n = 1, ..., N, t = 1, ..., T), i = 1, ..., I$$

Here different t-components of $S = NT$-dimensional demand represent different periods of time, from $t = 1$ until $t = T$. Then flows of arcs are $x_j = (xjnt, n = 1, ..., N, t = 1, ..., T), j = 1, ..., J$ and the scalar problem (3) can be directly extended to the vector case:

$$f(x) = \sum_{k=1}^{K} f_k(x_k) + \sum_{l=K+1}^{J} f_l(\sum_{k=1}^{K} b_{lk} x_k + x_{l0}) \tag{4}$$

where

$$x_k = (x_{knt}, n = 1, ..., N, t = 1, ..., T), k = 1, ..., K \tag{5}$$

The costs of arcs j representing transmission lines and transformers j directly depends not only on flows x_j but also on states y_j. So the cost of arc can be more conveniently expressed as $f_j(x_j, y_j)$. Here the state variable $y_j = (y_{j,t}, t = 1, ..., T)$ usually depends on time but not on n.

Each component y_{jt} of vector y_j is a non-negative integer defining the technical parameters of arc j, such as the number of parallel circuits of transmission line, the number and cross-section area of wires, the number and the power rating of transformers and so on. Assume that capacity of arc $X_{jt}(y_{jt})$ is an increasing function of its state yjt (this assumption will help us later to deal with capacity constraints). So we define the mixed integer programming problem:

$$f(x, y) = \sum_{k=1}^{K} f_k(x_k, y_k) + \sum_{l=K+1}^{J} f_l(\sum_{k=1}^{K} b_{lk} x_k + x_{l0}, y_l) \tag{6}$$

This problem can be reduced to continuous non-linear programming problem by choosing the cheapest state y_j for a fixed flow x_j, namely:

$$f_j(x_j) = \min_{y_j \in Y_j} f(x_j, y_j) \tag{7}$$

providing that the capacity constraints hold

$$|x_{jnt}| \leq X_{jnt}(y_{jt}) \tag{8}$$

The capacity constraints $|x_{jnt}| \leq X_{jnt}(y_{jt})$ are satisfied by increasing the state variable y_{jt} if inequality (8) does not holds. Notation $Y_j = Y_j(x_j)$ means a set of feasible states of arc j which can depend on flow x_j. So expression (7) defines the cost function, generally a multimodal one.

Expression (7) gives a convenient definition of S-dimensional cost function $f_j(x_j)$. However there remains the exponential complexity of minimization of non-convex cost function (6) depending on vector variables $x_k, k = 1, ..., K$. So we shall consider some other ways too.

An interesting way is to reduce the problem of mixed integer programming (6) to a problem of pure integer programming, regarding the states y_j as independent integer variables. Here for each fixed state $y = y_k$ the optimal value of flow x_k has to be calculated.

An advantage of pure integer programming approach is that the cost of arc $f_j(x_j, y_j)$ at a fixed state y_j is usually a convex function (compare it with nonconvex cost function (7) of non-linear programming approach). It is well known that a sum (6) of convex functions also is a convex function.

However we should carry out the optimization of convex function (6) many times, for each state y. It is a hard task, if N and T are not small. The practical experience shows, that the best computational results can be obtained by facing the mixed integer programming problem (6) directly. It means that we optimize states and flows together at each step of global line search.

4 Mixed Integer Global Line Search

We shall consider a special case

$$f_j(x_j, y_j) = \sum_{t=1}^{T}[g_{jt}(y_{jt-1}, y_{jt}) + \sum_{n=1}^{N} h_{jnt}(y_{jt})x_{jnt}^2] \tag{9}$$

Here the function $g_{jt}(y_{jt-1}, y_{jt})$ defines the cost of reconstruction of node j from state y_{jt-1} to the state y_{jt}. Expression $h_{jnt}x_{jt}^2$ means the power loss of flow x_{jnt} in arc j at state y_{jt}. It is well known that minimization of (6) under assumption (9) defines the natural distribution of power flows in a homogeneous electrical net for any fixed state y. The net is not homogeneous if it contains transmission lines of different voltages or if it includes lines and transformers together.

Suppose that $y_{jnt} = 0, if x_{jnt} = 0$ and that $y_{jnt+1} \geq y_{jnt}$. It means that zero state is feasible only for zero flows and that the state variable can't be decreased in time. The last assumption is usually true, but not always.

If the state of arc is zero from the time 1 until the time t we shall call it t-interrupted arc. It is supposed that after the time t the state of t-interrupted arc is non-zero. The optimization of each loop k is carried out in $T + 1$ stages.

At the 1-sth stage we compare all possible cases of T-interruption of arcs belonging to the loop k. For each fixed state we minimize sum (6) as a function of flow x_k. We do it by solving linear equations corresponding to the condition of zero first derivatives with regard to $x_{knt}, n = 1, \ldots, N, t = 1, \ldots, T$. In the most economical state of loop we replace zero state values $y_{jt} = 0$ by unit state values, that is by $y_{jt} = 1$. We accept it as an initial state for the next stage.

At the second stage we consider all cases of $T-1$ interruption and so on until the last $T+1$- sth stage. At $T+1$-sth stage we consider all non-zero states.

We compare sums of costs of arcs belonging to a loop k for all stages. The state which corresponds to the minimal cost function we accept as a result of global search along the "line" k. The enumeration of arcs may be changed after each step of global line search. The purpose of this change is to keep "most interrupted" arcs out of the tree.

We call an arc as "most interrupted" if it is interrupted for a longest time t_0. Here $t_0 = \arg\max_{l \in L_k} t_l$, where t_l we define as $y_{l,t} = 0, t = 1, ...t_l, y_{lt} \geq 1, t = t_l + 1, ..., T$ and L_k is a set of arcs belonging to a loop k.

If $T \leq 2$, then the optimization at each stage can be carried out by a simple comparison of all corresponding states. If T is greater, then some dynamic programming procedure usually is more efficient.

The optimization stops if the cost of net changes less then ϵ during K steps of global line search.

The CPU time τ of global line search software developed by J.Valeviciene can be estimated as

$$\tau = cKINM$$

Here c depends on computer, for PC approximately $c = 0.2 - 1.0 sec.$, K is a number of loops, I is a number of nodes, N is a number of flow components and M is an average number of states of arcs.

The algorithm was used since 1969 for the optimal planning of North-Western power system of the former USSR designing the new power transmission lines of 110 KV, 220 KV and 330 KV in the Leningrad branch of "Energosetprojekt", which was the leading institution in the country at the time.

Dynamic programming procedures developed in Riga were also used at the some place, with lesser success. The reason was that approximate global line search method was solving the problems up to 100 nodes and more. The exact dynamic programming procedures were directly applicable only to problems with tens of nodes.

Pardalos and Rosen (1987) present an approximation technique based on piecewise linear underestimation of concave cost functions $f_j(x_j)$. The resulting model is a linear, zero-one, mixed integer problem. A direct comparison of this approach and global line search techniques is an interesting problem of future research.

A complete review of western results in network optimization is given by Guisewite and Pardalos (1990).

References

[1] G.M. Guisewite and P.M.Pardalos, Minimum Concave-Cost Network Flow Problems: Applications, Complexity, and Algorithms, *Ann. of Operations Research*, **25** (1990) 75-100.

[2] J.Mockus , *Multimodal Problems in Engineering Design* (Nauka, Moscow, 1967) p. 216 (in Russian)

[3] P.M. Pardalos and J.B. Rosen J.B. *Constrained global optimization: Algorithms and applications*, Lecture Notes in Computer Science **268** (Springer, Berlin, 1987)

Network Optimization Problems, pp. 177-202
Eds. D.-Z. Du and P.M. Pardalos

Solving Nonlinear Programs with Embedded Network Structures

Mustafa Ç. Pınar
Institute for Numerical Analysis, The Technical University of Denmark, 2800 Lyngby, Denmark

Stavros A. Zenios
Decision Sciences Department, University of Pennsylvania, Philadelphia, PA 19104 USA

Abstract

We present an algorithm for solving large scale nonlinear programs with embedded network structures. It is based on a Linear-Quadratic Penalty (LQP) function that eliminates the non-network constraints from the problem. The resulting nonlinear and nonseparable problems are solved using a simplicial decomposition algorithm that induces separability in the objective function. As a result one can employ network simplex technology. At the same time the values of any side variables can be determined by inspection. The algorithm is implemented in the software system GENOS/LP. Extensive numerical results are reported for diverse application areas: multicommodity network flows, Naval personnel assignment, matrix balancing and some of the NETLIB test problems. Comparisons with general purpose optimizers, like MINOS and OB1, are included.

1 Introduction and Background

It is well documented in the optimization literature that network-structured optimization problems can be solved substantially faster than the general linear program.

This observation holds true even with the recent developments of Karmarkar's algorithm [1984] for linear programming, and the research that followed it on interior point methods. Furthermore, the superior performance of special purpose network algorithms has been documented for both pure and generalized networks, as well as for nonlinear programs. For example, in the mid-seventies, several studies established that codes based on the network simplex algorithm for pure network problems were 150-200 times faster than the state-of-the-art LP codes of the time. See, for example, Glover et al. [1979] and Mulvey [1978]. In the early eighties research concentrated on the generalized network problem. Once more the network simplex algorithm was shown to be approximately 50 times faster than LP codes. See, for instance, Brown and McBride [1985] and Mulvey and Zenios [1985]. This line of research was extended to the nonlinear network problem - see Dembo, Mulvey and Zenios [1989] - for a recent survey. Network specializations of nonlinear programming algorithms — like the primal truncated Newton or simplicial decomposition — were shown to be at least one order of magnitude faster than general purpose nonlinear programming solvers.

Every development in network algorithms was followed by research to use the new algorithms in solving linear programs with large embedded networks. These efforts were generally successful in solving linear programs where the majority of constraints and variables had a network structure. Such programs are known as *networks with side constraints and variables.* In this category are included several well-known classes of problems: the processing (or blending) problem of Koene [1982], the equal flow problem of Ali, Kennington and Shetty [1988], the multicommodity network flow problem, Kennington and Helgason [1980] and so on. The applications of these problems in management science are numerous and well documented in the above references.

In this paper we develop an algorithm for solving nonlinear networks with side constraints and variables. Our primary objective is to make special-purpose nonlinear network optimization software applicable to a broader class of problems. The technique we propose here can also solve linear network problems with side constraints and side variables. As such it fits in the line of research pursued in the past by several others: McBride [1985], Chen and Engquist [1986], Glover and Klingman [1981], Chen and Saigal [1977] and so on. (See, also, Kennington and Helgason [1980, Chapter 7].) Even in the case of linear networks, however, our approach differs significantly from the earlier studies. Most of the earlier work dealt with specializations of the simplex algorithm. These specializations aimed at developing basis partitioning techniques that would separate the network basis form the non-network component. These two separate components were treated using distinct computational procedures. Graph data structures were used to carry out operations on a tree (corresponding to the network basis). General sparse matrix factorizations were applied to the non-network component. When applied to networks with few side constraints these methods were proven very efficient.

The algorithm we propose here takes a different approach. We use an exact penalty function to move the side constraints into the objective function. We then introduce a smoothing of the penalty term in order to obtain a differentiable problem which we solve using a linearization procedure: simplicial decomposition. The use of penalty functions has been very effective in solving the multicommodity network flow problem. Ali, Kennington and Shetty [1988] use a relaxation approach together with subgradient optimization. Schultz and Meyer [1990] develop a barrier type algorithm and Zenios, Pınar and Dembo [1990] propose the use of a Linear-Quadratic Penalty (LQP) function. The last two algorithms were particularly successful in solving some very large problems from a Military Airlift Command Application. In this paper we extend the LQP algorithm of Zenios, Pınar and Dembo [1990] from the multicommodity network flow problem to the more general networks with side constraints and variables. We also discuss key features of the software system GENOS/LP that we develop based on the LQP algorithm. A comprehensive computational investigation provides information on the relative merits of the LQP special purpose algorithm compared to general purpose optimizers. Section 2 formulates the problem we are going to solve and develops the algorithm. Section 3 describes the GENOS/LP software system and reports its use on several diverse applications. Concluding remarks are given in Section 4.

2 The Linear-Quadratic Penalty Algorithm for Networks with Side Constraints and Variables

In this section we describe the Linear-Quadratic Penalty (LQP) algorithm for nonlinear programs with embedded network structures. We begin with a formulation of the problem and proceed with the main components of the LQP algorithm.

2.1 Problem Formulation

We consider the following nonlinear program:

[NLP]

$$\begin{array}{ll} \underset{x,z}{\text{minimize}} & f(x,z) \\ \text{subject to} & Ax = b \\ & Sx + Pz \leq d \\ & 0 \leq x \leq u \\ & 0 \leq z \leq r \end{array}$$

where:

f: $\Re^{n_1+n_2} \rightarrow \Re$ is the objective function, assumed to be convex and continuously differentiable,

$x \in \Re^{n_1}$ is the vector of decision variables which represent flows on a graph,

$z \in \Re^{n_2}$ is the vector of decision variables which represent the side (non-network) columns,

A is an $m \times n_1$ constraint matrix with network structure. It could be the node–arc incidence matrix of a network flow problem, or a block–diagonal matrix where each block is a node–arc incidence matrix as occurs in multicommodity network flows, stochastic networks and time-staged problems.

S is the $s \times n_1$ matrix of side (i.e., non-network) constraints imposed on the network flow variables,

P is the $s \times n_2$ matrix of side (i.e., non-network) constraints imposed on the side variables,

$u \in \Re^{n_1}$ are upper bounds on the flow variables x,

$r \in \Re^{n_2}$ are upper bounds on the side variables z,

$b \in \Re^m, d \in \Re^s$ are the right–hand side coefficients of the constraints.

Also, let

$$X = \{(x,z) | Ax = b, 0 \le x \le u, 0 \le z \le r\}.$$

Throughout the manuscript, transposition is indicated by a superscript ^{T}, $\nabla_x f$ and $\nabla_z f$ denote the gradient vector of the function f with respect to x and z, and all vectors are column vectors.

2.2 The Linear-Quadratic Penalty (LQP) Algorithm

To exploit the network structure we want to remove the side constraints and append them to the objective. To this end we use an exact penalty function

$$p(t) = \max\{0, t\}. \tag{1}$$

where t is a scalar variable. By placing the side constraints into the objective function using a penalty function we obtain a problem with network constraints. In particular, for multicommodity flows or stochastic networks, the penalty problem has a disjoint constraint set. Unfortunately, using an exact penalty function like (1) produces a non–differentiable problem.

To avoid the difficulties of non–differentiability we use a smoothing approximation to the exact penalty function. This approach has been proposed in the context of min-max optimization by Bertsekas [1975] and later by Zang [1980]. For the exact

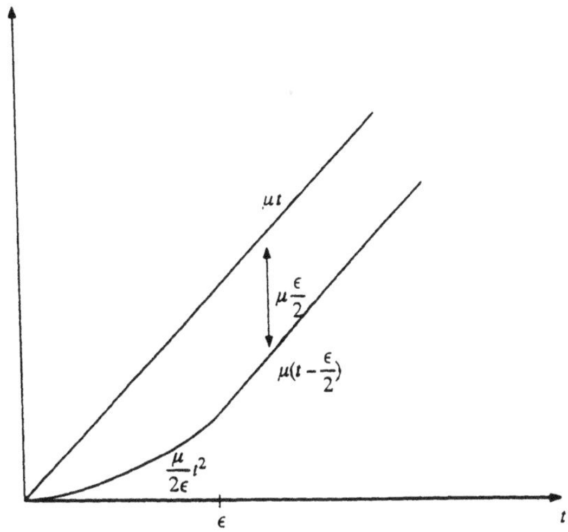

Figure 1: The linear-quadratic penalty function

penalty function (1), we consider the linear-quadratic penalty function of Zenios, Pınar and Dembo [1990]:

$$\phi(\epsilon, t) = \begin{cases} 0 & \text{if } t \leq 0 \\ \frac{t^2}{2\epsilon} & \text{if } 0 \leq t \leq \epsilon \\ t - \frac{\epsilon}{2} & \text{if } t \geq \epsilon \end{cases} \tag{2}$$

where t is a scalar real variable and ϵ is a positive real number. The linear-quadratic penalty function is depicted in Figure 1. The linear-quadratic penalty function is used to eliminate the side constraints by placing those in the objective function. The nonlinear network problem obtained by penalizing the side constraints $Sx + Pz \leq d$ is formulated as:

[NETNLP]

$$\begin{array}{ll} \underset{x,\,z}{\text{minimize}} & \Phi(x,z) = f(x,z) + \mu \sum_j \phi(\epsilon, \rho_j) \\ \text{subject to} & Ax = b \\ & 0 \leq x \leq u \\ & 0 \leq z \leq r \end{array}$$

where $\rho = Sx + Pz - d$, the linear-quadratic penalty function is given by (2) and μ is a positive scalar which determines the severity of the penalty. The resulting nonlinear network problem is solved repeatedly with adaptively changing parameters μ and ϵ until suitable stopping criteria are satisfied. The algorithm can be concisely stated as follows:

The Linear–Quadratic Penalty Algorithm

LQP–0 (Initialization.) Find an initial feasible solution to the network component of **NLP** ignoring the side constraints, i.e., solve the problem

$$\begin{array}{ll} \underset{x,z}{\text{minimize}} & f(x,z) \\ \text{subject to} & Ax = b \\ & 0 \leq x \leq u \\ & 0 \leq z \leq r \end{array}$$

If the solution to this problem satisfies all side constraints, stop. Otherwise choose initial values for penalty parameters μ and ϵ and go to **LQP–1**.

LQP–1 (Penalty Problem.) Solve – perhaps inexactly – the nonlinear network problem **NETNLP**. Go to **LQP–2**.

LQP–2 If the solution satisfies optimality criteria, stop. Otherwise, adjust the penalty parameters μ and ϵ and go to **LQP–1**.

There are three main components of the LQP algorithm which deserve special attention. The solution of the nonlinear network problem at step **LQP–1** demands the most computational effort. This problem is solved using the network specialized version of simplicial decomposition algorithm, see Mulvey, Zenios and Ahlfeld [1990]. The second component is the multiplier adjustment procedure which is crucial to the efficient performance of the LQP algorithm. Finally, we compute lower and upper bounds to the optimal value and test stopping criteria. We study these topics next.

2.3 Simplicial Decomposition for the Penalty Problem

Simplicial decomposition iterates by solving a sequence of linearized subproblems to generate extreme points of the feasible region of the network component and master problems which minimize the nonlinear objective function over the simplex spanned by the extreme points. For a more detailed treatment the reader is directed to Mulvey, Zenios and Ahlfeld [1990]. Here we discuss the specialization of the algorithm in solving the penalty problem **NETNLP**. In particular **NETNLP** is decomposed into a linear network problem that can be solved by the network simplex algorithm and a simple linear program that can be solved by inspection.

The Simplicial Decomposition Algorithm

SD–0 Set $\nu = 0$, and use $\binom{x^0}{z^0} \in X$ as the starting point. Let $Y = \emptyset$, and $v \leftarrow 0$ denote the set of generated vertices and its cardinality, respectively.

SD–1 (Linearized subproblem.) Compute the gradient of the penalty function Φ at the current iterate $\binom{x^\nu}{z^\nu}$ and solve a linear program to get a new vertex of the constraint set, i.e., solve for $y^{v+1} = \arg\min_{y \in X} y^T \nabla\Phi(x^\nu, z^\nu)$ and let $Y = Y \cup \{y^{v+1}\}$, $v \leftarrow v + 1$.

SD–2 (Nonlinear master problem.) Using the set of vertices Y to represent a simplex over the constraint set X, find an optimizer of the penalized objective function Φ over this subset of X. Let $w^* = \arg\min_{w \in W_v} \Phi(Bw)$ where $W_v = \{w_i | \sum_{i=1}^{v} w_i = 1, w_i \geq 0 \ \forall\ i = 1, 2, \ldots, v\}$ and $B = [y^1|y^2|\ldots|y^v]$ is the basis for the simplex generated by the set of vertices Y. The optimizer of Φ over the simplex is given by $\binom{x^{\nu+1}}{z^{\nu+1}} = Bw^*$.

SD–3 Let $\nu \leftarrow \nu + 1$, and return to Step 1.

The Subproblem. At step **SD–1** a new vertex $\binom{x}{z}$ is generated as the solution to the following subproblem:

$$\begin{aligned} \underset{x,\,z}{\text{Minimize}} \quad & x^T \nabla_x \Phi(x^\nu, z^\nu) + z^T \nabla_z \Phi(x^\nu, z^\nu) \\ \text{subject to} \quad & \\ & Ax = b \\ & 0 \leq x \leq u \\ & 0 \leq z \leq r \end{aligned}$$

where $\binom{x^\nu}{z^\nu}$ is the iterate at the ν-th iteration of simplicial decomposition. This problem decomposes into two independent linear programs as follows:

$$\begin{aligned} \underset{x}{\text{Minimize}} \quad & x^T \nabla_x \Phi(x^\nu, z^\nu) \\ \text{subject to} \quad & \\ & Ax = b \\ & 0 \leq x \leq u \end{aligned}$$

and

$$\begin{aligned} \underset{z}{\text{Minimize}} \quad & z^T \nabla_z \Phi(x^\nu, z^\nu) \\ \text{subject to} \quad & \\ & 0 \leq z \leq r \end{aligned}$$

The first problem is a linear network problem and is solved using the network simplex method. The second problem is solved trivially by assigning each component z_j of the vector z to its lower or upper bound depending on the sign of the gradient $\nabla_{z_j}\Phi(x^\nu, z^\nu)$, i.e.,

$$z_j = \begin{cases} r_j & \text{if } \nabla_{z_j}\Phi(x^\nu, z^\nu) \geq 0 \\ 0 & \text{if } \nabla_{z_j}\Phi(x^\nu, z^\nu) \leq 0 \end{cases} \tag{3}$$

However, when no upper bound is specified for the side variables, this procedure fails to produce an accurate approximation to the optimal value of the side variables. We consider an alternative scheme instead. Instead of taking a full step in the direction of either the lower or the upper bound as indicated by the sign of the gradient component, we choose a point between the current value of the side variable and the bound. This is allowed since the descent direction is not affected by this operation. Thus, the side variable portion of the new vertex is obtained as follows:

$$z_j = \begin{cases} \alpha(r_j - z_j^\nu) & \text{if } \nabla_{z_j}\Phi(x^\nu, z^\nu) \geq 0 \\ \alpha(z_j^\nu) & \text{if } \nabla_{z_j}\Phi(x^\nu, z^\nu) \leq 0 \end{cases} \tag{4}$$

where α is a positive scalar in the interval $(0, 1]$. This procedure is reminiscent of the trust region methods, Dennis and Schnabel [1983]. Using this procedure, the values of the side variables were computed to five digits of accuracy where this level of accuracy was not attained using the first procedure after an identical amount of computation time. The accuracy verification was made by comparing the value reported by the LQP algorithm and that of the general purpose code MINOS. We used $\alpha = 0.5$ in this study.

The Master Problem. At step **SD–2** a nonlinear master problem optimizes the objective function on the simplex specified by the extreme points generated by the subproblems. The master problem is formulated in the form:

$$\begin{array}{ll} \underset{w}{\text{Minimize}} & \Phi(Bw) \\ \text{subject to} & \\ & \displaystyle\sum_{i=1}^{v} w_i \;=\; 1 \\ & w_i \;\geq\; 0 \quad i = 1, \ldots, v \end{array}$$

where v is the number of extreme points generated by the subproblems, B is the matrix whose columns are the extreme points and $w = [w^1, w^2, \ldots, w^v]$ are the corresponding weights. The master problem, though nonlinear, is of significantly smaller size than the original problem since it is posed as a problem over the weights w. There are several standard methods that can be used for its solution, like, for example, Bertsekas's projected Newton method [1982]. If the simplicial decomposition algorithm drops vertices that carry zero weight at the optimal solution of the master problem, then subsequent master programs are locally unconstrained. Hence, methods of unconstrained optimization can be used to compute a descent direction. A simple ratio test determines the maximum feasible step length that will not violate the bounds. The master program can be rewritten in the form:

$$\min_{w \geq 0} \Phi(Dw) \tag{5}$$

where $D = [y_1 - y_v | y_2 - y_v | \dots | y_{v-1} - y_v]$ is the derived linear basis for the simplex generated by the vertices $y_1, y_2, \dots, y_v$. We denote by w the vector $[w_1, w_2, \dots, w_{v-1}]$ and the solution for w_v is computed as

$$w_v = 1 - \sum_{i=1}^{v-1} w_i \tag{6}$$

At the current iteration we have $v-1$ active vertices (i.e. $w_i > 0$, for $i = 1, \dots, v-1$) and the last vertex y_v lies along a direction of descent. Hence, given an iterate (x^ν, z^ν) a descent direction p to (5) can be obtained as the solution to

$$(D^T M D)p = -D^T \nabla \Phi(x^\nu, z^\nu), \tag{7}$$

The choice of the matrix M and alternative solution methods for system (7) are discussed in Zenios, Pınar and Dembo [1990].

2.4 Adjusting the Penalty Parameters

The procedure used to update the penalty parameters μ and ϵ consist of dynamically decreasing the value of ϵ to a small final tolerance and increasing the value of μ when certain criteria are met. Suppose $\binom{x^k}{z^k}$, μ^k, ϵ^k are given at iteration k of the LQP algorithm. Also let $\rho^k = Sx^k + Pz^k - d$ and define the set $\mathcal{V}(x, z) = \{j | \rho_j > \epsilon\}$ to be the set of violated constraints. The iterate $\binom{x^k}{z^k}$ is termed *ϵ–feasible* if the index set of violated constraints $\mathcal{V}(x^k, z^k)$ is empty. We distinguish between the following two cases when updating the penalty parameters:

Case 1: If $\mathcal{V}(x^k, z^k) = \emptyset$, this is an indication that the magnitude of the penalty parameter μ was adequate in the previous iteration since *ϵ–feasibility* is achieved. In this case the infeasibility tolerance ϵ should be reduced.

Case 2: If $\mathcal{V}(x^k, z^k) \neq \emptyset$, the current point is not *ϵ–feasible*, an indication that the penalty parameter μ should be increased. Let $\gamma = \eta\epsilon^k$ be a target degree of infeasibility where $\eta \in (0, 1]$. We consider the following update equation:

$$\mu^{k+1} = \mu^k \frac{\rho_j}{\gamma}, \tag{8}$$

or equivalently,

$$\mu^{k+1} = \mu^k \frac{\rho_j^k}{\eta\epsilon^k}. \tag{9}$$

And if $|\mathcal{V}(x^k, z^k)| \geq 1$, we get

$$\mu^{k+1} = \frac{\mu^k}{\eta\epsilon} \max_{j \in \mathcal{V}(x^k, z^k)} \rho_j^k. \tag{10}$$

In summary we have the following update procedure:

Pick η_1, $\eta_2 \in (0,1]$
  If $\mathcal{V}(x^k, z^k) = \emptyset$
    $\epsilon^{k+1} = \max\{\epsilon_{min}, \eta_1 \epsilon^k\}$
  Else
    $\mu^{k+1} = \frac{\mu^k}{\eta_2 \epsilon^k} \max_{j \in \mathcal{V}(x^k, z^k)} \rho_j^k$.

where ϵ_{min} is a suitable final feasibility tolerance. A suitable initial value for μ can be found through some preliminary experimentation. With the test problems we used in this study, the absolute maximum of objective function coefficients proved to be a good choice. The solution $\binom{x^0}{z^0}$ obtained by ignoring the side constraints can be used to provide an initial value for ϵ. A reasonable choice is to pick a value equal to a fraction of the maximum of the side constraint violations, i.e., in the interval $(0, \max_{j \in \mathcal{V}(x^0,z^0)} \rho_j^0)$. The value of parameters η_1 and η_2 was taken to be 0.5 for all computational tests reported in this study.

2.5 Bounds to the Optimal Value and Stopping Criteria

It is possible to compute lower bounds to the optimal objective function value during the course of the LQP algorithm. This computation is performed after the subproblem phase of the simplicial decomposition and is based on a first-order Taylor series expansion of the function Φ around the current iterate.

Let v^* be the optimal value of **NLP** and $\binom{\hat{x}}{\hat{z}}$ an optimal solution of **NETNLP** for given penalty parameters μ and ϵ and $\mathbf{x} = \binom{x}{z}$ for notational simplicity. Also, let

$$X = \{\mathbf{x} | Ax = b, 0 \leq x \leq u, 0 \leq z \leq r\}.$$

Then

$$\Phi(\hat{x}, \hat{z}) \leq v^* \tag{11}$$

since **NETNLP** is a relaxation of **NLP**. Therefore the optimal solution of **NETNLP** is a lower bound for the optimal objective value. But in the presence of inexact minimizations of the penalized objective function Φ, this is not always guaranteed to be a lower bound. Hence, we consider the first order Taylor series expansion of Φ around a point $\mathbf{y}$

$$\Phi(\mathbf{x}) = \Phi(\mathbf{y}) + (\mathbf{x} - \mathbf{y})^T \nabla \Phi(\mathbf{y}) + o(||\mathbf{y}||^2) \tag{12}$$

Ignoring the second order term define the function h

$$h(\mathbf{x}) = \Phi(\mathbf{y}) + (\mathbf{x} - \mathbf{y})^T \nabla \Phi(\mathbf{y}) \tag{13}$$

By convexity of Φ, $\min_{x \in X} h(x) \leq \Phi(\hat{x})$, and hence, $\min_{x \in X} h(x) \leq v^*$. This bound is readily computed by the simplicial decomposition algorithm that generates extreme points of X by minimizing a linearized approximation to the objective function over X; see step **SD–1**. However, it is possible to obtain tighter lower bounds to the optimal value as follows. We slightly change notation for expositional simplicity. We denote the side constraint matrix by E and temporarily ignore the distinction between network and side variables. Let $\mathbf{x}^k$ be an arbitrary iterate and denote by $Q(\epsilon, \rho^k) = \sum_{j=1}^{s} \phi(\epsilon, \rho_j^k)$ where $\rho^k = E\mathbf{x}^k - d$. Recall the subproblem objective function of the simplicial decomposition algorithm:

$$\nabla\Phi(\mathbf{x}^k)^T \cdot \mathbf{x} = (\nabla f(\mathbf{x}^k) + \mu\nabla_\rho Q(\epsilon, \rho^k)^T E) \cdot \mathbf{x} \tag{14}$$

We define the following lower bound function

$$V(\hat{u}) = (\nabla f(\mathbf{x}^k)^T - \hat{u}E)\mathbf{x}' + \hat{u}d \tag{15}$$

where the vector $\hat{u}$ is given by

$$\hat{u} = -\mu\nabla_\rho Q(\epsilon, \rho^k) \tag{16}$$

and $\mathbf{x}'$ is the solution to the linearized subproblem, i.e., $\mathbf{x}' = \arg\min_{\mathbf{x}\in X} x^T \nabla\Phi(\mathbf{x}^k)$. Next we show that $V(.)$ provides a lower bound superior to the linearized subproblem bound. The analysis is a generalization of the result given in Brown et al. [1989]. To proceed we need the following intermediate result.

Lemma 1. Let $g : \Re^n \longmapsto \Re$ be a convex and at least once continuously differentiable function with the property that

$$g(\mathbf{0}) = 0 \tag{17}$$

where $\mathbf{0}$ is the zero vector. Then

$$g(y) - y^T\nabla g(y) \leq 0 \qquad \forall \;\; y. \tag{18}$$

Proof. Consider a first-order Taylor series expansion of g. By convexity

$$g(x) \geq g(y) + (x - y)^T \nabla g(y) \qquad \forall \;\; x, y \; \in \; \Re^n$$

In particular,

$$g(\mathbf{0}) \geq g(y) - y^T\nabla g(y)$$

However, by hypothesis $g(\mathbf{0}) = 0$, and the result follows. ∎

The assumption in Lemma on f holds for example for linear objective functions and quadratic objective functions of the form $\sum_j a_j x_j^2$.

Proposition 2. Let f be as in Lemma 1 and $\mathbf{x}' = \arg\min_{\mathbf{x}\in X} x^T \nabla\Phi(\hat{\mathbf{x}})$ where $\hat{\mathbf{x}}$ is the current iterate. Then

$$h(\mathbf{x}') \leq V(\hat{u})$$

where h is given by (13).

Proof. We will show that

$$h(\mathbf{x}') - V(\hat{u}) \leq 0$$

Equivalently we want to show

$$\Phi(\hat{\mathbf{x}}) + (\mathbf{x}' - \hat{\mathbf{x}})^T \nabla\Phi(\hat{\mathbf{x}}) - (\nabla f(\hat{\mathbf{x}})^T - \hat{u}E)\mathbf{x}' - \hat{u}d \leq 0$$

Consider the left hand side. Let $\hat{\rho} = E\hat{\mathbf{x}} - d$. By algebraic manipulation and using the definition of $\hat{u}$,

$$\begin{aligned}
&\Phi(\hat{\mathbf{x}}) + (\mathbf{x}' - \hat{\mathbf{x}})^T \nabla\Phi(\hat{\mathbf{x}}) - (\nabla f(\hat{\mathbf{x}})^T - \hat{u}E)\mathbf{x}' - \hat{u}d \\
&\quad = f(\hat{\mathbf{x}}) + \mu Q(\epsilon, \hat{\rho}) - \nabla f(\hat{\mathbf{x}})^T \hat{\mathbf{x}} + \hat{u}(E\hat{\mathbf{x}} - d) \\
&\quad = f(\hat{\mathbf{x}}) - \nabla f(\hat{\mathbf{x}})^T \hat{\mathbf{x}} + \mu Q(\epsilon, \hat{\rho}) - \mu \nabla_\rho Q(\epsilon, \hat{\rho})^T (E\hat{\mathbf{x}} - d) \\
&\quad = \underbrace{f(\hat{\mathbf{x}}) - \nabla f(\hat{\mathbf{x}})^T \hat{\mathbf{x}}}_{1} + \underbrace{\mu(Q(\epsilon, \hat{\rho}) - \nabla_\rho Q(\epsilon, \hat{\rho})^T \hat{\rho})}_{2}
\end{aligned}$$

By the assumption imposed on f, the first term is nonpositive following Lemma 1. The nonpositivity of the second term follows from the fact that $\mu > 0$ and

$$Q(\epsilon, \mathbf{0}) = 0$$

and by invoking Lemma 1. Hence the claim is established. ∎

We now describe the procedure for generating upper bounds in the linear-quadratic penalty algorithm. For a general discussion on bounding exterior penalty function algorithms see Fiacco and McCormick [1968]. Define the set $R^0 = \{\mathbf{x} \in X | Sx + Pz > r\}$ and assume that R^0 is non–empty. Let $\mathbf{x} = \binom{x}{z} \in R^0$ and $\binom{\hat{x}}{\hat{z}}$ be an — perhaps approximate — optimal solution of **NETNLP**. Then a new interior point $\binom{\underline{\mathrm{x}}}{\underline{\mathrm{z}}}$ is generated as follows: let $y = S\hat{x} + P\hat{z} - d$, and $I = \{i | y_i > 0\}$.

$$\beta = \max_{i \in I} \frac{y_i}{y_i + y_i'} \tag{19}$$

and define

$$\underline{\mathrm{x}} = (1 - \beta)\hat{x} + \beta x \tag{20}$$

$$\underline{\mathrm{z}} = (1 - \beta)\hat{z} + \beta z \tag{21}$$

It is easily seen that $\binom{\underline{\mathrm{x}}}{\underline{\mathrm{z}}}$ is feasible for **NLP** and thus provides an upper bound, Fiacco and McCormick [1968, Theorem 29, p. 107]. The same result also states that the upper bound converges monotonically to the optimal objective value. Obviously this procedure requires an interior point to be generated at the beginning of the algorithm. For example, in Zenios, Pınar and Dembo [1990], a solution satisfying the mutual capacity constraints of the multicommodity flow problem is computed based on the solution to the network relaxation. We were also able to generate initial feasible solutions for the Naval personnel assignment problems used in this study due to a special property of the problem. This is detailed in the forthcoming section on numerical experiments.

Therefore the LQP algorithm generates both upper and lower bounds for the optimal objective value during the execution of the algorithm for problems where an initial feasible solution can be computed. The algorithm terminates when both of the following error measures are within acceptable tolerance:

1. Absolute error in side constraint feasibility

$$\|Sx + Pz - d\|_\infty \leq \epsilon_{min}$$

2. Bound gap

$$\frac{f(\underline{x}, \underline{z}) - V(\hat{u})}{V(\hat{u})} \leq \epsilon_{gap}$$

where $\mathbf{x} = \binom{x}{z}$ is the current iterate and $\binom{\underline{x}}{\underline{z}}$ is obtained from (20)–(21). The values of ϵ_{min} and ϵ_{gap} used in this study are 10^{-5} and 10^{-2} respectively. The ability to compute improving upper bounds is an important feature of our approach since computation can be stopped as soon as a reasonable improvement in the upper bound is achieved.

3 Numerical Experience

The LQP algorithm was implemented to solve problems of the form [**NLP**]. The code was written in Fortran 77. We refer to the code as the GENOS/LP system. The computational testing was performed on DEC stations 3100 and 5100/200 running Ultrix and for large problems on a CRAY Y-MP. On the DEC stations, the code was compiled with the default compiler optimization option. For the CRAY experiments the code was tailored to take advantage of vectorization capabilities of the CRAY architecture. Before we report the results of computational testing, we discuss briefly the main components of the vectorized code.

3.1 Vector Computing

The simplicial decomposition algorithm is particularly rich in dense linear algebra computations which can be efficiently vectorized. We mention here the main components of the linear algebra involved in the LQP method.

Computing Descent Directions. The following system of linear equations is solved to compute a descent direction during the course of the simplicial decomposition algorithm:

$$(D^T MD)p = -D^T \nabla\Phi(x, z), \tag{22}$$

where D is a projection matrix, $\nabla\Phi$ denotes the gradient of the objective function, the pair (x, z) is an arbitrary iterate and M is a matrix which usually approximates the second derivatives of the function Φ. The matrix D tends to be very large depending on problem size. Typically for D can be $100000 \times N$ where N is the number of extreme points used in the master problem solution. N varies from 1 to 100. The computation of the product

$$D^T MD$$

can be very efficiently vectorized.

Function and Gradient Evaluations. Having computed a descent direction, a one-dimensional search is executed to compute the next iterate. The time spent in the search procedure is dominated by the computation of function values and the gradient vector. The function and gradient evaluation of the original linear objective function can be vectorized trivially as it involves a simple DO-loop over all variables in the problem. However, the function and gradient values contributed by the nonseparable penalty function requires the evaluation of the side constraints. These computations are also vectorized.

Other Linear Algebra. The solution of the system

$$(D^T MD)p = -D^T \nabla\Phi(x^\nu, z^\nu),$$

at every step of the master problem also requires the computation of the right-hand side reduced gradient vector $D^T \nabla\Phi(x^\nu, z^\nu)$. This is a fully dense matrix-vector product suitable for vector architectures.

To illustrate the impact of vectorization on the performance of the above components, we give in Table 1 below the time spent in these master problem components during execution of the LQP algorithm with problem PDS3 both with and without vectorization. Compiler vectorization refers to automatic vectorization of the code using compiler options whereas user vectorization refers to restructuring of code segments and use of library subroutines such as the BLAS (Basic Linear Algebra) subroutines as explained in Pınar and Zenios [1990].

Opt. level	Descent Dir.	Other Lin. Alg.	Func. and Grad. Evals.
no vectorization	35.2	17.6	36.1
compiler vectorization	3.7	13.2	12.3
user vectorization	3.4	0.8	12.0

Table 1:Reduction in CPU spent in the main master problem components due to vectorization with PDS3.

As evidenced by the results, significant gains are realized in the master problem phase with vectorization. It is not possible to improve the subproblem solution time through vectorization due to the inherently scalar nature of the graph data structures used in implementing the network simplex algorithm.

3.2 Solving Multicommodity Network Flow Problems

The multicommodity network flow problem can be seen as a special case of the networks with side constraints model. The side constraints in this case have the following simple form: commodity flows on all or a fraction of the arcs compete for a joint arc capacity. This special structure of the side constraints considerably simplifies the computer representation and evaluation of these constraints in the context of the LQP algorithm. Extensive computational experience with multicommodity network flow problems using the LQP algorithm is reported in Zenios, Pınar and Dembo [1990]. Further work with a parallel decomposition of the algorithm is given in Pınar and Zenios [1990]. We only give a summary of the results here. The first set of test problems is a collection of linear multicommodity network flow problems derived from a Military Airlift Command (MAC) application. They are referred to as the Patient Distribution System (PDS) problems. The second set of test problems are randomly generated linear multicommodity network problems communicated to us by J.L. Kennington, see Ali and Kennington [1977]. The characteristics of the test problems are given in Table 2.

Problem	No. of arcs	No. of nodes	No. of commodities	No. of rows	No. of columns
PDS1	339	126	11	1473	3816
PDS3	1117	390	11	4593	12590
PDS5	2149	686	11	7546	23639
PDS10	4433	1399	11	15389	48763
PDS15	7203	2125	11	23375	79233
PDS20	10116	2447	11	31427	105728
KEN11	176	121	121	14694	21349
KEN13	225	169	169	28632	42659

Table 2: Characteristics of multicommodity network flow problems.

Computational results with multicommodity flow problems using the LQP algorithm are given in Table 3. For each problem we report the total number of simplicial decomposition iterations, number of extreme points retained upon completion, and the CPU time consumed during the subproblem and master problem phase of the simplicial decomposition algorithm. These two components comprise the total CPU usage during execution of the algorithm.

Test Problem	GENOS/LP				OB1
	Simplicial iters	Subproblem time	Master time	Total time	time
PDS1	23	1.01	0.85	1.86	
PDS3	41	12.04	6.73	18.77	
PDS5	71	55.12	37.97	93.09	
PDS10	103	232.31	175.82	408.13	1530
PDS15	121	559.35	381.08	940.43	
PDS20	145	1225.83	720.06	1945.89	16000
KEN11	16	7.4	8.1	15.5	21
KEN13	87	66.8	70.6	137.5	67

Table 3: Multicommodity flow problems solution statistics on the CRAY Y–MP.

With all the test problems, both the infeasibility tolerance $\epsilon_{min} = 10^{-5}$ and the bound gap tolerance $\epsilon_{gap} = 10^{-2}$ were achieved. PDS1 and PDS3 were also solved with the general purpose package MINOS of Murtagh and Saunders [1987]. The optimal values reported by MINOS matched the LQP optimal values to 5 digits. Some of these problems were solved on the same computer by Marsten et al. [1990] using the code OB1 based on interior point methods. The results are also given in Table 3. It is clear that the LQP algorithm outperforms substantially state-of-the-art implementations of interior point methods. By virtue of the linearization in the subproblem phase of simplicial decomposition, the linear network flow problems for individual commodities can be solved on parallel processors. The results of this study are reported elsewhere; see Pınar and Zenios [1990].

3.3 Solving the Naval Personnel Assignment Problem

In this section we report numerical results obtained using GENOS/LP on two Naval Personnel Assignment problems. Each year thousands of decisions are made to (re)allocate the Navy Enlisted Personnel to a fleet of combat units and to mission areas within these units. Allocations are made in such a manner as to provide the best defense at the lowest cost. All mission areas within a combat unit require personnel with different skills to support operational capabilities. A unit's capability to perform its functions in all its mission areas is referred to as "readiness". Readiness is measured based on the skills of personnel assigned. A shortage of skilled personnel would decrease the level of readiness of a mission area and thus degrade the capabilities of the unit. Clearly maximizing the level of readiness is a complex decision making problem given the large number of mission areas and personnel to be matched. This problem can be formulated as a network optimization problem with side constraints and variable(s). The reader is directed Krass and Thompson [1990] for more details on the model. Expressed in matrix notation the model has the following form:

$$\begin{array}{lrcl}\underset{x,z}{\text{minimize}} & cx - z & & \\ \text{subject to} & Ax & = & b \\ & Sx + Pz & \geq & d \\ & 0 \leq x & \leq & u \\ & 0 \leq z & \leq & r\end{array}$$

where A is a node-incidence matrix for the network and represents the flow conservation conditions and S and P are matrices used to capture the non-network requirements. The variables x denote the flow variables and z is the side variable which represents the level of readiness to be maximized. The objective is to maximize the level of readiness for all units considered and minimize the cost of the assignment.

For the Naval assignment problems used in this study, an initial feasible solution was readily computed since the solution to the network relaxation satisfied the side constraints when the side variable was ignored, i.e., let x^0 be a solution of the network relaxation

$$\begin{array}{lrcl}\underset{x}{\text{minimize}} & cx & & \\ \text{subject to} & Ax & = & b \\ & 0 \leq x & \leq & u\end{array}$$

If x^0 is such that

$$Sx^0 \geq d$$

then z^0 is computed as

$$z^0 = \min_i \frac{(\sum_{ij} s_{ij} x_{ij} - d_i)}{p_{ij}} \tag{23}$$

where s_{ij} and p_{ij} denote the entries at the i-th row and j-th column of the matrices S and P respectively.

The first model — NAVY — is a simplified version of the complete model which we we call HUGENAVY. The size and characteristics of both problems are given in Table 4. In addition the problem NAVY has a nonzero assignment cost vector c whereas the larger problem HUGENAVY has an assignment cost vector which is identically zero. The objective in problem HUGENAVY is solely to minimize the readiness level.

Problem	LP form		Network		No. of side const.	Opt. value
	Rows	Columns	Nodes	Arcs		
NAVY	4144	6842	3457	6841	687	-2.72347×10^5
HUGENAVY	36013	64542	30639	64541	5374	-0.5340

Table 4: Problem Characteristics of Naval Personnel Assignment.

Both problems were solved on the CRAY Y–MP. We give in Table 5 the solution statistics of the LQP algorithm. All times are stated in CPU seconds exclusive of input/output. Major iterations refer to the total number of times step 1 of the LQP algorithm is executed. It is interesting to note that the larger Navy problem is solved in a time very close to the solution time of the smaller problem. This can be attributed to the larger number of major iterations the algorithm took in the case of the problem NAVY because the smaller problem is more tightly constrained than the larger. Since the iterates generated by the LQP algorithm become only feasible on termination, the previous observation leads to the conclusion that, though much larger in size, HUGENAVY is a relatively easier problem for our method.

Problem	**GENOS/LP**				**MINOS time**	**OB1 time**
	Simpl. iters	**Subproblem time**	**Master time**	**Total time**		
NAVY (CRAY Y-MP)	6	45	149	194	NA	NA
NAVY (DEC 5100)	6	132	1428	1560	600	NA
HUGENAVY(CRAY Y-MP)	2	157	181	276	NA	150

Table 5: Performance of the LQP algorithm on the Naval personnel assignment problems.

We also report the solution time by MINOS for NAVY. The LQP algorithm was outperformed by MINOS on this problem. However, MINOS was not able to produce a feasible solution to HUGENAVY after one hour of CPU time on a CRAY Y-MP whereas this problem was solved within 5 minutes using the LQP method. On the other hand, the same problem was solved in less than 3 minutes using the OB1 code based on interior point methods. This indicates that the LQP algorithm based on nonlinear programming technology is competitive with current linear programming technology while it outperforms the state-of-the-art simplex based linear programming code MINOS. We also note that the value of the side variable which represents the readiness level was computed to 5 digits of accuracy by the LQP method as confirmed for both problems by the OB1 and MINOS solutions.

We also experimented with nonlinear versions of NAVY and HUGENAVY problems. We refer to these problems as NAVYQ and HUGENAVYQ where the objective function is a separable quadratic function of the form $\sum_j a_j x_j^2$. For the problem NAVYQ, the coefficients a_j are precisely the coefficients given in the linear model. For the HUGENAVYQ problem the coefficient vector was taken to be identically unity. We report the solution statistics in Table 6. Both results were obtained on a CRAY Y-MP.

Problem	Major iters	Subprob. time	Master time	Total time	Lower bound	Upper bound
NAVYQ	18	89	367	456	−273025.94	−252403.
HUGENAVYQ	8	270	867	1137	0.7220×10^8	0.7320×10^8

Table 6: Performance of the LQP algorithm on the nonlinear Naval personnel assignment problems.

With both problems, the infeasibility tolerance $\epsilon_{min} = 10^{-5}$ was attained on termination. Using MINOS to solve NAVYQ a feasible solution with objective function value −269515.4 was obtained in 485 CPU seconds on the CRAY Y-MP. This solution is 1% better than the best feasible solution produced by the LQP algorithm in 456 seconds. However, the LQP algorithm produced a more accurate solution on HUGENAVYQ. MINOS was not used for this problem due to the anticipated CPU time and memory requirements.

3.4 Solving Constrained Matrix Estimation Problems

In this section we report results with constrained versions of two matrix estimation problems from the World Bank. The matrix estimation problem is that of adjusting the entries of Social Accounting Matrices (SAM) for an economy and can be formulated as a nonlinear network optimization problem, see Zenios, Drud and Mulvey [1989]. The first problem SAMKE is a SAM model for Kenya and the second problem SAMBO is a SAM model for Botswana. Both problems were derived from econometric studies conducted at the World Bank. Both problems have separable objective functions. The problem SAMKE has a weighted entropy objective function of the form $ax \cdot (\log \frac{x}{b} - 1.0)$ for each flow variable and the problem SAMBO has a quadratic objective function of the form $a \cdot (x - b)^2$. We note that these functions do not satisfy the assumption of Lemma 1 of section 2.5 and therefore we rely on the subproblem lower bound for these problems. Since no initial feasible solution can be readily obtained, we do not compute upper bounds.

We constructed side constraints for these problems as follows. A number of arcs were randomly chosen and a fraction of the sum of the optimal flow values on these arcs was taken as the right-hand side of the inequality. This was repeated as many times as the number of side constraints we added to the problem. Therefore, the side constraints have the following form:

$$\sum_{(ij)\in\mathcal{E}} x_{ij} \geq \beta \sum_{(ij)\in\mathcal{E}} x^*_{ij}$$

where $\mathcal{E}$ is an arbitrary subset of the arcs of the underlying graph and x_{ij} is the flow variable for arc (i,j), x^*_{ij} are the optimal flows obtained by solving the original matrix estimation problem and $\beta \in (0,1]$. Characteristics of the problems are summarized in Table 7.

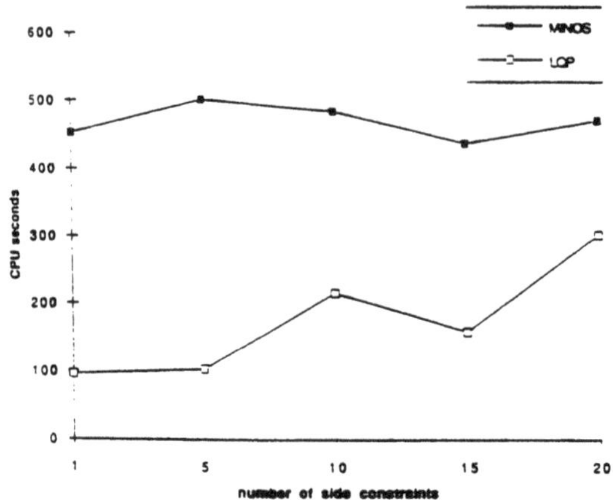

Figure 2: Variation of the solution time for constrained matrix estimation problem SAMBO as a function of the number of side constraints with GENOS/LP and MINOS.

Problem	Network		Optimal value
	Nodes	Arcs	
SAMKE	50	202	−7768.11
SAMBO	128	662	10.71

Table 7: Characteristics of matrix estimation problems.

Starting with one side constraint, both problems were solved with an increasing number of side constraints using the LQP algorithm. Tests were performed on a DEC station 3100. All times are given in CPU seconds. We provide in Figure 2 the variation of the CPU time taken by GENOS/LP algorithm to solve SAMBO and the CPU time taken by MINOS on the same problem as a function of the increasing number of side constraints added to the problem. We observe that while MINOS time does not vary considerably, GENOS/LP outperforms MINOS by a significant margin. However the advantage of GENOS/LP is reduced as the number of side constraints increase. This is not surprising since the LQP method is more sensitive to the size of the network component which gets smaller in percentage as more side constraints are added. In Figure 3, we plot the optimal values reported by GENOS/LP and MINOS with the problem SAMBO as a function of the number of side constraints. In all experiments, GENOS/LP was able to produce reasonably accurate solutions to the problem.

We provide in Table 8 a summary of the LQP algorithm statistics for both problems. The problems are referred to as SAMKE25 and SAMBO20 to indicate the

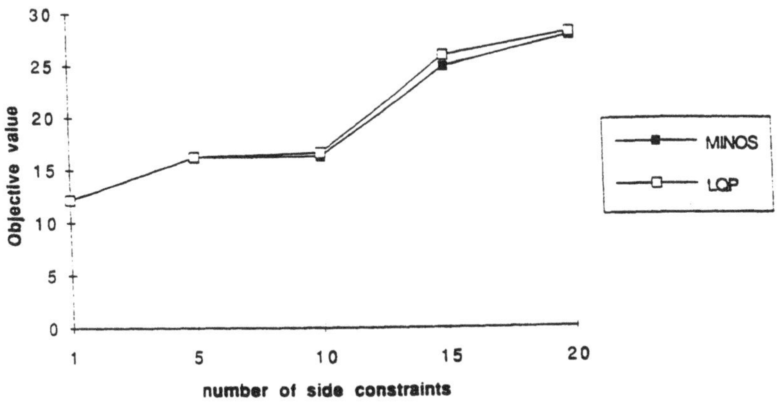

Figure 3: Variation of the optimal value for constrained matrix estimation problem SAMBO as a function of the number of side constraints with GENOS/LP and MINOS.

number of side constraints present in the problem.

Problem	Major iters.	Infeasibility	Total time	Objective value	Lower bound
SAMKE25	3	5.82×10^{-4}	14.	−7624.85	−7678.11
SAMBO20	7	6×10^{-4}	303.	28.10	21.89

Table 8: The LQP statistics with the matrix estimation problems.

The lower bounds in both cases are not very tight. The solution statistics with MINOS for problems SAMKE25 and SAMBO20 are given in Table 9.

Problem	Number of iterations	Optimal value	CPU time
SAMKE25	475	−7630.94	7
SAMBO20	1804	27.73	472

Table 9: Performance of MINOS on the matrix estimation problems.

Although MINOS provides more accurate solutions and was faster with the smaller SAMKE25, the LQP method outperforms MINOS on the larger problem with respect to CPU usage for different number of side constraints while it produced an acceptable level of accuracy.

3.5 Analyzing the NETLIB Test Problems

Using the network extracting heuristics of Bixby and Fourer [1988], we analyzed a subset of NETLIB linear programming test problems. A summary of the characteristics of the test problems and the associated network formulations are given in Table 10. As can be observed from the statistics, two of the problems have a large network component to warrant some attention.

Problem	Linear Programming			Network		Side	Side
	Rows	Columns	Opt. Val.	Nodes	Arcs	Const.	Vars.
RECIPE	92	180	-2.6661×10^2	54	140	30	14
GREENBEA	2400	5443	-7.2462×10^7	895	4641	1423	606
GIFFPINC	617	1092	6.9022×10^6	523	1071	69	1
SCAGR25	472	500	-1.4753×10^7	200	372	147	127
SCRS8	491	1169	9.9429×10^2	301	1096	156	78
SHIP12L	1165	5427	1.4701×10^6	735	5321	104	
SIERRA	1228	9252	1.5394×10^7	878	2726	349	
STANDATA	468	3686	1.2576×10^3	96	331	226	789

Table 10: The NETLIB problems in LP and network forms.

Our experience with the NETLIB problems using the LQP algorithm revealed that solving these problems as general linear programs is more efficient. We report results with two problems GIFFPINC and SHIP12L. The LQP statistics are given below in Table 11. The number of major iterations refer to the number of executions of step 1 of the LQP algorithm. We also report the final objective function value reported on termination and the best lower bound computed thus far. Infeasibility refers to the maximum degree of violation of the side constraints. It was not possible to provide an initial feasible solution for these problems and hence no feasible iterates and upper bounds to optimal value were computed. The tests were performed on a DEC station 3100. SHIP12L was solved in 110 CPU seconds using MINOS and GIFFPINC was soved in 25 seconds using the same code.

Problem	Major iters.	Infeasibility	Total time	Objective value	Lower bound
GIFFPINC	10	0.242×10^{-2}	8240.9	7.581×10^6	6.766×10^6
SHIP12L	15	2.37	7200	1.768×10^6	NA

Table 11: The LQP statistics with the NETLIB problems.

The statistics clearly indicate that these problems proved to be extremely hard for the LQP algorithm. In the case of GIFFPINC, although an acceptable level of infeasibility and a reasonable lower bound was achieved, the objective value is off the known optimal value by a considerable margin. The case of SHIP12L was more problematic.

The computation was stopped after two CPU hours and the iterate was still far from reaching the absolute infeasibility tolerance $\epsilon_{min} = 10^{-5}$. It was also impossible to assess the quality of the solutions to nonlinear network (penalty) problems due to the poor quality of the lower bounds. We also note that both problems had a dense side constraint matrix structure, a factor which affects the LQP algorithm negatively.

To conclude, we remark that the LQP technology was not effective in dealing with the NETLIB problems although the experience contributed to the robustness testing of the GENOS/LP code.

3.6 Integration with MINOS

The LQP method delivers quickly an approximate solution to the problem. When higher accuracy is needed a linear programming solver may be used. The linear program NAVY is solved with the general purpose linear programming solver MINOS of Murtagh and Saunders [1987]. The statistics are given in Table 12.

Number of Phase-I pivots	1247
Total number of pivots	2423
CPU time(DEC 5100)	10 mins.

Table 12: Performance of MINOS on NAVY.

Interfacing the GENOS/LP system with MINOS may provide MINOS with an advanced starting point. However, since the LQP algorithm is essentially based on an exterior point penalty function, no basis for the problem is readily available. The optimal network basis produced as a result of solving linear network subproblems is input to MINOS. This idea produced a significant reduction in the number of pivots taken by MINOS to reach optimality. A comparison is given below in Table 13.

	MINOS	MINOS with advanced start
Number of Phase-I pivots	1247	1750
Total number of pivots	2423	1782

Table 13: Performance of MINOS on NAVY using advanced start.

As can be observed from Table 13 the total number of iterations were reduced significantly. Due to the anticipated CPU usage this strategy was not applied to HUGE-NAVY.

4 Conclusions

We presented in this paper a solution method suitable for large scale optimization problems with embedded network structures, and results of extensive computational testing with various test problems. The LQP method is an exterior point method based on a smooth penalty function and can produce feasible iterates if an initial feasible point is available. For applications where several problem instances need to be solved such as the Patient Distribution System and the Naval Personnel Assignment, there is a high payoff in exploiting the network structure and developing a specialized algorithm. Particularly as the problem size gets bigger, the benefits of using the LQP algorithm become more accentuated. In this paper we presented strong evidence to support this claim. For smaller problems it is more beneficial to use general purpose algorithms as evidenced by the analysis with the NETLIB problems. Another important factor which affects the performance of the LQP algorithm is the sparsity pattern of the side constraint matrix. With the PDS problems and Naval personnel assignment problems the side constraint matrix had a favourable sparsity pattern whereas the NETLIB problems had a very dense structure. However, the LQP algorithm may still be a viable alternative in smaller problems with relatively few side constraints as observed in the case of constrained matrix estimation problems.

In summary the LQP algorithm is able to provide quickly approximate solutions to the problem. For large problems it outperforms state-of-the-art general purpose optimization software while it remains competitive with the more recent interior point based optimization technology. To achieve higher accuracy, the LQP solution can be used as an advanced start for a general purpose linear programming solver.

Acknowledgments. This research was partially supported by NSF grants SES-91-00216 and CCR-91-0402 and AFOSR grant 91-0168. The assistance of Mr. Ted Thompson and Mr. Iosif Krass with supplying the data for the Navy Personnel Assignment problem is gratefully acknowledged. Mr. John Gregory kindly provided assistance with the CRAY experiments and OB1. Professor Bob Fourer kindly made his network extraction program available.

References

[1] A.I. Ali and J.L. Kennington, MNETGN Program Documentation, Technical Report IEOR 77003. Department of Industrial Engineering and Operations Research, Southern Methodist University, Dallas (1985).

[2] A.I. Ali, J.L. Kennington and B. Shetty, The Equal Flow Problem, *European Journal of Operational Research* **36** (1988) 107–115.

[3] D.P. Bertsekas, Nondifferentiable Optimization via Approximation, *Mathematical Programming Study* **3** (1975) 1–25.

[4] D.P. Bertsekas, Projected Newton Methods for Optimization Problems with Simple Constraints, *SIAM Journal on Control and Optimization* **20** (1982) 221–246.

[5] R.E. Bixby and R. Fourer, Finding Embedded Network Rows in Linear Programs I. Extraction Heuristics, *Management Science* **34** (1988) 342–376.

[6] G.G. Brown, G.W. Graves, H. Lange, C. Staniec and R.K. Wood Dual Decomposition Methods for Solving Multicommodity Flow Problems, Technical Report, Naval Postgraduate School (1989).

[7] G.G. Brown and R.D. McBride, Solving Generalized Networks, *Management Science* **30** (1984) 1497–1523.

[8] C.H.J. Chen and M. Engquist, A Primal Simplex Approach to Pure Processing Networks, *Management Science* **32** (1986) 1582–1598.

[9] S. Chen and R. Saigal, A Primal Algorithm for Solving a Capacitated Network Flow Problem with Additional Linear Constraints, *Networks* **7** (1977) 59–79.

[10] R.S.Dembo, J.M. Mulvey and S.A. Zenios, Large-Scale Nonlinear Network Models and Their Application. *Operations Research* **37** (1989) 353–372.

[11] J.E. Dennis,Jr. and R.B. Schnabel, *Numerical Methods for Unconstrained Optimization and Nonlinear Equations* (Prentice–Hall, New Jersey, 1983).

[12] A.V. Fiacco and G.P. McCormick, *Nonlinear Programming: Sequential Unconstrained Minimization Techniques* (John Wiley, New York, 1968).

[13] A.B. Gamble, A.R. Conn and W.R. Pulleyblank, A Network Penalty Problem, *Mathematical Programming* **50** (1991) 53–74.

[14] F. Glover, J. Hultz, D.Klingman and J.Stutz, Generalized Networks: A Fundamental Computer Based Planning Tool, *Management Science* **24** (1978) 1209–1220.

[15] F. Glover and D. Klingman, The Simplex SON Algorithm for LP/Embedded Network Problems, *Mathematical Programming Study* **15** (1981) 148–176.

[16] N. Karmarkar, A New Polynomial Time Algorithm for Linear Programming, *Combinatorica* **4** (1984) 373–395.

[17] J.L. Kennington and R.V. Helgason, *Algorithms for Network Programming* (John Wiley and Sons, New York, 1980).

[18] J. Koene, Minimal Cost Flow in Processing Networks, A Primal Approach, Ph.D. Thesis, Eindhoven University of Technology, Eindhoven, The Netherlands (1982).

[19] I.A. Krass and T.J. Thompson, Mathematical Formulation of EDPROJ - Readiness Connection, Technical Report, Navy Personnel Research and Development Center, San Diego CA (1990).

[20] R. Marsten, R. Subramanian, M. Saltzman, I. Lustig and D. Shanno, Interior Point Methods for Linear Programming, *Interfaces* **20** (1990) 105–116.

[21] R.D. McBride, Solving Embedded Generalized Network Problems, *European Journal of Operational Research* **21** (1985) 82–92.

[22] J.M. Mulvey, S.A. Zenios and D.P. Ahlfeld, Simplicial Decomposition for Convex Generalized Networks, *Journal of Information and Optimization Sciences* **11** (1990) 359–387.

[23] J.M. Mulvey and S.A. Zenios, Solving Large Scale Generalized Networks, *Journal of Information and Optimization Sciences* **6** (1985) 95–112.

[24] J.M. Mulvey, Testing of a Large Scale Network Optimization Program, *Mathematical Programming* **15** (1978) 291–315.

[25] B.A. Murtagh and M.A. Saunders, MINOS 5.1 User's Guide, Report SOL 83–20R, December 1983 (revised January 1987), Stanford University.

[26] M.Ç. Pınar, Decomposition and Parallel Solution of Network Structured Optimization Problems, Ph.D. Thesis, University of Pennsylvania, Philadelphia PA 19104 (1992).

[27] M.Ç. Pınar and S.A. Zenios, Parallel Decomposition of Multicommodity Network Flows using Smooth Penalty Functions, *ORSA Journal on Computing* **4** (1992) (forthcoming).

[28] G.L. Schultz and R.R. Meyer., An Interior Point Method for Block Angular Optimization, *SIAM Journal on Optimization* **1** (1991) .

[29] I. Zang, A Smoothing-out Technique for Min-Max Optimization, *Mathematical Programming* 19 (1980) 61–77.

[30] S.A. Zenios, A. Drud and J.M. Mulvey, Balancing Large Social Accounting Matrices with Nonlinear Network Programming. *Networks* 19 (1989) 569–585.

[31] S.A. Zenios, M.Ç. Pınar and R.S. Dembo, A Smooth Penalty Function Algorithm for Network Structured Problems, Department of Decision Sciences Report 90–12–05, The Wharton School, University of Pennsylvania, Philadelphia, PA. 19104 (1990).

Network Optimization Problems, pp. 203-231
Eds. D.-Z. Du and P.M. Pardalos

On Algorithms for Nonlinear Dynamic Networks

Warren B. Powell
Elif Berkkam
Irvin J. Lustig
Department of Civil Engineering and Operations Research, School of Engineering and Applied Science, Princeton University, Princeton, NJ 08544 USA

Abstract

We consider the problem of minimizing costs over a dynamic, acyclic network with convex, separable link cost functions. The standard approach is to formulate the problem as a convex, separable optimization problem subject to flow conservation constraints, where the decision variable is the flow x_{ij} on link (i, j). We show that standard network algorithms applied to dynamic problems exhibit surprisingly poor performance for networks with as few as 10 or 20 time periods, suggesting that dynamic networks are intrinsically much harder to solve than static networks of comparable size. The problem can be reformulated using decision variables θ_{ij}, which gives the fraction of the total flow passing through node i that should be routed over link (i, j). This formulation has been used by other researchers in the development of parallel algorithms which take advantage of the simple constraint structure. We show that this reformulation produces substantially faster execution times for Frank-Wolfe type methods than the same methods applied to the standard formulation.

1 Introduction

We consider the problem of optimizing flows over a dynamic network with separable nonlinear cost functions. The problem can be motivated by problems arising in network models of dynamic fleet management for common carriers in freight transportation (truck, rail, containers). In these models, supplies of vehicles enter the network in the first few time periods but then flow through the network and exit via a supersink. An important characteristic of these problems is that most of the nodes are pure transshipment nodes and only the supersink is a deficit node. Powell *et al.* [11] presents a model with this structure to manage a fleet of trucks over time under uncertain demands. Gallagher [5] introduces a similar model with multicommodity flows to optimize the routing of messages over communication networks.

We show in this paper that dynamic networks are intrinsically much more difficult to solve with first-order methods than static networks of comparable size. Algorithms applied to standard formulations of nonlinear dynamic networks can perform very poorly. Even special packages for nonlinear networks such as GENOS (Mulvey and Zenios [8]) require exceptionally long run times for relatively small networks with 10 or 20 time periods. By contrast, even relatively simplistic algorithms such as Frank-Wolfe are shown to work quite well when applied to a different formulation of the same problem.

There are two approaches that can be used to solve this problem. The first is to view it as a standard minimum cost flow problem with nonlinear cost functions and link flows, x_{ij}, as decision variables. The second approach does not require the use of any network flow algorithms but takes advantage of the acyclic structure of the network (Figure 1) to describe the impact of decisions made now on future costs. This new approach uses flow fractions, θ_{ij}, as decision variables. The decision variable θ_{ij} represents the fraction of total flow passing through node i that is to be routed over link (i, j). We refer to the classical formulation, which uses x_{ij} as decision variables, as NDN-X (Nonlinear Dynamic Network with X variables), and refer to the new formulation, which uses θ_{ij}, as NDN-T.

The formulation NDN-T appears to have been first introduced by Gallagher [5] for multicommodity flow problems arising in telecommunications. The motivation for the formulation was the development of distributed algorithms for routing in telecommunications. As shown below, the NDN-T formulation uses a very simple constraint structure that lends itself easily to parallel computation. The same formulation was developed independently by Powell *et al.* [11] in the context of managing fleets of vehicles under uncertainty. However, neither of these papers really investigate the behavior of solution algorithms for dynamic networks. Researchers have long realized that the structure of dynamic networks could be used to develop specialized algorithms (see, for example, Aronson [1]). By contrast, there has been relatively little recognition of the challenges posed by dynamic networks.

There are relatively few algorithms specialized for dynamic networks (see, for example, the extensive review in Aronson and Chen [2]). Aronson [1] presents a specialization of the network simplex algorithm that takes advantage of breaks in the tree that limit the effect of pivots on earlier time periods. This result, however, appears to be restricted to pure dynamic networks that arise in inventory planning problems. White and Bomberault [12] offer a specialization of a primal dual algorithm for dynamic networks motivated by empty railcar models. Powell *et al.* [11] propose a stochastic formulation of the dynamic vehicle allocation problem which produces a nonlinear, dynamic network with the structure considered here. A flow splitting algorithm of the type presented above is introduced and shown to exhibit good performance in limited tests. Bertsekas [3] and Bertsekas *et al.* [4] explore in depth enhancements of the flow splitting formulation we refer to as NDN-T. Taking advantage of the simple structure of the constraint set, these papers develop projection algorithms and second order algorithms.

In this paper we use this formulation to expose the impact of the dynamic structure of the network on algorithmic performance. Section 2 presents the formulation of the nonlinear dynamic network problem in the traditional form NDN-X. Section 3 presents a more detailed description of the NDN-T formulation in terms of dynamic networks, and states a simple backward recursion for calculating derivatives, with a derivation left to the appendix. Section 4 outlines several standard algorithms that can be used with the new formulation, taking advantage of the simple structure of the constraint set. Finally, section 5 compares the NDN-X and NDN-T formulations.

2 The NDN-X Formulation

We begin by presenting the basic problem in a form that explicitly reflects the dynamic structure of the problem. Let i and j refer to cities (points in space) and let a node in the network be denoted by (i,t). For notational simplicity only, we assume any links emanating from (i,t) terminate in period $t+1$. Define

$$\begin{aligned} R_i^t &= \text{net surplus } (R_i^t > 0) \text{ or deficit } (R_i^t < 0) \text{ at node } i \text{ at time } t \\ x_{ij}^t &= \text{total flow from } i \text{ to } j \text{ departing at time } t \text{ (and arriving at time } t+1) \\ f_{ij}^t(x_{ij}^t) &= \text{cost of sending flow on the link from node } (i,t) \text{ to node } (j,t+1) \end{aligned}$$

The formulation NDN-X is then

$$\begin{aligned} \underset{x}{\text{minimize}} \quad F(x) &= \sum_t \sum_i \sum_j f_{ij}^t(x_{ij}^t) \\ \text{subject to} \quad Ax &= R \qquad (1) \\ x &\geq 0 \qquad (2) \end{aligned}$$

where A is the node-arc incidence matrix for the network. It is assumed that the functions f_{ij}^t are convex and that R is the vector of surpluses and deficits at each node. This formulation produces a problem with a separable, nonlinear objective function with nonseparable network constraints.

The relative ease with which derivatives can be calculated, due to separability, makes this traditional formulation favorable. In addition, the conservation of flow constraints can be handled in the context of first order nonlinear programming algorithms, such as the Frank-Wolfe algorithm, since the linearized subproblems are pure networks. The gradient g of F is defined by

$$g_{ij}^t(x_{ij}^t) = \frac{\partial f_{ij}^t(x_{ij}^t)}{\partial x_{ij}^t}. \tag{3}$$

If a first-order algorithm is applied to NDN-X, then the linearized subproblem is based on the gradient at some point $\hat{x}$. This problem is then a pure network problem of the form

$$\begin{aligned} \underset{y}{\text{minimize}} \quad & \sum_t \sum_i \sum_j g_{ij}^t(\hat{x}_{ij}^t) \cdot y_{ij}^t \\ \text{subject to} \quad & Ay = R \\ & y \geq 0 \end{aligned} \tag{4}$$

The Frank-Wolfe algorithm applied to NDN-X generates iterates $x^{(k)}$ for $k \geq 0$ using the following steps:

Step 1. Set $x^{(0)} = 0$ and find $g(x^{(0)})$. Solve (4) with $\hat{x} = x^{(0)}$ and set k=1, so that our initial solution is $x^{(1)} = y^{(0)}$.

Step 2. Evaluate $g_{ij}^t(x^{(k)})$. Solve the linear network problem (4) and let $y^{(k)}$ be the optimal solution.

Step 3. Find the optimum step size α^* by solving

$$\min_{0 \leq \alpha \leq 1} F(x^{(k)} + \alpha(y^{(k)} - x^{(k)})).$$

Step 4. Update $x^{(k+1)} = x^{(k)} + \alpha^*(y^{(k)} - x^{(k)})$.

Step 5. If $\frac{|F(x^{(k+1)}) - F(x^{(k)})|}{F(x^{(k)})} < \epsilon$, then stop, otherwise set $k \leftarrow k+1$ and go to step 2.

The problem with a standard application of Frank-Wolfe is the nature of the extreme solution, as depicted in Figure 2. The structure of the problem, where flows typically enter the network in the first few time periods and leave through the supersink, produces a pattern of flows from the subproblem that follows a tree. In fact, we are actually just solving a shortest path problem into the supersink over an uncapacitated network with linear costs. For dynamic networks with at least five or ten time periods, the result is a set of flows that converges on a single path several time periods into the problem. This extreme point solution is an unusually poor approximation to the optimal solution, and hence is the cause of extremely slow rates of convergence. In fact, standard algorithms, applied to even relatively small problems with a large (greater than 20) number of time periods converge so slowly that they may stop well short of optimality. In the next section, we show how a simple transformation takes advantage of the dynamic structure of the problem, and allows us to develop very simple and efficient algorithms for solving these networks.

3 The Transformed Problem NDN-T

In this section we give a different formulation of the same problem, where the decision variables are fractions of total supply at a node instead of link flows. The transformation from the original formulation NDN-X into this new formulation is performed using

$$x_{ij}^t = \theta_{ij}^t \cdot S_i^t \tag{5}$$

where

$$\begin{aligned}
\theta_{ij}^t &= \text{fraction of total supply at node } i \text{ at time } t \text{ to be sent to node } j \text{ at time } t+1 \\
\theta^t &= \{\ldots, \theta_{ij}^t, \ldots\} = \text{values of } \theta \text{ at a fixed time } t \\
P &= \text{total number of time periods} \\
\theta &= \{\theta^1, \theta^2, \ldots, \theta^P\} \\
\hat{\theta}^t &= \{\theta^1, \theta^2, \ldots, \theta^t\} = \text{all } \theta \text{ values up to time } t \\
S_i^t &= \text{total endogenous flow through node } i \text{ at time } t \\
S_i^t + R_i^t &= \text{total available supply at node } i \text{ at time } t
\end{aligned}$$

For simplicity of notation, we assume all flows pass from period t to $t+1$. Flows from period t to $t+m$, $m \geq 1$, are easily handled, whereas flows between regions within a time period would require us to solve certain systems of linear equations, adding substantially to the computational effort.

The supply at a node S_i^t is defined not as a constraint but is actually an implicit function of θ. This is because S_j^{t+1} can be calculated as

$$S_j^{t+1} = R_j^{t+1} + \sum_i \theta_{ij}^t \cdot S_i^t \tag{6}$$

As a result, the total flow S_i^t through a node depends on the partial vector $\hat{\theta}^t$ that gives the decisions made in earlier time periods. We will henceforth use the notation S_i^t and $S_i^t(\theta)$ interchangeably, using the latter when the stress on the functional nature of S_i^t is needed.

Our objective is to minimize the total cost of flow along the links over the entire network using the link cost functions

$$\hat{f}_{ij}^t(\theta) = f_{ij}^t(\theta_{ij}^t \cdot S_i^t(\theta)). \tag{7}$$

The optimization problem can be stated as follows :

$$\begin{aligned}
\underset{\theta}{\text{minimize}} \quad \hat{F}(\theta) &= \sum_t \sum_i \sum_j \hat{f}_{ij}^t(\theta) && (8) \\
\text{subject to} \quad \sum_j \theta_{ij}^t &= 1 \qquad \forall\, i, t && (9) \\
\theta_{ij}^t &\geq 0 \qquad \forall\, i, j, t && (10)
\end{aligned}$$

The first constraint ensures that the sum of all the fractions of total supply at a particular node i add up to unity and the second one is the nonnegativity constraint of the flow fractions.

By (7) and (6), we have that the new formulation produces a nonseparable objective function, but separable constraints. For the special case of dynamic networks we show in the appendix that the derivatives of $\hat{F}(\theta)$ can be calculated using only nominally more effort than required for the original NDN-X case, producing subproblems that are trivial to solve when using a Frank-Wolfe algorithm.

Through a straightforward application of the chain rule, we can obtain the following recursions for the derivatives:

$$\frac{\partial \hat{F}}{\partial \theta_{ij}^t} = \left(\frac{\partial f_{ij}^t}{\partial x_{ij}^t} + \frac{\partial \hat{F}}{\partial S_j^{t+1}(\theta)}\right) \cdot S_i^t(\theta) \tag{11}$$

$$\frac{\partial \hat{F}}{\partial S_i^t(\theta)} = \sum_j \left(\frac{\partial f_{ij}^t}{\partial x_{ij}^t} \cdot \theta_{ij}^t + \frac{\partial \hat{F}}{\partial S_j^{t+1}(\theta)} \cdot \theta_{ij}^t\right) \tag{12}$$

To simplify the notation, let the gradients be denoted as

$$g_{ij}^t = \frac{\partial \hat{F}(\theta)}{\partial \theta_{ij}^t} \tag{13}$$

$$\text{and} \quad \hat{g}_i^t = \frac{\partial \hat{F}(\theta)}{\partial S_i^t(\theta)} \tag{14}$$

It will be clear that g_{ij}^t represents the gradient of $\hat{F}$ evaluated at some point θ.

These derivatives can be found by a backward pass, where the main idea is to start at time $t = P$ and move backward in time over the entire network. The **Backward Pass** procedure works as follows where it is assumed that θ_{ij}^t and S_i^t are known for all i, j, and t.

Step 1. Set $g_{ij}^{P+1} = 0$ for all i, j and $\hat{g}_i^{P+1} = 0$ for all i.

Step 2. For each $t = P, P-1, \ldots, 1$, compute

$$g_{ij}^t = \left(\frac{\partial f_{ij}^t}{\partial x_{ij}^t} + \hat{g}_j^{t+1}\right) \cdot S_i^t \tag{15}$$

$$\text{and} \quad \hat{g}_i^t = \sum_j \theta_{ij}^t \left(\frac{\partial f_{ij}^t}{\partial x_{ij}^t} + \hat{g}_j^{t+1}\right). \tag{16}$$

The loop is derived from the recursions (11) and (12). Note that

$$\frac{\partial f_{ij}^t}{\partial x_{ij}^t}$$

is the gradient of the original objective function evaluated at the value $x_{ij}^t = \theta_{ij}^t \cdot S_i^t$.

For the overall algorithm we need to calculate the supplies S_i^t and the gradients g_{ij}^t for all nodes and all time periods using a given θ value. The values S_i^t can be found by a forward pass, where the main idea is to start at time $t = 1$ and move forward in time over the entire network. This **Forward Pass** procedure works as follows:
Step 1. Set $S_i^1 = R_i^1$ for all i to determine the initial supplies for period 1.

Step 2. For each $t = 1, 2, \ldots, P-1$, update for each j, $S_j^{t+1} = R_j^{t+1} + \sum_i \theta_{ij}^t \cdot S_i^t$.

4 Solution Algorithms for NDN-T

In this section we outline three standard algorithms that can be used with the new formulation: the Frank-Wolfe algorithm, a gradient projection algorithm, and an active set strategy. All of these algorithms take advantage of the special structure of the constraint set. In the gradient projection method, the required projection operator is particularly simple (see Bertsekas [3] for a discussion of projection methods using this problem formulation). The active set method also uses the structure of the constraints to find a basis. These algorithms are only used to illustrate the general performance of the transformation, and do not represent a comprehensive study of algorithms for this class of problems.

4.1 Frank-Wolfe

A standard application of the Frank-Wolfe algorithm involves solving the following subproblem determined by computing the gradients g_{ij}^t at some point θ:

$$\underset{\beta}{\text{minimize}} \quad \sum_t \sum_i \sum_j g_{ij}^t \beta_{ij}^t \tag{17}$$

$$\text{subject to} \quad \sum_j \beta_{ij}^t = 1 \quad \forall i, t \tag{18}$$

$$\beta_{ij}^t \geq 0 \quad \forall i, j, t \tag{19}$$

The solution to this subproblem is the vector β^*. The problem decomposes by each node (i, t). Hence, for each pair (i, t), we simply choose the index j corresponding to the most negative value g_{ij}^t. We then set $\beta_{ij}^t = 1$ and the values $\beta_{ij}^t = 0$ for $k \neq j$.

Using the backward and forward passes, the complete algorithm can be built up very efficiently, which is as follows:
Step 1. Let $\theta^{(0)} = 0$ and calculate g_{ij}^t at the point $\theta = 0$. Solve

$$\underset{\beta}{\text{minimize}} \sum_t \sum_i \sum_j g_{ij}^t \cdot \beta_{ij}^t$$

subject to (18) and (19). The solution β^* becomes the initial solution $\theta^{(1)} = \beta^*$. Set $k = 1$.

Step 2. Use the Forward Pass to find the vector $(S_i^t)^{(k)}$. .

Step 3. Compute the derivatives g_{ij}^t evaluated at $\theta^{(k)}$ using the Backward Pass.

Step 4. Solve the linearized subproblem (17) subject to (18) and (19) to get the solution $\beta^{(k)}$.

Step 5. Find the optimum step size α^* by solving:

$$\underset{0\leq\alpha\leq 1}{\text{minimize}}\, \hat{F}(\theta^{(k)} + \alpha(\beta^{(k)} - \theta^{(k)}))$$

Step 6. Update $\theta^{(k+1)} = \theta^{(k)} + \alpha^*(\beta^{(k)} - \theta^{(k)})$.

Step 7. If $\frac{|\hat{F}(\theta^{(k+1)}) - \hat{F}(\theta^{(k)})|}{\hat{F}(\theta^{(k)})} < \epsilon$, then stop. Otherwise set $k=k+1$ and go back to Step 2.

The linearized subproblem is solved easily when compared with the classical linearized subproblem since it needs to look only for the most negative gradient out of each node to decide on which arc to put flow. By contrast, the classical formulation using link flows produces a subproblem that requires the solution of a linear network.

4.2 Gradient Projection Method

An attraction of Frank-Wolfe is that it produces a feasible descent vector, but the cost is the use of an extreme point solution. Gradient methods generally offer a better search direction, but require a projection operation to regain feasibility. The simple constraint structure of the transformed problem allows this projection to be performed with relative ease.

Let $\theta_{i\cdot}^t$ be the vector of flow fractions out of node (i,t). We scale the gradient out of node (i,t), $g_{i\cdot}^t$, by dividing all entries out of that node by the most negative element. Let the scaled gradient be denoted as $\bar{g}_{i\cdot}^t$. The updated flow fractions out of node (i,t) are obtained by

$$\beta_{i\cdot}^t = \theta_{i\cdot}^t + \bar{g}_{i\cdot}^t, \tag{20}$$

which will be infeasible. Thus we project $\beta_{i\cdot}^t$ back onto the feasible region using the method presented by Held *et al.* [7] and determine the feasible direction d^k by

$$d^k = \beta^{\text{proj}} - \theta \tag{21}$$

This direction is then used in the main iterate

$$\theta^{(k+1)} = \theta^{(k)} + \alpha_k^* \cdot d^k \tag{22}$$

4.3 Active Set Method

An alternative use of the simple constraint set is the active set method (see Gill *et al.* [6]). Having a single separable constraint for each node (i,t) makes it easy to find a basis by eliminating one variable using each constraint. In doing so we distinguish between the nonnegativity constraints that hold exactly (active) and those that do not (inactive). At each node, apart from the convexity constraint which is always active, whenever $\theta^t_{ij} = 0$, we set it as active and whenever $\theta^t_{ij} > 0$, it is referred to as inactive. Out of all the inactive variables we eliminate one and make it basic. Let

$$\begin{aligned} E^t_i &= \text{the set of variables } \theta^t_{ij} \text{ that are eliminated (basic),} \\ \text{and} \quad N^t_i &= \text{the set of variables } \theta^t_{ij} \text{ that are not eliminated} \\ &\quad\ \text{and that are inactive (nonbasic),} \end{aligned}$$

where the sets are mutually exclusive.

The basic iteration is

$$\theta^{(k+1)} = \theta^{(k)} + \alpha^* \cdot d^k \tag{23}$$

We start by eliminating one of the inactive flow fractions, which is strictly positive, out of each node (i,t) such that

$$\theta^t_{il} = 1 - \sum_{j:\theta^t_{ij} \in N^t_i} \theta^t_{ij} \qquad \text{for} \quad \theta^t_{il} \in E^t_i \tag{24}$$

If θ^t_{ij} is nonbasic then the direction of movement is the negative gradient out of node i evaluated at θ^t_{ij}:

$$d^t_{ij} = -g^t_{ij} \tag{25}$$

If θ^t_{ij} is basic then

$$d^t_{ij} = \sum_{k:\theta^t_{ik} \in N^t_i} g^t_{ik} \tag{26}$$

Since the elements with $d^t_{ij} < 0$ are going to decrease from their current values in order to maintain feasibility $((\theta^t_{ij})^{k+1}) \geq 0)$, a ratio test must be performed that evaluates the distance of the variables to zero:

$$\alpha_{\max} = \min_{i,j,t} \left\{ \left| \frac{\theta^t_{ij}}{d^t_{ij}} \right|, d^t_{ij} < 0 \right\} \tag{27}$$

The variables are then updated with respect to a stepsize α^* such that $0 \leq \alpha^* \leq \alpha_{\max}$ is obtained through a one dimensional search.

5 Numerical Results

5.1 Experimental Design

Experiments were run to provide an indication of the important properties of each formulation and to contrast the performance of the different algorithms on each formulation. Relatively little formal experimentation has been reported specifically for dynamic networks, with the notable exception of the work by Aronson and Chen [2]. This work focussed on dynamic networks that featured potentially unbalanced transportation problems in each time period with inventory carry-over arcs from one period to the next. Aside from the fact that the networks are linear, these networks exhibit a basically different structure from the deep networks that we consider.

Given the exploratory nature of our experiments, we used randomly generated networks to test the algorithms. The network generator, however, was designed in the context of dynamic fleet management problems arising in truckload trucking. In this application, the nonlinear "cost" functions arise from an attempt to capture the uncertainity in the demand for transportation from city i to city j (see Powell *et al.* [9] and Powell [10] for complete details of the model). Let D be the uncertain demand for transportation in a particular market, and let x be the flow of vehicles. Then $\min\{D, x\}$ vehicles will move loaded, generating revenue r, while $x - \min\{D, x\}$ will move empty, at a cost c. Let $p(x)$ be the expected cost (negative profit) generated on this link, given by:

$$p(x) = E_D[\ c(x - \min\{D, x\}) - r\min\{D, x\}] \tag{28}$$

If D is described by a simple density function $f_D(x) = \lambda e^{-\lambda x}$, where $\lambda = 1/E[D]$, then

$$p(x) = cx - \frac{1}{\lambda}(r + c)(1 - e^{-\lambda x}) \tag{29}$$

Of course, any nonlinear cost function could be used, but we felt that it helped our design of the network generator to use parameters and assumptions that were motivated by an application. For example, we were able to choose input supplies (representing the supplies of vehicles in the fleet distributed among the set of cities) in a manner consistent with the demand for the vehicles.

In order to generate the parameters r, c, and λ for each market, we generated random coordinates for cities, uniformly over a 1000 by 2000 mile rectangle, from which we could calculate distances d_{ij} for each city pair (i, j). Given these distances, we used

$$r_{ij} = 1.2 d_{ij} \tag{30}$$

$$c_{ij} = 0.6 d_{ij} \tag{31}$$

$$\lambda_{ij} = \frac{1}{d_{ij}} \tag{32}$$

where 1.2 and 0.6 are typical per mile revenues and costs for the trucking industry. The external supplies that enter the network at the first time period are generated by :

$$R_i(1) = \gamma \sum_j \frac{1}{\lambda_{ij}} \tag{33}$$

where $\gamma \in [0.3, 0.7]$. By choosing the fraction γ within this range, we made sure that flows stayed on the curved part of the nonlinear cost function, which corresponds to the hatched region in Figure 3. Inconsistency between the flows (representing the supply of vehicles) and the market demands (which determines the shape of the cost function) may result in landing on the linear portion of the function by being too far off to the right or to the left. The last parameter required is the link density. We generated each link with probability α, using $\alpha = 0.5$ for most problems.

In addition to using our algorithms, certain sets of experiments were run using GENOS (Mulvey and Zenios [8]). GENOS includes implementations of the primal truncated Newton method and simplicial decomposition, both specialized for nonlinear generalized networks. To accomodate the predefined functions in GENOS, the experiments which involved comparisons against GENOS used cost functions of the form ae^{-bx}. In terms of the parameters of our problem, the functions were given by:

$$p(x) = \frac{1}{\lambda}(r + c)e^{-\lambda x} \tag{34}$$

The test problems are designed to compare the classical formulation against the new one using Frank-Wolfe, and to see how well the new algorithm performs against GENOS. Beyond that there is the question of how different methods like Frank-Wolfe, projected gradient, and the active set method compare with each other when they are applied to the new formulation. Toward this goal, a series of experiments were run to test the effect of the number of regions and time periods on networks with various densities. Of particular interest is the effect of longer planning horizons on the rate of convergence.

We used a simple stopping rule based on the relative change in the objective function from one iteration to the next, given by:

$$\frac{\left|\hat{F}(\theta^{(k+1)}) - \hat{F}(\theta^{(k)})\right|}{\hat{F}(\theta^{(k)})} < \epsilon \tag{35}$$

where ϵ is a parameter that we set to 0.001. For some of the experiments, we solved the problem with one algorithm and then measured how long a competing algorithm required to produce the same objective function value.

The codes that implement the solution algorithms are written using the C programming language. Computational tests are performed on a Silicon Graphics 4D/70 workstation running SGI Unix V3.2 with code compiled with the MIPS cc compiler using the default optimization level.

5.2 Results and Conclusions

We began by experimenting with different algorithms for NDN-T to investigate the properties of the transformation. Six test problems were randomly generated. Each problem is characterized by the number of cities, number of time periods, and network density. Other parameters (such as r_{ij} and c_{ij}) were fixed as described earlier. Initial experiments indicated that the projected gradient algorithm was superior to others. To obtain fair comparisons between the three algorithms for NDN-T, we ran the projected gradient algorithm until it satisfied the ϵ-optimality test. The other two algorithms were then run until they produced an objective value that met or was closest to the result obtained using the projected gradient algorithm. Figure 4 illustrates the rate of convergence of the three methods.

Table 2 gives the results of a side-by-side comparison of the Frank-Wolfe algorithm for NDN-T and NDN-X. Here the results of the Frank-Wolfe run for NDN-T were taken from table 1. Then, Frank-Wolfe was used on NDN-X until it produced an objective value that met or came closest to the results for NDN-T. The results show a dramatic deterioration in performance as the number of time periods is increased. For 20 time periods, the Frank-Wolfe algorithm applied to NDN-X could not even reach the result obtained using NDN-T within a reasonable time.

This behavior is explained by the nature of the extreme point solution given by the Frank-Wolfe algorithm for NDN-X, as illustrated in figure 2. By contrast, figure 5 illustrates the Frank-Wolfe solution for NDN-T using the same network and link flows as figure 2. In the X-formulation, many nodes have no flow moving through them in the extreme solution, especially in the later time periods. As a result, the one dimensional search uniformly decreases the flow on all the links of these nodes. The links emanating from the same nodes in the T-formulation will also experience a net reduction in flow. However, this formulation also allows the algorithm to shift flow between the links emanating from these nodes, further refining the solution. It must be acknowledged that Frank-Wolfe is not the best algorithm for either of the formulations.

In table 3, we used the best available algorithm for each formulation. For NDN-T, we used the projected gradient algorithm, and for NDN-X we used GENOS, a package designed for nonlinear (generalized) networks. GENOS includes specialized implementations of both the primal truncated Newton algorithm and simplicial decomposition. These experiments revealed some limitations of the system. First, we were unable to run our larger problems due to restrictions in the GENOS software. Second, initial experiments produced almost pathologically slow execution times using the primal truncated Newton algorithm. We concluded that additional work was needed on this algorithm and that the run times were probably not an accurate measure of the performance of the algorithm. As a result, we only report the results using simplicial decomposition. Recall that we were forced to use different link cost functions to accomodate GENOS. A new set of test problems were generated using at most 10 cities and 10 time periods. In each case GENOS was run until its internal optimality conditions were satisfied. Then, the projected gradient algorithm was run for NDN-T until the objective function value met or came the closest to that produced by GENOS. The execution times reported in table 3 indicate the dramatic improvements in run times over GENOS, especially in the larger problems. This result should not be too surprising when we consider that simplicial decomposition is just a generalization of Frank-Wolfe, and must still use extreme points with the qualities depicted in figure 2.

A final set of experiments was run to investigate the possibility that the results are sensitive to the structure of the input flows. All the networks generated up to now exhibit the property that $R_i(t) = 0$ for $t \geq 2$. While this is fairly realistic for dynamic fleet management problems, it produces the sparse structure of the extreme point solution exhibited in Figure 2. The final set of tests was conducted using a set of input flows that satisfied $R_i(t) > 0$ for all cities and time periods. The values for the external supply vector are obtained by:

$$R_i(t) = \rho[\delta^{t-1}(1-\delta)\sum_j \frac{1}{\lambda_{ij}}] \tag{36}$$

where $\rho = 0.5$ and $\delta = 0.3$. This expression puts a declining amount of flow into later time periods, with the total amount of flow entering all time periods comparable to that used in earlier experiments. Again, the reason is to insure that link flows stay on the "interesting" part of the nonlinear cost functions. In this case, a plot of the nonzero flows in an extreme point solution of Frank-Wolfe looks more like figure 5 for both formulations. The results, shown in table 4, indicate that the relative results do not change significantly as a result of all positive input flows. The explanation is that while a plot of nonzero flows looks more like a dense tree, a plot of the flow volumes in the extreme point solution would still show a noticeable funneling of flows onto a single path.

6 Appendix: Calculation of the Derivatives

In this section we wish to use the acyclic structure of network to develop backward recursions for the derivatives $\partial \hat{F}/\partial\theta^t_{ij}$, in order to minimize the function $\hat{F}(\theta)$. For this we need to evaluate its derivative with respect to θ^t_{ij}. To make the calculations easier to follow, we can split the objective function into three parts by considering a particular time t. The objective function is then viewed as a combination of the earlier time periods $t' < t$, the present time t, and the future $t' > t$. For any value t, let

$$\hat{F}(\theta) = H^t_1(\theta) + H^t_2(\theta) + H^t_3(\theta) \tag{37}$$

where

$$H^t_1(\theta) \quad = \quad \sum_{t'<t}\sum_k\sum_l \hat{f}^{t'}_{kl}(\theta) \tag{38}$$

$$H_2^t(\theta) = \hat{f}_{ij}^t(\theta) + \sum_{k \neq i} \sum_{l \neq j} \hat{f}_{kl}^t(\theta) \tag{39}$$

$$H_3^t(\theta) = \sum_{t'>t} \sum_k \sum_l \hat{f}_{kl}^{t'}(\theta) \tag{40}$$

There are some important facts about these functions stated in the following two lemmas, the first of which is stated without proof:

Lemma 6.1 *For any time $t < P$, $H_3^t(\theta) = H_2^{t+1}(\theta) + H_3^{t+1}(\theta)$.*

Lemma 6.2 *For $s \geq t$, the partial derivative*

$$\frac{\partial H_1^t(\theta)}{\partial \theta_{ij}^s} = 0.$$

Proof. For $t' < t$, $\hat{f}_{kl}^{t'}(\theta) = f_{kl}^{t'}(\theta_{kl}^{t'} S_i^{t'}(\theta))$ by equation (7). Since $S_i^{t'}(\theta)$ does not depend on any values $\theta_{ij}^{\hat{t}}$ for $\hat{t} > t'$, and H_1^t is a sum of the functions $\hat{f}_{kl}^{t'}$ with $t' < t$, it follows that H_1^t has no dependence on any value of θ_{ij}^s for $s \geq t$. □

To find $\partial \hat{F} / \partial \theta_{ij}^t$, we first need to evaluate $\partial H_3^t / \partial \theta_{ij}^t$, which corresponds to differentiating the function of future time periods with respect to a change in the flow fraction at the present time t. Then we need to evaluate $\partial H_2^t / \partial \theta_{ij}^t$ which corresponds to differentiating the function of the present time period with respect to a change at the present time t. Lemma 6.2 has shown that $\partial H_1^t / \partial \theta_{ij}^t = 0$, i.e., the derivative of the function of the past time periods with respect to the current time period is zero. So we have the following two propositions:

Proposition 6.3 *The partial derivative*

$$\frac{\partial H_3^t}{\partial \theta_{ij}^t} = \frac{\partial \hat{F}}{\partial S_j^{t+1}(\theta)} \cdot S_i^t(\theta).$$

Proof. From Lemma 6.1, it follows that

$$\frac{\partial H_3^t}{\partial \theta_{ij}^t} = \frac{\partial}{\partial \theta_{ij}^t} \left(H_3^{t+1} + \sum_k \sum_l \hat{f}_{kl}^{t+1} \right). \tag{41}$$

Since the derivatives for $k \neq j$ are zero due to the structure of the dynamic network, the expression (41) is simplified to

$$\frac{\partial H_3^t}{\partial \theta_{ij}^t} = \frac{\partial}{\partial \theta_{ij}^t} \left(H_3^{t+1} + \sum_l \hat{f}_{jl}^{t+1} \right). \tag{42}$$

Using the chain rule and equation (7),

$$\frac{\partial \hat{f}_{jl}^{t+1}(\theta)}{\partial \theta_{ij}^{t}} = \frac{\partial f_{jl}^{t+1}(x_{jl}^{t+1})}{\partial \theta_{ij}^{t}} \tag{43}$$

$$= \frac{\partial f_{jl}^{t+1}(x_{jl}^{t+1})}{\partial x_{jl}^{t}} \cdot \frac{\partial x_{jl}^{t+1}}{\partial \theta_{ij}^{t}}. \tag{44}$$

It follows from the chain rule that

$$\frac{\partial x_{jl}^{t+1}}{\partial \theta_{ij}^{t}} = \frac{\partial x_{jl}^{t+1}}{\partial S_{j}^{t+1}(\theta)} \cdot \frac{\partial S_{j}^{t+1}(\theta)}{\partial \theta_{ij}^{t}} \tag{45}$$

where

$$\frac{\partial S_{j}^{t+1}(\theta)}{\partial \theta_{ij}^{t}} = \frac{\partial}{\theta_{ij}^{t}} \sum_{k} S_{k}^{t}(\theta) \cdot \theta_{kj}^{t} \tag{46}$$

$$= \frac{\partial}{\theta_{ij}^{t}} S_{i}^{t}(\theta) \cdot \theta_{ij}^{t} \tag{47}$$

$$= S_{i}^{t}(\theta). \tag{48}$$

This last expression is obtained by using the relationship (6) and realizing that $S_i^t(\theta)$ is an implicit function of $\hat{\theta}(t-1)$, and hence is constant with respect to θ_{ij}^t. Using the chain rule once again helps build the recursive structure by

$$\frac{\partial H_{3}^{t+1}}{\partial \theta_{ij}^{t}} = \frac{\partial H_{3}^{t+1}}{\partial S_{j}^{t+1}(\theta)} \cdot \frac{\partial S_{j}^{t+1}(\theta)}{\partial \theta_{ij}^{t}} \tag{49}$$

$$= \frac{\partial H_{3}^{t+1}}{\partial S_{j}^{t+1}(\theta)} \cdot S_{i}^{t}(\theta) \tag{50}$$

Substituting (48) into (45) and then into (44), and finally together with (50) into the original expression (42) we obtain

$$\frac{\partial H_{3}^{t}}{\partial \theta_{ij}^{t}} = \frac{\partial H_{3}^{t+1}}{\partial S_{j}^{t+1}(\theta)} \cdot S_{i}^{t}(\theta) + \sum_{l} \frac{\partial f_{jl}^{t+1}}{\partial x_{jl}^{t}} \cdot \frac{\partial x_{jl}^{t+1}}{\partial S_{j}^{t+1}(\theta)} \cdot S_{i}^{t}(\theta) \tag{51}$$

$$= \left(\frac{\partial H_{3}^{t+1}}{\partial S_{j}^{t+1}(\theta)} + \sum_{l} \frac{\partial f_{jl}^{t+1}}{\partial x_{jl}^{t}} \cdot \frac{\partial x_{jl}^{t+1}}{\partial S_{j}^{t+1}(\theta)} \right) \cdot S_{i}^{t}(\theta) \tag{52}$$

$$= \frac{\partial \hat{F}}{\partial S_{j}^{t+1}(\theta)} \cdot S_{i}^{t}(\theta). \tag{53}$$

The last equality follows from the fact that $\hat{F} = H_1^{t+1} + H_2^{t+1} + H_3^{t+1}$ and that Lemma 6.2 implies that $(\partial H_1^{t+1}(\theta)/\partial S_j^{t+1}(\theta)) = 0$. □

It is useful to note that

$$\frac{\partial x_{jl}^{t+1}}{\partial S_j^{t+1}(\theta)} = \frac{\partial}{\partial S_j^{t+1}(\theta)}(\theta_{jl}^{t+1} \cdot S_j^{t+1}) \quad (54)$$

$$= \theta_{jl}^{t+1} \quad (55)$$

Proposition 6.4 *The derivative of the present cost function $H_2^t(\theta)$ with respect to θ_{ij}^t is*

$$\frac{\partial H_2^t(\theta)}{\partial \theta_{ij}^t} = \frac{\partial f_{ij}^t(x_{ij}^t)}{\partial x_{ij}^t} \cdot S_i^t(\theta).$$

Proof. Here we are evaluating the relative change in the objective function due to a change in the flow fraction θ_{ij}^t at the present time period t. Since this change is only along the arc from node i in period t to node j in period $t+1$,the derivative is not effected by changes along arcs other than this specific arc. From the definition of $H_2^t(\theta)$ given in (39),

$$\frac{\partial H_2^t(\theta)}{\partial \theta_{ij}^t} = \frac{\partial \hat{f}_{ij}^t(\theta)}{\partial \theta_{ij}^t} \quad (56)$$

$$= \frac{\partial f_{ij}^t(\theta_{ij}^t \cdot S_i^t(\theta))}{\partial \theta_{ij}^t} \quad (57)$$

$$= \frac{\partial f_{ij}^t(x_{ij}^t)}{\partial x_{ij}^t} \cdot S_i^t(\theta). \quad (58)$$

□

Using Proposition 6.3 and Proposition 6.4, we are now ready to state

Theorem 6.5 *The partial derivative of the function $\hat{F}$ with respect to θ_{ij}^t is*

$$\frac{\partial \hat{F}(\theta)}{\partial \theta_{ij}^t} = \left(\frac{\partial f_{ij}^t(x_{ij}^t)}{\partial x_{ij}^t} + \frac{\partial \hat{F}}{\partial S_j^{t+1}(\theta)}\right) \cdot S_i^t(\theta).$$

Proof. Differentiating (37) and using Lemma 6.2 gives

$$\frac{\partial \hat{F}(\theta)}{\partial \theta_{ij}^t} = \frac{\partial H_2^t(\theta)}{\partial \theta_{ij}^t} + \frac{\partial H_3^t(\theta)}{\partial \theta_{ij}^t}.$$

The first expression is presented as Proposition 6.4 and the second one as Proposition 6.3. Combining the two proves the theorem. □

The next step is to find $\partial\hat{F}/\partial S_i^t(\theta)$. Hence we need to evaluate $\partial H_3^t(\theta)/\partial S_i^t(\theta)$ and $\partial H_2^t/\partial S_i^t(\theta)$. We present the following two propositions:

Proposition 6.6 *The partial derivative of the future cost function $H_3^t(\theta)$ with respect to the flow $S_i^t(\theta)$ through node i at time t is*

$$\frac{\partial H_3^t(\theta)}{\partial S_i^t(\theta)} = \sum_j \frac{\partial \hat{F}(\theta)}{\partial S_j^{t+1}(\theta)} \cdot \theta_{ij}^t .$$

Proof. From Lemma 6.1 and the definitions of $H_2^{t+1}(\theta)$ and $H_3^{t+1}(\theta)$ given by equations (39) and (40), one can write

$$\frac{\partial H_3^t(\theta)}{\partial S_i^t(\theta)} = \frac{\partial}{\partial S_i^t(\theta)} \left(\left(\sum_j \sum_l \hat{f}_{jl}^{t+1}(\theta) \right) + H_3^{t+1}(\theta) \right) \tag{59}$$

Using the chain rule,

$$\begin{aligned} \frac{\partial \hat{f}_{jl}^{t+1}(\theta)}{\partial S_i^t(\theta)} &= \frac{\partial f_{jl}^{t+1}(\theta_{jl}^{t+1} \cdot S_j^{t+1}(\theta))}{\partial S_i^t(\theta)} & (60) \\ &= \frac{\partial f_{jl}^{t+1}(x_{jl}^{t+1})}{\partial x_{jl}^{t+1}} \cdot \frac{\partial x_{jl}^{t+1}}{\partial S_i^t(\theta)} . & (61) \end{aligned}$$

Now the last derivative in (61) can be written as

$$\frac{\partial x_{jl}^{t+1}}{\partial S_i^t(\theta)} = \frac{\partial x_{jl}^{t+1}}{\partial S_j^{t+1}(\theta)} \cdot \frac{\partial S_j^{t+1}(\theta)}{\partial S_i^t(\theta)} . \tag{62}$$

where it follows from equation (6) that

$$\begin{aligned} \frac{\partial S_j^{t+1}(\theta)}{\partial S_i^t(\theta)} &= \frac{\partial}{S_i^t(\theta)} \sum_k S_k^t(\theta) \cdot \theta_{kj}^t & (63) \\ &= \frac{\partial}{S_i^t(\theta)} (S_i^t(\theta) \cdot \theta_{ij}^t) & (64) \\ &= \theta_{ij}^t . & (65) \end{aligned}$$

Since for each j, $S_j^{t+1}(\theta)$ is a function of $S_i^t(\theta)$, it follows from the chain rule that

$$\frac{\partial H_3^{t+1}(\theta)}{\partial S_i^t(\theta)} = \sum_j \frac{\partial H_3^{t+1}(\theta)}{\partial S_j^{t+1}(\theta)} \cdot \frac{\partial S_j^{t+1}(\theta)}{\partial S_i^t(\theta)} \tag{66}$$

$$= \sum_j \frac{\partial H_3^{t+1}(\theta)}{\partial S_j^{t+1}(\theta)} \cdot \theta_{ij}^t. \tag{67}$$

Substituting (65) into (62), then into (61), and finally together with (67) into the original expression (59) we obtain:

$$\frac{\partial H_3^t(\theta)}{\partial S_i^t(\theta)} = \left(\sum_j \sum_l \frac{\partial f_{jl}^{t+1}(x_{jl}^{t+1})}{\partial x_{jl}^{t+1}} \cdot \frac{\partial x_{jl}^{t+1}}{\partial S_j^{t+1}(\theta)} \cdot \theta_{ij}^t\right) + \sum_j \frac{\partial H_3^{t+1}(\theta)}{\partial S_j^{t+1}(\theta)} \cdot \theta_{ij}^t \tag{68}$$

$$= \sum_j \left(\left[\sum_l \frac{\partial f_{jl}^{t+1}(x_{jl}^{t+1})}{\partial x_{jl}^{t+1}} \cdot \frac{\partial x_{jl}^{t+1}}{\partial S_j^{t+1}(\theta)}\right] + \frac{\partial H_3^{t+1}(\theta)}{\partial S_j^{t+1}(\theta)}\right) \cdot \theta_{ij}^t \tag{69}$$

$$= \sum_j \frac{\partial \hat{F}}{\partial S_j^{t+1}(\theta)} \cdot \theta_{ij}^t. \tag{70}$$

The last equality follows from the fact that the expression in parentheses in equation (69) is precisely

$$\frac{\partial H_2^{t+1}(\theta)}{\partial S_j^{t+1}(\theta)} + \frac{\partial H_3^{t+1}(\theta)}{\partial S_j^{t+1}(\theta)}$$

and the fact that

$$\frac{\partial H_1^{t+1}(\theta)}{\partial S_j^{t+1}(\theta)} = 0,$$

which follows from Lemma 6.2. □

Proposition 6.7 *The derivative of the present cost function $H_2^t(\theta)$ with respect to the flow $S_i^t(\theta)$ through node i at time t is*

$$\frac{\partial H_2^t(\theta)}{\partial S_i^t(\theta)} = \sum_j \frac{\partial f_{ij}^t(x_{ij}^t)}{\partial x_{ij}^t} \cdot \theta_{ij}^t.$$

Proof. ¿From the definition of $H_2^t(\theta)$ in equation (39),

$$\frac{\partial H_2^t(\theta)}{\partial S_i^t(\theta)} = \sum_j \frac{\partial \hat{f}_{ij}^t(\theta)}{\partial S_i^t(\theta)} \tag{71}$$

$$= \sum_j \frac{\partial f_{ij}^t(x_{ij}^t)}{\partial S_i^t(\theta)} \tag{72}$$

$$= \sum_j \frac{\partial f_{ij}^t(x_{ij}^t)}{\partial x_{ij}^t} \cdot \theta_{ij}^t, \tag{73}$$

with the last equality derived in a similar fashion as in equations (62) through (65). □

We now can compute the derivative of $\hat{F}$ with respect to $S_i^t(\theta)$, stating the following theorem.

Theorem 6.8 *The derivative of the cost function $\hat{F}$ with respect to the flow $S_i^t(\theta)$ through node i at time t is*

$$\frac{\partial \hat{F}}{\partial S_i^t(\theta)} = \sum_j \left(\frac{\partial f_{ij}^t(x_{ij}^t)}{\partial x_{ij}^t} \cdot \theta_{ij}^t + \frac{\partial \hat{F}}{\partial S_j^{t+1}(\theta)} \cdot \theta_{ij}^t \right).$$

Proof.

Differentiating (37) and using $(\partial H_1^t(\theta)/\partial S_i^t(\theta)) = 0$, gives

$$\frac{\partial \hat{F}}{\partial S_i^t(\theta)} = \frac{\partial H_2^t(\theta)}{\partial S_i^t(\theta)} + \frac{\partial H_3^t(\theta)}{\partial S_i^t(\theta)}.$$

The first expression is presented as Proposition 6.4 and the second one as Proposition 6.3. By combining the two we obtain the desired final derivative. □

References

[1] J. E. Aronson, A survey of dynamic network flows, *Annals of Operations Research*, **20** (1989) 1–66.

[2] J. E. Aronson and B. D. Chen, A forward network simplex algorithm for solving multiperiod network flow problems, *Naval Research Logistics Quarterly*, **33** (1986) 445–467.

[3] D. P. Bertsekas, Algorithms for nonlinear multicommodity flow problems, in *International Symposium on Systems Optimization and Analysis*, (Springer-Verlag, 1979) p. 210–224.

[4] D. P. Bertsekas, E. M. Gafni, and R. G. Gallagher, Second derivative algorithms for minimum delay distributed routing in networks, *IEEE Transactions on Communications*, **COM-32** (1984) 911–919.

[5] R. G. Gallagher, A minimum delay routing algorithm using distributed computation, *IEEE Transactions on Communications*, **COM-25** (1977) 73–85.

[6] P. E. Gill, W. Murray, and M. H. Wright, *Practical Optimization* (Academic Press, London, 1981).

[7] M. Held, P. Wolfe, and H. P. Crowder, Validation of subgradient optimization, *Mathematical Programming* **6** (1974) 62–88.

[8] J. M. Mulvey and S. A. Zenios, GENOS 1.0 user's guide: A generalized network optimization system, Tech. Rep. 87-12-03, Department of Decision Sciences, The Wharton School, University of Pennsylvania (1987).

[9] W. B. Powell, A stochastic model of the dynamic vehicle allocation problem, *Transportation Science* **20** (1986) 117–129.

[10] W. B. Powell, A Comparative Review of Alternative Algorithms for the Dynamic Vehicle Allocation Problem, in *Vehicle Routing: Methods and Studies* (North Holland, New York, 1988), p. 249–292.

[11] W. B. Powell, Y. Sheffi, and S. Thiriez, The dynamic vehicle allocation problem with uncertain demands, in *Ninth International Symposium on Transportation and Traffic Theory* (1984) p. 357–374.

[12] W. W. White and A. M. Bomberault, A network algorithm for empty freight car allocation, *IBM Systems Journal* **8** (1969) 147–171.

Numerical results of different methods for NDN-T

No.	Cities	Periods	Dens.	Objective Function			CPU time (sec.)			Iterations		
				FW	PG	AS	FW	PG	AS	FW	PG	AS
1	20	5	0.5	-189,887	-189,708	-190,170	6.12	2.80	2.82	37	15	22
2	40	5	0.5	-766,618	-766,766	-768,092	29.52	10.55	16.83	47	15	36
3	20	10	0.5	-367,280	-366,568	-367,788	10.32	6.58	6.43	31	18	26
4	40	10	0.5	-1,597,279	-1,598,139	-1,599,719	80.94	33.69	43.07	65	24	46
5	20	20	0.5	-710,476	-710,377	-710,687	30.44	19.98	23.38	47	28	47
6	40	20	0.5	-3,268,208	-3,268,165	-3,268,859	190.85	107.49	110.48	76	39	58

Table 1

Comparison of NDN-T and NDN-X using Frank-Wolfe

No.	Cities	Periods	Dens.	Objective Function		CPU time (sec.)		Iterations	
				NDN-T	NDN-X	NDN-T	NDN-X	NDN-T	NDN-X
1	20	5	0.5	-189,887	-189,815	6.12	22.15	37	103
2	40	5	0.5	-766,618	-766,851	29.52	152.01	47	181
3	20	10	0.5	-367,280	-367,560	10.32	70.74	31	171
4	40	10	0.5	-1,597,279	-1,597,414	80.94	898.82	65	529
5	20	20	0.5	-710,476	-706,348*	30.44	246.54	47	297
6	40	20	0.5	-3,268,208	-2,065,743	190.85	999.88**	76	268

Table 2

* result up to stopping criteria of 0.000001

** result up to CPU Time of 1000

Comparison of GENOS verses NDN-T*

No.	Cities	Periods	Dens.	Objective Function *		CPU time (sec.)	
				GENOS	NDN-T	GENOS	NDN-T
1	8	2	0.5	13,981	13,981	2.20	0.68
2	8	4	0.5	29,604	29,594	5.76	0.46
3	8	8	0.5	63,373	63,348	10.94	0.98
4	10	2	0.2	10,607	10,609	1.85	1.85
5	10	5	0.2	35,554	35,545	5.59	0.47
6	10	10	0.2	62,485	62,472	10.78	0.74
7	10	2	0.5	43,180	43,179	4.63	0.36
8	10	5	0.5	90,342	90,237	10.86	0.69
9	10	10	0.5	164,096	163,883	25.12	1.60

Table 3

* same exponential function is used in both GENOS and NDN-T
* objective function values reflect the minimum cost attained

Comparison of NDN-T and NDN-X using Frank-Wolfe
(with external supplies at each period)

No.	Cities	Periods	Dens.	Objective Function		CPU time (sec.)		Iterations	
				NDN-T	NDN-X	NDN-T	NDN-X	NDN-T	NDN-X
1	20	5	0.5	-176,807	-177,107	8.24	13.31	46	62
2	40	5	0.5	-608,662	-606,669	43.69	88.20	67	105
3	20	10	0.5	-328,328	-328,095	13.55	46.31	40	110
4	40	10	0.5	-1,228,038	-1,228,730	78.96	398.80	61	232
5	20	20	0.5	-603,379	-601,000	15.71	140.48	23	164
6	40	20	0.5	-2,764,947	-932,789	170.15	998.72*	65	268

Table 4

* result up to CPU Time of 1000

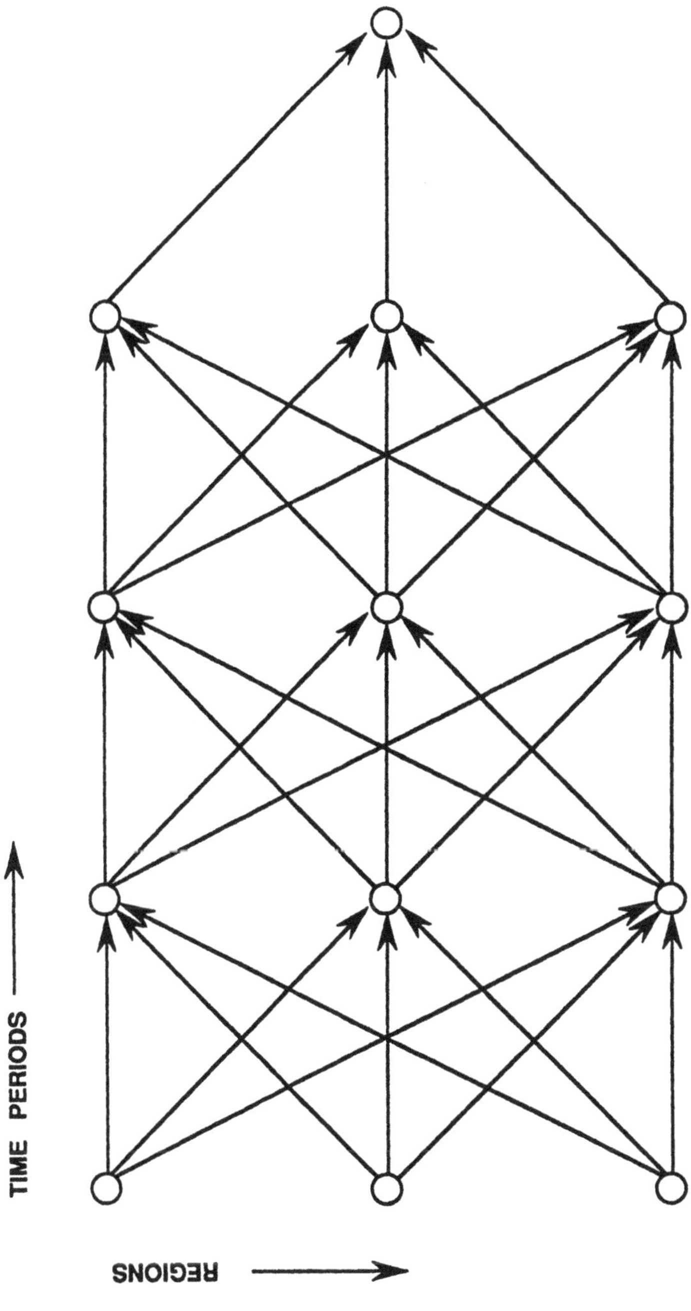

Figure 1

Space-time representation of a network

Example problem of 10 Regions and 10 Time periods

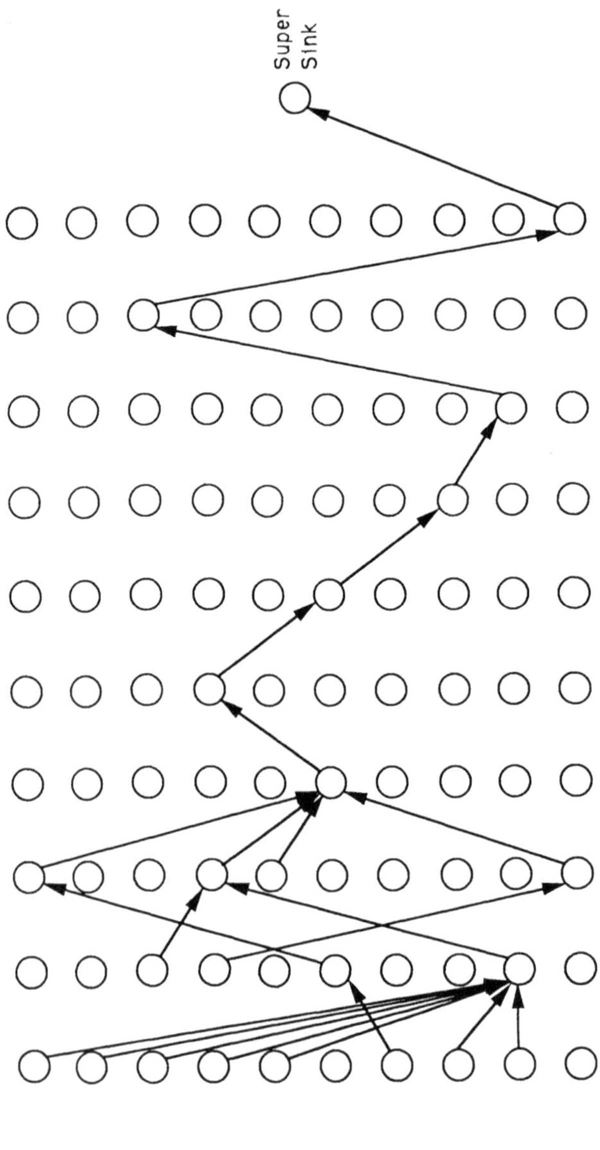

Figure 2

Solution of classical linearized subproblem using Frank-Wolfe

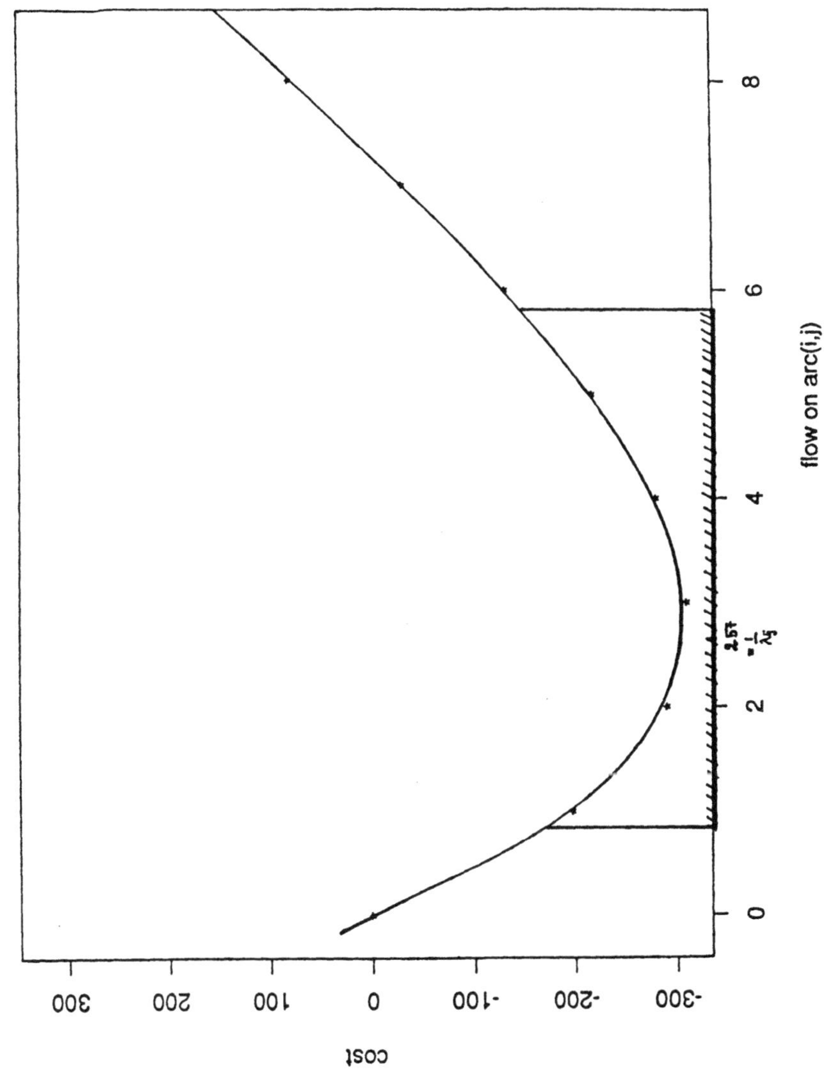

Figure 3

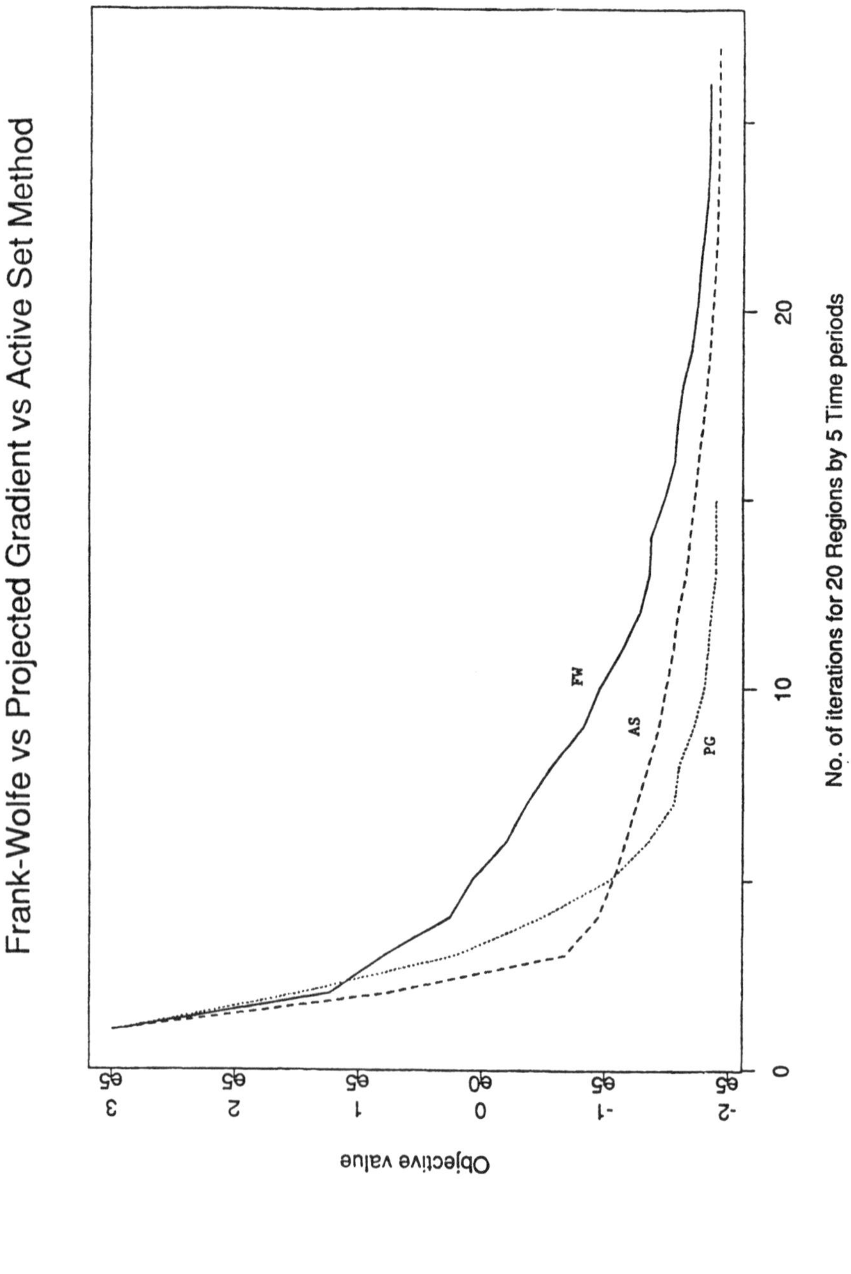
Frank-Wolfe vs Projected Gradient vs Active Set Method
Objective value
3e5
2e5
1e5
0e0
-1e5
-2e5
0
10
20
FW
AS
PG
No. of iterations for 20 Regions by 5 Time periods

Figure 4

Example problem of 10 Regions and 10 Time periods

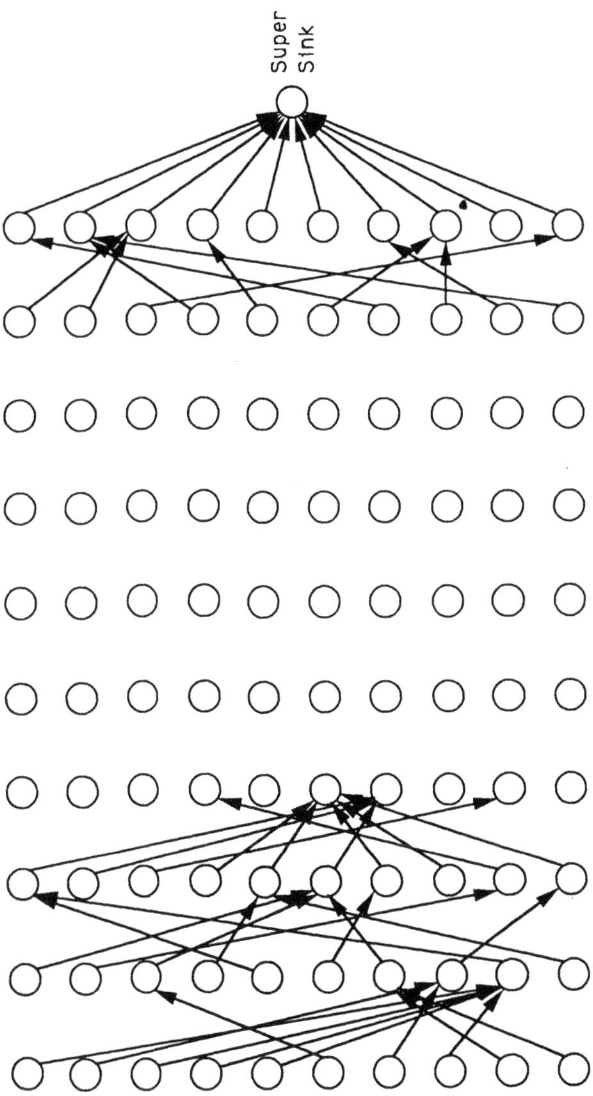

Figure 5

Solution of linearized subproblem for the transformed problem using Frank-Wolfe

Network Optimization Problems, pp. 233-262
Eds. D.-Z. Du and P.M. Pardalos

Strategic and Tactical Models and Algorithms for the Coal Industry Under the 1990 Clean Air Act†

Hanif D. Sherali
Department of Industrial and Systems Engineering, Virginia Polytechnic Institute and State University, Blacksburg, Virginia 24061-0118

Quaid J. Saifee
Association of American Railroads, 26555 Evergreen Road, Suite 1120, Southfield, Michigan 48076-4251

Abstract

This paper is concerned with a study of the effect of the Acid Rain Provision of the 1990 Clean Air Act on the investment, production, and distribution operations in the coal industry, with concentration on the development of new mines, shutting down inefficient strips of existing mines, and on blending and distribution problems. The problem here is to determine which new mines to open and when, and what decisions and schedules to make for the shipment of coal from the mines to silos, cleaning and blending operations at silos, and subsequent shipment of coal to customers over a multi-period time horizon, so as to satisfy the demand at a minimum total operational cost. A long-term strategic model is developed to meet this objective. The final product is a computer based decision tool which will serve as a mechanism for implementing cost effective decisions in light of complex variations in the production levels of existing and potential mines, ore quality, and demand and quality requirements. The strategic model will play a useful role in planning future growth and in making capital investment decisions. The model can also be used to study the effect of various policies, by testing the sensitivity, feasibility, and the cost of system operations under different perturbations of system configuration, data

†**Acknowledgement:** This research has been supported by the Department of the Interior's Mineral Institute program administered by the Bureau of Mines under allotment grant # G1114151

and demand specifications. Real operational data from the Westmoreland Coal Company and hypothetical test data are used for testing purposes. We also present a related short-term tactical model in the appendix, that can be used to assist in decisions regarding day-to-day operations.

1 Introduction

The research conducted in this study was motivated by the Acid Rain Provision of the 1990 Clean Air Act, passed by Congress on October 28, and signed into law by President Bush on November 15, 1990. This Provision in the Act places stringent restrictions on the sulfur dioxide and nitrous oxide emissions, particularly on electric utilities operating coal-fired generators, and it promotes the role of Southwest Virginia (our area of study) as becoming a central player in the coal industry due to its low sulfur coal resources, as compared with other coal mined east of the Mississippi River. However, in order for different companies in the industry to remain solvent and competitive in light of anticipated sudden changes in quality requirements of future coal demand, they will need to judiciously plan for the usage of their reserves, and the operation of their cleaning and distribution facilities.

This paper addresses the combined problems of choosing new mines to develop, determining mine production, ore purification, and blending of different grades of coal, along with the problem of distributing the coal from mines to silos to customers, in order to satisfy customer demands placed over several time periods, each having specified quality requirements as driven by the Clean Air Act. A long-term strategic model is developed to address this problem from a planning perspective, and is designed to help coal companies in making strategic decisions such as the development of new mines or the development of new production units at mines. Also, this model is intended to assist coal companies by providing guidelines for the usage of their existing and potential reserves, and in the operation of their cleaning and distribution facilities, along with a possible investment in new technologies to offset the development of new mines. Additionally, a modified formulation of Sherali and Puri's [14] short-term tactical model is presented in the appendix. This model can be used to aid in the day-to-day operational decision making process.

Information and data for these models have been provided by the Westmoreland Coal Company, one of the largest coal mining companies in Southwest Virginia. The company owns several coal mines, some of which are currently in use (existing mines), and some of which can be developed later as needs arise. Each mine can produce coal at a specific rate and has certain quality specifications that vary over time. This coal needs to be appropriately shipped to silo facilities, where it is subjected to a beneficiation process in order to be partially cleaned to a desirable degree. The different grades of coal at individual silo facilities then need to be blended and shipped to customers in order to satisfy demands for various quantities having stipulated quality specifications.

In our case study, the Westmoreland Coal Company owns two large silo facilities, known as the Bullitt facility and the Wentz facility. The Bullitt facility has six silos, each with a capacity of 14,000 tons. Silo 1 and Silo 2 store coal that does not require to be cleaned. Silo 3 contains high sulfur coal and Silo 4 contains low sulfur coal. After beneficiation, coal from Silo 3 is transferred to Silo 5 and coal from Silo 4 is transferred to Silo 6 for storage. The Wentz facility has five silos, three with a capacity of 6,000 tons and two with a capacity of 9,000 tons. One of them is used to store coal which does not need to be cleaned, two of them are used to store the remaining coal prior to cleaning, and the remaining two are used for storing coal after it has been cleaned. This being a typical structure in the industry, our model partitions the silos within each facility into two categories based on two different kinds of coal stored. These categories are denoted as J_1 and J_2 types of silos. The J_1 silos are "run-of-mine" (ROM) blend silos, where a shipment is received, stored, and is directly blended with the other coal in its run-of-mine form itself. Each J_2 type of silo, on the other hand, constitutes a pair of silos, one of which receives the run-of-mine coal, while the other stores the cleaned coal following the beneficiation process, holding it ready for the blending operation. These silo pairs are indexed by the sets J_{2W} and J_{2B} for the Wentz and Bullitt facilities, respectively.

Besides the conventional mine to silo to customer shipments, there might also be a coal transfer between the silo facilities at two different geographical sites. Such a transfer occurs due to the "stoker" customers. In the example of the Westmoreland Coal Company, "stoker" customers are provided with cleaned, sifted coal from the Wentz facility. Simultaneously, an almost equivalent amount of sifted by-product is shipped to the Bullitt facility. The normal blending processes resume at this stage.

The problem is to find out which production units of potential mines should be opened and at what time periods, and to determine optimal schedules for shipping coal from mines to silos, for cleaning and blending at the silos, and for the distribution of coal to the customers. This has to be accomplished subject to restrictions involving storage at the silos, production capacity limits, and material flow balance constraints, so that customer demands having required quality specifications are satisfied at minimal total cost. The long-term strategic model developed to address this problem has the structure of a fixed-charge, mixed-integer, zero-one programming problem. Zero-one integer variables are used to model the decision of opening a production unit at each particular mine. In addition to the cleaning and shipping costs, the objective function incorporates a fixed-charge component to reflect the cost of opening a new production unit at a potential mine. Also, a storage cost component is added to the objective function in order to penalize the underutilization of active resources. The constraints include production capacity, storage, material flow balance, and quality requirement restrictions, along with restrictions on the sequence in which units can be opened, and on the maximum number of units that can be opened in any time period without any substantial surcharge penalties.

2 Related Models in the Literature

The problems related to the coal industry have been analyzed in a number of different ways. Different techniques varying over linear, nonlinear, and mixed integer 0-1 programming problems have been used to approach different problems related to the coal industry.

Young et al. [18], Faulkner [5] and Johnson [7] provide few of the pioneering papers in the area. Many practical case studies of Operations Research in the coal mining industry have been surveyed by Tomlinson [16]. Knight and Manula [9] have developed a long-term simulation model to study potential coal production and utilization systems in Pennsylvania. Gershon [6] has formulated a mixed-integer model for a mine scheduling problem. Lietaer [10] presents a linear programming model with an objective of preparing a combination of mining works at a minimum total operational cost subject to different restrictions including operational restrictions on the mines and the concentrators, among other constraints. The problem of allocating coking coals from collieries to washeries and blending plants is also modeled as a linear programming problem by Williams and Haley [17]. Sherali and Puri [14] develop three short-term tactical models for analyzing day-to-day coal flow operations in the coal industry, with a concentration on the blending and distribution problems. A modified formulation of their most accurate model is presented in the Appendix of this paper.

Steinmann and Schwinn [15] have formulated a zero-one programming model to minimize the total resources necessary for balancing the capacity structure of the coal mine, subject to capacity constraints for a particular mine, and have reported on computational experience with different algorithms used to solve this problem. Two link models, a transportation–transshipment model defined for the coal distribution network, and a location–allocation model defined for the potential location of coal handling facilities within receiving and shipping nodes of the network, have been developed by Osleeb and Ratick [12] to determine the optimal capacity, placement, and railroad and marine interface of coal handling facilities within and between the New England ports and converting power plants.

A mixed-integer location-allocation problem has also been formulated by Osleeb et al. [13] to evaluate the potential for reducing water-borne coal transportation costs, and concomitantly, the cost of delivering coal to the European markets. Candler [4] formulates a blending problem with integer constraints on a minimum usage of each coal type in a mix, and recommends a sampling plan along with penalty function and rejection options.

As evident in the literature, different papers have considered different detailed aspects of specific problems faced by the coal industry. However, the imminent problem of developing a strategic plan for a time-staged resource management along with coal-blending and distribution operations in order to comply with stringent future quality requirements as prompted by the 1990 Clean Air Act, has not been addressed for the coal industry. However, strategic models developed for other contexts do share a philosophical structure with our model. For example, Aboudi et al. [1] develop a long-term planning model for the petroleum production and distribution problem using mixed-integer programming techniques, and Khan [8] also presents a mixed-integer model to minimize the total disposal costs in an urban solid waste disposal problem.

Section 3 below presents a formulation of our strategic planning model. Exact and heuristic solution procedures are described in Section 4, and Section 5 presents computational test results using real and hypothetical data sets. The Appendix contains a related formulation that deals with the tactical day-to-day operational problem.

3 Formulation of a Long-Term Strategic Model

As described earlier, the problem at hand is to determine which production units of potential mines should be opened and at what time periods, along with optimal schedules for shipping coal from mines to silos, for cleaning and blending coal at the silos, and for the distribution of coal to the customers, subject to various constraints. These constraints include production capacity, silo capacity, material flow balance, demand satisfaction for customers, and quality requirement restrictions, along with restrictions on the sequence in which units can be opened, and on the maximum number of units that can be opened in any time period, without any substantial surcharge penalties.

The model requires specific data pertaining to the existing and potential mines, silos, and customers. This model considers periods equal to six months in duration, with a horizon of three to five years. As this is a long-term model, all costs such as cleaning and shipping costs, revenues for shipping better quality coal, and penalties for storage and underutilization at mines, are present values using a certain rate of return. The subscript t attached to these cost factors reflects this representation. In this regard, note that the fixed-charge cost to open a production unit at a potential mine is the present value of the semi-annualized payments over the life of the mine that are to be made over the horizon of the model.

Real data from the Westmoreland Coal Company and nine other similar, hypothetical data sets have been used for making the model runs. Given below are the data requirements, along with the notation used.

$t = 1, \ldots, T \equiv$ number of time periods (≤ 8).

$i = 1, \ldots, I \equiv$ number of existing and potential mines (≤ 23).

$q = 1, \ldots, h_i \equiv$ number of units at a potential mine $i (\leq 30)$.

$j = 1, \ldots, J \equiv$ number of silo units (≤ 10).

$k = 1, \ldots, K \equiv$ number of customers (≤ 12).

K_{st} = "stoker customers" served by Wentz silos.

p_{it} = production (tons) at existing mine i, in period t.

$p_{iq}(t)$ = a function which represents production (tons) at unit q of potential mine i, in period t of its life. (Note that t denotes the period of its life after this unit has been opened, and not the period of the model.)

a_{it} = ash content (%) in coal produced at existing mine i, in period t.

s_{it} = sulfur content (%), in coal produced at existing mine i, in period t.

a_{iqt} = ash content (%), in coal produced at unit q of potential mine i, in period t. (Note that t denotes the period of horizon, and not the period of the unit's life after it has been opened.)

s_{iqt} = sulfur content (%), in coal produced at unit q of potential mine i, in period t. (The same comment applies here as for a_{iqt})

$I_1 \equiv$ existing mines.

$I_2 \equiv$ potential mines.

$J_1 \equiv$ ROM silo storage units.

$J_{2W} \equiv$ cleaned silo units at the Wentz facility.

$J_{2B} \equiv$ cleaned silo units at the Bullitt facility.

$J_2 \equiv$ cleaned silo units ($J_{2W} \cup J_{2B}$).

SC_{jt} = storage capacity of silo unit j in period t. (For $j \in J_2$, this is taken as the sum of the two associated silo storage capacities.)

c_{ijt}^{sh1} = present value of shipping cost (per ton), from existing mine i to silo j in period t.

c_{iqjt}^{sh1} = present value of shipping cost (per ton), from a unit q of a potential mine i to silo j in period t.

c_{ijt}^{cl} = present value of cleaning cost (per ton), at silo j for coal from existing mine i in period t.

c_{iqjt}^{cl} = present value of cleaning cost (per ton), at silo j for coal from a unit q of a potential mine i in period t.

$\alpha_{ij} \in (0,1]$ = total weight attenuation factor (output per ton input) at silo $j \in J_2$ for coal from existing mine i.

$\alpha_{iqj} \equiv$ defined similar to α_{ij} for production unit q at potential mine i.

$\beta_{ij} \in (0,1]$ = ash content attenuation factor at silo $j \in J_2$ for coal from existing mine i.

$\beta_{iqj} \equiv$ defined similar to β_{ij} for production unit q at potential mine i.

$\gamma_{ij} \in (0,1]$ = sulfur content attenuation factor at silo $j \in J_2$ for coal from existing mine i.

$\gamma_{iqj} \equiv$ defined similar to γ_{ij} for production unit q at potential mine i.

Note: α_{ij}, α_{iqj}, β_{ij}, β_{iqj}, γ_{ij}, and γ_{iqj} are assumed to be 1 for $j \in J_1$.

A_1 = flow arcs (i,j) from mine i to silo j; $F_i^1 = \{ j : (i,j) \in A_1 \}$, $R_j^1 = \{ i : (i,j) \in A_1 \}$.

A_2 = flow arcs (j,k) from silo j to customer k; $F_j^2 = \{ k : (j,k) \in A_2 \}$, $R_k^2 = \{ j : (j,k) \in A_2 \}$.

A_3 = arcs (j,j') representing by-product flow from Wentz silo $j \in J_{2W}$ to Bullitt silo $j' \in J_{2B}$, corresponding to stoker customer shipments; $F_j^3 = \{ j' : (j,j') \in A_3 \}$, $R_j^3 = \{ j : (j',j) \in A_3 \}$.

c_{jkt}^{sh2} = present value of shipping cost (per ton), from silo j to customer k in period t.

$c_{jj't}^{sh3}$ = present value of shipping cost (per ton), from Wentz silo $j \in J_{2W}$ to Bullitt silo $j' \in J_{2B}$ in period t.

d_{kt} = demand placed by customer k during period t.

ASH_{ukt} = upper limit on ash percentage in coal to be delivered to customer k in period t.

SUL_{ukt} = upper limit on sulfur percentage in coal to be delivered to customer k in period t.

R_{kt} = present value of additional revenue earned per percentage point below the maximum specification of ash content, in coal delivered to customer k in period t.

f_{iqt} = fixed charge (equivalent present value as described above) to open a unit q at a potential mine i in period t.

$f_t^{\max} = \text{maximum}\{ f_{iqt} : q = 1, \ldots, h_i,\ i \in I_2 \}$ for all $t = 1, \ldots, T$.

PE_{it} = present value of penalty per ton of coal unused at existing mine i in period t.

PE_{iqt} = present value of penalty per ton of coal unused at unit q of potential mine i in period t.

U_t = upper limit on the number of units that can be opened at a potential mine i in period t.

Mathematical Formulation

A detailed mathematical formulation is given below, followed by an explanation and motivation of the objective function and the various constraints. The decision variables track the flow of coal from mines through silos (within the blending process) to the different customers. Specifically, these variables correspond to (i) the amount (tons) shipped from existing mine i through silo j to customer k in time period t, given by y_{ijkt} (ii) stoker customer by-product amount (tons) shipped from existing mine to i to silo $j \in J_{2W}$, which is then shipped to silo $j' \in J_{2B}$, and then to customer k in period t, given by $Y_{ijj'kt}$ (iii) amount (tons) shipped from unit q of potential mine i through silo j to customer k in time period t, given by w_{iqjkt} (iv) stoker customer by-product amount (tons) shipped from unit q of potential mine i to silo $j \in J_{2W}$, which is then shipped to silo $j' \in J_{2B}$, and then to customer k in a period t, given by $W_{iqjj'kt}$ (v) a binary variable that takes on a value of one if unit q of a potential mine $i \in I_2$ is opened or initiated in a period t, and is zero otherwise, given by x_{iqt}. Note that the aggregate shipments of coal from mines to silos, the blending process at the silos, and the aggregate shipments of blended coal from the silo facilities to the customers can be readily computed via these variables.

Auxiliary Decision Variables

1. z_{it} = slack variable equal to the amount (tons) of coal produced at existing mine i during period t that remains unused in that period, for $i \in I_1$, $t = 1, \ldots, T$.

2. z_{iqt} = slack variable equal to the amount (tons) of coal produced at a unit q of potential mine i during period t that remains unused in that period, for $i \in I_2$, $q = 1, \ldots, h_i$, $\delta = 1, \ldots, T$.

3. ASH_{kt} = % ash content in blended coal delivered to customer k in period t, for $k = 1, \ldots, K$, $t = 1, \ldots, T$.

4. SUL_{kt} = % sulfur content in blended coal delivered to customer k in period t, for $k = 1, \ldots, K$, $t = 1, \ldots, T$.

5. θ_{1t} = number of mines opened within the set limit U_t in period t, for $t = 1, \ldots, T$.

6. θ_{2t} = number of mines opened beyond the set limit U_t in period t, for $t = 1, \ldots, T$.

Objective Function:

$$\begin{aligned}
\text{Minimize} \quad & \sum_{j=1}^{J} \sum_{i \in R_j^1 \cap I_1} \sum_{k \in F_j^2} \sum_{t=1}^{T} (c_{ijt}^{sh1} + c_{ijt}^{cl} + c_{jkt}^{sh2})\, y_{ijkt} \\
+ & \sum_{j \in J_{2W}} \sum_{i \in R_j^1 \cap I_1} \sum_{j' \in F_j^3} \sum_{j' \in F_{j'}^2} \sum_{t=1}^{T} (c_{ijt}^{sh1} + c_{j'kt}^{sh2} + c_{jj't}^{sh3} + c_{ijt}^{cl})\, Y_{ijj'kt} \\
+ & \sum_{j=1}^{J} \sum_{i \in R_j^1 \cap I_2} \sum_{q=1}^{h_i} \sum_{k \in F_j^2} \sum_{t=1}^{T} (c_{iqjt}^{sh1} + c_{iqjt}^{cl} + c_{jkt}^{sh2}) w_{iqjkt} \\
+ & \sum_{j \in J_{2W}} \sum_{i \in R_j^1 \cap I_2} \sum_{q=1}^{h_i} \sum_{j' \in F_j^3} \sum_{k \in F_{j'}^2} \sum_{t=1}^{T} (c_{iqjt}^{sh1} + c_{j'kt}^{sh2} + c_{jj't}^{sh3} + c_{iqjt}^{cl}) W_{iqjj'kt} \\
+ & \sum_{i \in R_j^1 \cap I_2} \sum_{q=1}^{h_i} \sum_{t=1}^{T} f_{iqt} x_{iqt} + \sum_{i \in R_j^1 \cap I_1} \sum_{t=1}^{T} PE_{it} z_{it} + \sum_{i \in R_j^1 \cap I_2} \sum_{q=1}^{h_i} \sum_{t=1}^{T} PE_{iqt} z_{iqt} \\
+ & \sum_{t=1}^{T} 0.2 f_t^{\max} \theta_{2t} - \sum_{k=1}^{K} \sum_{t=1}^{T} (ASH_{ukt} - ASH_{kt}) d_{kt} R_{kt}
\end{aligned}$$

Constraints:

1. *Flow balance at the mines:*

$$p_{it} - \left[\sum_{j \in F_i^1} \sum_{k \in F_j^2} y_{ijkt} + \sum_{j \in F_i^1} \sum_{j' \in F_j^3} \sum_{k \in F_{j'}^2} Y_{ijj'kt} \right] = z_{it}$$

for $i \in I_1$, and $t = 1, \ldots, T$

$$\sum_{t \le \delta} p_{iq}(\delta - t + 1) x_{iqt} - \left[\sum_{j \in F_i^1} \sum_{k \in F_j^2} w_{iqjk\delta} + \sum_{j \in F_i^1} \sum_{j' \in F_j^3} \sum_{k \in F_{j'}^2} W_{iqjj'k\delta} \right] = z_{iq\delta}$$

for $i \in I_2$, $q = 1, \ldots, h_i$ and $\delta = 1, \ldots, T$

2. *Capacity constraints on silos:*

$$\begin{aligned}
& \sum_{i \in R_j^1 \cap I_1} \sum_{k \in F_j^2} \frac{1}{2}(1 + \alpha_{ij}) y_{ijkt} + \sum_{i \in R_j^1 \cap I_1} \sum_{j' \in J_{2W}} \sum_{k \in F_j^2} \frac{1}{2}(1 + \alpha_{ij}) Y_{ij'jkt} \\
+ & \sum_{i \in R_j^1 \cap I_2} \sum_{q=1}^{h_i} \sum_{k \in F_j^2} \frac{1}{2}(1 + \alpha_{iqj}) w_{iqjkt} + \sum_{i \in R_j^1 \cap I_2} \sum_{q=1}^{h_i} \sum_{j' \in J_{2W}} \sum_{k \in F_j^2} \frac{1}{2}(1 + \alpha_{iqj}) W_{iqj'jkt} \\
\le & \; SC_{jt}
\end{aligned}$$

for $j \in (J_{2B} \cup J_{2W}) \cap J_2$, and $t = 1, \ldots, T$

$$\sum_{i \in R_j^1 \cap I_1} \sum_{k \in F_j^2} y_{ijkt} + \sum_{i \in R_j^1 \cap I_2} \sum_{q=1}^{h_i} \sum_{k \in F_j^2} w_{iqjkt} \leq SC_{jt}$$

for $j \in J_1$, and $t = 1, \ldots, T$

3. *Wentz to Bullitt transfer:*

$$\sum_{j' \in F_j^3} \sum_{k \in F_{j'}^2} Y_{ijj'kt} \leq \sum_{k \in F_j^2 \cap K_{st}} y_{ijkt} \leq 1.1 \sum_{j' \in F_j^3} \sum_{k \in F_{j'}^2} Y_{ijj'kt}$$

for $j \in J_{2W} \cap F_i^1$, $i \in R_j^1 \cap I_1$, and $t = 1, \ldots, T$

$$\sum_{j' \in F_j^3} \sum_{k \in F_{j'}^2} W_{iqjj'kt} \leq \sum_{k \in F_j^2 \cap K_{st}} w_{iqjkt} \leq 1.1 \sum_{j' \in F_j^3} \sum_{k \in F_{j'}^2} W_{iqjj'kt}$$

for $j \in J_{2W} \cap F_i^1$, $i \in R_j^1 \cap I_2$, $q = 1, \ldots, h_i$, and $t = 1, \ldots, T$

4. *Demand constraints for customers:*

$$\sum_{j \in R_k^2} \sum_{i \in R_j^1 \cap I_1} \alpha_{ij}\, y_{ijkt} + \sum_{j \in J_{2W} \cap R_{j'}^3,} \sum_{i \in R_j^1 \cap I_1} \sum_{j' \in F_j^3} \alpha_{ij}\, Y_{ijj'kt} + \sum_{j \in R_k^2} \sum_{i \in R_j^1 \cap I_2} \sum_{q=1}^{h_i} \alpha_{iqj}\, w_{iqjkt}$$
$$+ \sum_{j \in J_{2W} \cap R_{j'}^3,} \sum_{i \in R_j^1 \cap I_2} \sum_{q=1}^{h_i} \sum_{j' \in F_j^3} \alpha_{iqj}\, W_{iqjj'kt} = d_{kt}$$

for $k = 1, \ldots, K$, and $t = 1, \ldots, T$

5. *Customer product quality constraints:*

$$ASH_{kt} = \frac{1}{d_{kt}} \left[\sum_{j \in R_k^2} \sum_{i \in R_j^1 \cap I_1} y_{ijkt}\, a_{it}\, \beta_{ij} + \sum_{j \in J_{2W} \cap R_{j'}^3,} \sum_{i \in R_j^1 \cap I_1} \sum_{j' \in F_j^3} Y_{ijj'kt}\, a_{it}\, \beta_{ij} \right.$$
$$+ \sum_{j \in R_k^2} \sum_{i \in R_j^1 \cap I_2} \sum_{q=1}^{h_i} w_{iqjkt}\, a_{iqt}\, \beta_{iqj} + \sum_{j \in J_{2W} \cap R_{j'}^3,} \sum_{i \in R_j^1 \cap I_2} \sum_{q=1}^{h_i} \sum_{j' \in F_j^3} W_{iqjj'kt} a_{iqt}\, \beta_{iqj}$$

for $k = 1, \ldots, K$, $t = 1, \ldots, T$

$$SUL_{kt} = \frac{1}{d_{kt}} \left[\sum_{j \in R_k^2} \sum_{i \in R_j^1 \cap I_1} y_{ijkt} s_{it} \gamma_{ij} + \sum_{j \in J_{2W} \cap R_{j'}^3,} \sum_{i \in R_j^1 \cap I_1} \sum_{j' \in F_j^3} Y_{ijj'kt}\, s_{it}\, \gamma_{ij} \right.$$
$$\left. + \sum_{j \in R_k^2} \sum_{i \in R_j^1 \cap I_2} \sum_{q=1}^{h_i} w_{iqjkt}\, s_{iqt}\, \gamma_{iqj} + \sum_{j \in J_{2W} \cap R_{j'}^3,} \sum_{i \in R_j^1 \cap I_2} \sum_{q=1}^{h_i} \sum_{j' \in F_j^3} W_{iqjj'kt}\, s_{iqt}\, \gamma_{iqj} \right]$$

for $k = 1, \ldots, K$, $t = 1, \ldots, T$

$0 \leq ASH_{kt} \leq ASH_{ukt}$, for $k = 1, \ldots, K$, and $t = 1, \ldots, T$

$0 \leq SUL_{kt} \leq SUL_{ukt}$, for $k = 1, \ldots, K$, and $t = 1, \ldots, T$

6. *Restricting a unit of mine to be opened only once:*

$$\sum_{t=1}^{T} x_{iqt} \leq 1$$

for $i \in I_2$, $q = 1, \ldots, h_i$.

7. *Upper limit on the number of units that can be opened in a time period:*

$$\sum_{i \in I_2} \sum_{q=1}^{h_i} x_{iqt} \leq \theta_{1t} + \theta_{2t} \text{ for } t = 1, \ldots, T$$

$$0 \leq \theta_{1t} \leq U_t, \quad \theta_{2t} \geq 0, \text{ for } t = 1, \ldots, T$$

8. *Sequencing of units at a potential mine (with some implied constraints added to derive a tighter relaxation.)*

$$x_{iqt} \leq \sum_{\delta \leq t} x_{iq'\delta}$$

for $i \in I_2$, $q' = 1, \ldots, q-1$, $q = 2, \ldots, h_i$, $t = 1, \ldots, T$

9. *Disaggregated constraints (further restrictions added to derive a tighter relaxation)*

$$\sum_{j \in F_i^1} w_{iqjk\delta} + \sum_{j \in F_i^1 \cap J_{2W}} \sum_{j' \in F_j^3} W_{iqjj'k\delta} \leq d_{k\delta} \sum_{t \leq \delta} x_{iqt}$$

for $i \in I_2$, $q = 1, \ldots, h_i$, $k = 1, \ldots, K$, $\delta = 1, \ldots, T$

10. *Nonnegativity and binary constraints:*

$y_{ijkt} \geq 0$, for $i \in I_1$, $j = 1, \ldots, J$, $k = 1, \ldots, K$, $t = 1, \ldots, T$

$y_{iqjkt} \geq 0$, for $i \in I_2$, $q = 1, \ldots, h_i$, $j = 1, \ldots, J$, $k = 1, \ldots, K$, $t = 1, \ldots, T$

$Y_{ijj'kt} \geq 0$, for $i \in I_1$, $j \in J_{2W}$, $j' \in J_{2B}$, $k = 1, \ldots, K$, $t = 1, \ldots, T$

$Y_{iqjj'kt} \geq 0$, for $i \in I_2$, $q = 1, \ldots, h_i$, $j \in J_{2W}$, $j' \in J_{2B}$, $k = 1, \ldots, K$, $t = 1, \ldots, T$

$z_{it} \geq 0$, for $i \in I_1$, $t = 1, \ldots, T$

$z_{iqt} \geq 0$, for $i \in I_2$, $q = 1, \ldots, h_i$, $t = 1, \ldots, T$

$x_{iqt} = 0$ or 1, for $i \in I_2$, $q = 1, \ldots, h_i$, $t = 1, \ldots, T$

Comments on the Formulation:

Given below is an explanation of some of the finer points related to the objective function and the constraints.

As this is a long-term model, all the cost coefficients in the objective function are present values. A fixed charge cost is incurred whenever a unit at a potential mine is opened. Shipment of better ash quality coal is rewarded, and storage and underutilization at a mine is discouraged by accommodating a penalty term in the objective function. Also, in any time period t, if the number of units that can be opened exceeds the specified upper limit on the number of units that can be opened given the available budgetary resources, then a surcharge equal to 20% of the maximum fixed cost over the different units of the potential mines is applied to the number of units that exceed the stated limit. This surcharge (its value being governed by finance considerations) reflects the burden for acquiring additional capital for developing units beyond what the available budget permits.

In the flow balance constraints (1) for the mines, we require that all the coal produced within each six month period should account for flow within this duration, without any carryover of inventory. If the flow falls far shorter than the production, despite the underutilization penalty in the objective function, then this mine is a strong candidate for a shutdown. Also, in the flow balance constraint for a new unit of a potential mine, the production rate function at that unit is multiplied by a binary variable that takes on a value of one when this unit is opened and is zero otherwise, such that the constraint reflects the appropriate rate of production according to its age in the given period.

As each J_2 type of silo constitutes a pair of silos, one of which receives the run-of-mine coal, while the other stores the cleaned coal following the beneficiation process, we have used the coefficient $\frac{(1+\alpha_{ij})}{2}$ with the flow variables in the capacity constraints (2) for the J_2 type of silos. Furthermore, the normal storage or handling capacity of the silo is multiplied by the number of working days in a time period to derive the parameter SC_{jt}.

The Wentz to Bullitt transfer constraints (3) are interval constraints, and they reflect a transfer of coal from the Wentz silos to the Bullitt silos as actually practiced by the Westmoreland Coal Company. These constraints can easily be generalized for any coal company, and can be removed from the formulation if no such type of transfer is practiced in a particular company.

In the demand constraints (4) for customers, the flow variables are multiplied by coefficients α_{ij} to take into account the total weight attenuation during the benefici-ation process at the silos.

In the customer product quality constraints (5), we have simply used coefficients a_{iqt} and s_{iqt} instead of using $\sum_{\delta \leq t} a_{iq}(t-\delta+1)x_{iq\delta}$ and $\sum_{\delta \leq t} s_{iq}(t-\delta+1)x_{iq\delta}$ respectively, as it would lead to nonlinear constraints that would have to be linearized. (These latter functions are of the same nature as the production rate function used in the flow balance constraints for a unit of a given potential mine.) However, since the variation in the ash/sulfur content over the horizon is not too significant, we can approximate these nonlinear terms by a_{iqt} and s_{iqt}, which in effect assumes that the ash/sulfur content fraction varies as if the unit was opened at time period 1.

In the constraints (6), every unit of a given potential mine is restricted to be opened only once. Each constraint (7) has a right-hand side equal to the sum of θ_{1t} and θ_{2t}, where θ_{1t} is bounded from above by the maximum number of units from all potential mines that can be opened in time period t. This upper limit depends on the availability of resources such as capital, equipment, and manpower. The variable θ_{2t}, is simply nonnegatively restricted, but is penalized as described in the comment pertaining to the objective function.

The constraints (8) enforce the sequencing of units at a potential mine. Here, it is sufficient to use only the constraints having $q' = q-1$ in order to obtain a valid representation. However, additional constraints have been added for $1 \leq q' < q-1$ as well, in order to obtain a tighter relaxation. Given below is an example to illustrate this feature.

Consider $i = 1$, $q = 1,2,3$ and $t = 2$. For $q' = q-1$, we would obtain the following constraints: $x_{122} \leq x_{111} + x_{112}$ and $x_{132} \leq x_{121} + x_{122}$. Along with constraints (6), these restrictions are sufficient to require the sequencing of units 1,2, and 3 in this order up to period 2. However, if we let $1 \leq q' < q-1$, we would include an additional constraint: $x_{132} \leq x_{111} + x_{112}$. Although this is implied in the integer sense, it is not implied in the continuous sense, and hence it assists in obtaining a tighter linear programming relaxation for our problem. Such a tighter relaxation can enhance the performance of both exact and heuristic solution procedures (see Nemhauser and Wolsey [11]).

In a similar spirit, the disaggregated constraints (9) are included in the formulation to further tighten the continuous relaxation. Analogous to the standard disaggregated fixed-charge problem formulation (see Nemhauser and Wolsey [11]), these constraints could be written as

$$\sum_{j \in F_i^1} w_{iqjk\delta} + \sum_{j \in F_i^1 \cap J_{2W}} \sum_{j' \in F_j^3} W_{iqjj'k\delta} \leq \min \left[\sum_{t \leq \delta} p_{iq}(\delta - t + 1)x_{iqt},\ d_{k\delta} \sum_{t \leq \delta} x_{iqt} \right]$$

for $i \in I_2$, $q = 1, \ldots, h_i$, $k = 1, \ldots, K$, $\delta = 1, \ldots, T$.

However, as we already have flow balance constraints of the form

$$\sum_{j \in F_i^1} w_{iqjk\delta} + \sum_{j \in F_i^1 \cap J_{2W}} \sum_{j' \in F_j^3} W_{iqjj'k\delta} \leq \sum_{t \leq \delta} p_{iq}(\delta - t + 1) x_{iqt}$$

for $i \in I_2$, $q = 1, \ldots, h_i$, $k = 1, \ldots, K$, $\delta = 1, \ldots, T$, we have simply added the constraints

$$\sum_{j \in F_i^1} w_{iqjk\delta} + \sum_{j \in F_i^1 \cap J_{2W}} \sum_{j' \in F_j^3} W_{iqjj'k\delta} \leq d_{k\delta} \sum_{t \leq \delta} x_{iqt}$$

for $i \in I_2$, $q = 1, \ldots, h_i$, $k = 1, \ldots, K$, $\delta = 1, \ldots, T$ in our formulation. The computational results given in Section 5 exhibit the effect of constraints (8) and (9) in tightening the continuous relaxation of the problem.

4 Solution Procedures

This section presents methodologies for solving the long-term strategic model, using both exact and heuristic techniques aimed at deriving near optimal solutions.

Solving the Model Using Commercial Softwares

The long-term model is a mixed-integer 0-1 programming problem. We first tried solving the various test problems generated using ZOOM, which is the default solver in GAMS (see Brook et al., [3]). However ZOOM could only find an optimal solution for two out of the ten test problems even with a 10% optimality tolerance, and did not even find an integer feasible solution for the remaining problems except for test problem 4. Hence, we had to resort to more sophisticated, state-of-the-art mixed-integer solvers, namely, OSL (developed by IBM) and CPLEX (MIP option, developed by CPLEX Optimization Inc.). Different links to connect these softwares with GAMS are available through the GAMS Development Corporation. After coding the problem in GAMS, these solvers can be called to execute the solution process, provided of course, they have been installed within the system. The optimality tolerance while solving these test problems was kept at 10%. As the results of Section 5 indicate, these options, if available, are viable alternatives. CPLEX was able to solve seven out of ten problems, while OSL solved all the test problems. However, CPLEX found better quality solutions than did OSL for five of the seven instances it solved.

If these options are unavailable and one has access only to a linear programming code, or if a robust procedure is required for solving larger sized problems, we propose the following linear programming based heuristic procedure.

We remark here that we tried to derive approximate solutions for our test problems using the Pivot and Complement Heuristic of Balas and Martin [2], but it failed to solve all but the smallest of our test problems. Hence, we designed our own heuristic procedures, exploiting the nature of our problem. (Some of the ideas below (see Step 2 in particular) are portable to other 0-1 mixed-integer programming contexts as well.)

Linear Programming Based Heuristic (LPH)

This heuristic procedure employs a sequential rounding scheme based on a series of continuous linear programming relaxations, exploiting the structure of the problem in determining which variables to round up to 1 at each step of the process. The following are the steps involved in this heuristic procedure:

Step 1: Solve the linear programming (LP) relaxation of the problem. If the solution obtained has all the x-variables at binary values, then stop; the optimal solution obtained for the LP relaxation solves the mixed-integer programming problem. Otherwise, go to Step 2.

Step 2: (Optional, if package such as MINOS is available to handle nonlinear objective functions.) Replace the objective function by

$$\text{Maximize} \sum_{i \in R_j^1 \cap I_2} \sum_{q=1}^{h_i} \sum_{t=1}^{T} (x_{iqt} - 1/2)^2 \tag{11}$$

and incorporate the following additional constraint in the problem

$$z \leq \nu_{\text{LP}}(1 + \lambda) \tag{12}$$

where z represents the objective function of the strategic model, ν_{LP} is the value of its linear programming relaxation, and 100λ is a specified % deviation permitted from ν_{LP}. Solve the continuous relaxation of this linearly constrained nonlinear programming problem. (We used MINOS 5.2 available with GAMS for this purpose.) As the form of the objective function indicates, a solution to this problem encourages all the binary variables to take a value of 0 or 1, in order to maximize the objective function, while satisfying the problem constraints, including the additional constraint (12). In fact, a solution is integer feasible to the original strategic model and has an objective value within the specified tolerance $\nu_{\text{LP}}(1+\lambda)$ if and only if it solves the above problem. However, this problem is nonconvex, and so MINOS might stall at a nonoptimal local maximum, which is not integer valued. Hence, a range of values of $\lambda \in [0, 0.1]$ say, may be attempted in a sequential set of runs, and the best solution obtained may be recorded. (Note that if $\lambda = 0$, and if the solution obtained has all the x-variables at binary values, then an alternative optimal solution to the LP relaxation has been obtained which solves the original mixed-integer program.) Proceed to Step 3.

Step 3: Define $F = \{$ a set of mines i for which some unit q has a fractional variable x_{iqt} for some period $t\}$. For each $i \in F$, let $q(i)$ be the *imminent unit* defined as the smallest index q for which x_{iqt} is fractional for some t. Note that by the constraints of the problem, and the definition of $q(i)$, we must have $x_{iq\delta} = 1$ for each $q < q(i)$, for some $\delta \leq t$, where t is the smallest time period for which the x-variable for $q(i)$ is fractional. Now, for each $i \in F_i$, let $t_{i1} < t_{i2} < \ldots < t_{in_i}$ be the time periods t for which the variables $x_{iq(i)t}$ are fractional, and let $\nu_{i1}, \ldots, \nu_{in_i}$ be the fractional values of $x_{iq(i)t_{ik}}$, $k = 1, \ldots, n_i$. Accordingly, compute

$$T_i = \sum_{k=1}^{n_i} \nu_{ik} t_{ik} + T(1 - \sum_{k=1}^{n_i} \nu_{ik}) \text{ for each } i \in F$$

where T is number of time periods in the horizon. Note that the smaller the value of T_i, the relatively earlier is the tendency for the imminent unit of mine i to be opened. Hence, find

$$T^* = \underset{i \in F}{\text{minimum}} (T_i) \tag{13}$$

Instead of selecting a mine that is determined simply by (13) as the one for which the imminent unit should be opened, we examine a band of T_i values by finding

$$\overline{F} = \{ i \in F : T_i \leq T^* + 1 \}$$

For mines within this band, identify a mine r according to

$$r \in \underset{i \in \overline{F}}{\text{argmax}} \{\nu_{ik(i)}\}, \text{ where } \nu_{ik(i)} = \underset{k=1,\ldots,n_i}{\text{maximum}} \{\nu_{ik}\} \text{ for } i \in \overline{F}$$

If the fractional x-variables represent only one unit of a potential mine at different time periods, then select $k(r) = 1$, for fixing $x_{rq(r)t_{r1}} = 1$.

Step 4: Fix $x_{rq(r)t_{rk(r)}} = 1$, along with all the other binary variables which turned out to be 1 in the most recent linear programming relaxation solved. Re-solve the linear programming relaxation after fixing the above binary variables. If the problem is infeasible, go to Step 5. If the solution obtained has all the x-variables at binary values, then stop; the optimal solution obtained is prescribed as a heuristic solution. If the solution obtained still has fractional values for some of the x-variables, then return to Step 3.

Step 5: Replace $k(r)$ by $k(r) - 1$. (Note that by the structure of the problem and the nature of the solution to the last feasible linear programming relaxation, it must be that the revised $k(r) \geq 1$.) Return to Step 4.

5 Computational Experience

Real data from the Westmoreland Coal Company and nine other similar, hypothetical data sets are generated to test different heuristics and the commercial softwares used to solve the problem. The real data provided by Westmoreland Coal Company has 11 existing mines and 8 potential mines. Of the latter, one has four units, two of them have two units each, and the remaining five have one unit each. There are 8 silos and 6 customers. The horizon considered in running the model with this data is 6 periods (3 years) in duration.

Nine hypothetical test problems are created using a different number of existing mines, potential mines, units in each potential mine, silo units, and customers. Also, various combinations and connections between existing mines, potential mines, units in a potential mine, silo units, and customers are incorporated into these test problems. These problems vary in the range of 388 constraints, 27 binary variables, and 481 total number of variables, to 4155 constraints, 240 binary variables and 6565 total number of variables. Table 1 provides a list of specifications for all the test problems generated.

In all these data sets, we have included a special hypothetical potential mine with a single production unit that has a very high rate of production, and has a zero ash and sulfur percentage content in the coal produced by it. An inordinately large fixed-charge is required to open this mine. Also, the coal produced by it is given very high shipping and cleaning costs. In effect, it is ensured that this special hypothetical unit is used only when the model would otherwise have been infeasible. In other words, whenever the model uses this particular mine, it implies that the model is infeasible and that the infeasiblity lies where the flow is satisfied using the coal from this special mine. (While this should be included in practice, none of our test problems needed to resort to this hypothetical mine.)

Effect of Implied Constraints

After coding the test problems in GAMS, we also performed an investigation on the effect of using constraints that are implied in the integer sense, though not in the continuous sense, for the purpose of derving a tighter relaxation of the model, as discussed in Section 3. Tighter relaxations of a discrete programming problem help in improving the performance of any exact or heuristic solution procedure. To ascertain the effect of the implied constraints in (8) and (9) on the model, linear programming relaxations of all the test problems were run using MINOS 5.2 as a solver, both with and without these implied constraints. The results in Table 2 exhibit that the implied constraints used in the formulation produced a tighter relaxation for 6 out of the 10 test problems.

Comparison of Different Commercial Softwares

The test problems were first solved by ZOOM (Zero/One Optimization Method), which is a default solver in GAMS for mixed-integer programming problems. As it turned out, ZOOM faced significant difficulties even in obtaining a feasible solution to the linear programming relaxation of the original mixed-integer programming problem. After trying different combinations of options, like CHEAT, DIVE, EXPAND, FACTOR, GAP, PARTIAL, QUIT (see Brooke, Kendrick and Meeraus [3]) in the GAMS/ZOOM options file, we were able to find an integer solution for only three out of the ten test problems using an optimality tolerance of 10%. (It should be noted that if OSL, CPLEX, or ZOOM does not find an optimal solution within the specified range, then it gives the best solution found up to the point of termination, mentioning that there is no optimal solution within the specified range.)

Next, we solved the ten problems using OSL (Optimization Subroutine Library) and CPLEX, again with a 10% optimality tolerance. Separate GAMSLINKS, developed by the GAMS Development Corporation, were used to link GAMS with OSL and with CPLEX. The OSL runs were made on an IBM RS/6000 workstation model 320H, running AIX 3.1. The runs with CPLEX as a solver were made on a SUN Sparc 1, running SUNOS 4.1.1. The runs using ZOOM as the solver were made on an IBM 3090. (Different computers have been used since the OSL and CPLEX runs were made at the GAMS Development Corporation.) The results obtained after solving the problems with ZOOM/GAMS, OSL/GAMS and CPLEX/GAMS are tabulated in Table 3.

OSL was able to solve all the test problems as shown in Table 3 while CPLEX was unable to solve three of these problems due to memory limitations on the SUN Sparc 1 computer. On the other hand, ZOOM could only solve three test problems, two of which were solved within 10% of optimality, while for the third problem (number 4), only a feasible integer solution within about 43% of the linear programming lower bound could be found. No feasible solution was found for the remaining problems. Comparing CPLEX and OSL in terms of the number of iterations and the relative gap from the linear programming relaxation value, the results in the table show that for 4 out of the 7 problems that CPLEX solved, it required more iterations than did OSL, but for 5 out of these 7 problems, it obtained a better solution than did OSL. (Blank spaces in the table specify those problems which could not be solved either by CPLEX due to shortage of memory or by ZOOM due to different kinds of numerical difficulties.)

Comparison of Results Obtained Using Commercial Softwares and Heuristic LPH

Table 4 presents the results using Heuristic LPH to solve the ten test problems. In all the test problems except for problem number 2, the proposed heuristic consumes more iterations to solve the problems than do the commercial softwares OSL and

Prob. Num.	Num. of existing mines	Num. of potential mines	Num. of units in potential mines	Num. of customers	Num. of time periods	Num. of silos	Num. of Constr.	Num. of vars.	Num. of binary vars.	CPU time (sec)	Iters.
1	8	5	9	7	3	8	388	481	27	3.7	560
2	11	8	13	6	6	8	950	1145	78	31.2	2047
3	5	8	14	7	6	8	1195	2309	84	57.6	2599
4	4	8	15	5	6	8	776	1337	90	25.5	1885
5	7	8	15	10	6	4	1310	2987	90	93.3	3367
6	5	8	15	6	6	4	1053	863	96	36.6	2238
7	7	8	13	6	8	8	1290	1957	104	56.7	2634
8	4	8	14	6	8	6	1251	1752	112	105.8	5016
9	6	13	24	10	8	8	2981	4149	192	497.8	9850
10	7	16	30	12	8	10	4155	6565	240	584.6	6947

Table 1: **Test problem specifications for the long-term strategic model**

Legend: CPU time = CPU seconds for solving the linear programming (LP) relaxation on an IBM 3090 computer Iters. = Number of iterations required to solve the linear programming relaxation of the problem. Vars. = Variables, Num. = Number, Prob. = Prcblem, Constr. = Constraints.

Test problem number	Solved without implied constraints (objective value)	Solved with implied constraints (objective value)
1	258368136	258368136
2	296801038	316045551
3	76531010	76531010
4	31379965	31379965
5	77480004	77740986
6	328998463	351037130
7	391664242	392528791
8	56816347	56816347
9	655812636	673834448
10	409673011	412477641

Table 2: **Comparison of solutions obtained with and without implied constraints**

CPLEX, given the 10% optimality tolerance used within the latter methods. It should be noted, however, that while solving the problems using our heuristic, we did not use any advanced bases from one LP run to the next. The effort for the proposed heuristic can be substantially reduced by using the optimal basis obtained for one run in the subsequent problem solved, since only one additional fractional binary variable is fixed at 1 from one run to the next in the sequence of problems solved. This automation can be done by incorporating the MPS file generated through GAMS 2.25 within a FORTRAN or C program, and updating this file from one call to the next of the solver MINOS 5.2, OSL, or CPLEX.

As far as the solution quality is concerned, Heuristic LPH obtains a better solution than does OSL for 3 of the test problems and the same solution for 2 of them. It also obtains a better solution than does CPLEX for a single case. In this case, Heuristic LPH actually identifies an alternative optimal linear programming solution that happens to be integer feasible at Step 2 of the procedure, using $\lambda = 0.0$. Overall, as can be seen by comparing Tables 3 and 4, for the most part, the solutions obtained via the different procedures are comparable in quality. Finally, it should be noted that because OSL and CPLEX employ a branch-and-bound technique, it can be expected that as the problem size increases, the computer memory requirements would increase

Test Prob.	OSL				CPLEX				ZOOM			
	MIP solution value	RG %	CPU1 time (secs)	Iters	MIP solution value	RG %	CPU2 time (secs)	Iters	MIP solution value	RG %	CPU3 time (secs)	Iters
1	259637230	0.49	1.8	178	259637230	0.49	6.0	426	259637230	0.49	5.4	1121
2	368434780	16.5	1335.9	47348	373343025	18.12	414.2	8765	–	–	–	–
3	86359033	12.84	27.9	1419	83701728	9.36	230.1	4436	–	–	–	–
4	33697853	7.38	20.6	1455	33642804	7.21	243.0	6825	44930032	43.18	355.7	49077
5	78278455	1.03	32.6	2126	78278455	1.03	83.9	2045	79326334	2.38	139.2	7554
6	446829414	27.28	196.0	13046	–	–	–	–	–	–	–	–
7	448381433	14.22	121.9	5912	440208987	12.14	179.9	4271	–	–	–	–
8	60458788	6.41	41.1	2238	60584061	6.63	127.7	2634	–	–	–	–
9	965484334	43.28	166.9	3994	–	–	–	–	–	–	–	–
10	449239319	8.91	221.5	4004	–	–	–	–	–	–	–	–

Table 3: **Comparison of solutions obtained and effort required for solving the test problems with OSL/GAMS, CPLEX/GAMS, and ZOOM/GAMS**

Legend: CPU1 time = Resource usage in CPU seconds on an IBM RS/6000 workstation model 320H
CPU2 time = Resource usage in CPU seconds on a SUN Sparc 1
CPU3 time = Resource usage in CPU seconds on an IBM 3090
Iters = Total number of simplex iterations required to solve the problem
RG = % Relative gap of the MIP solution obtained from the initial linear programming relaxation value

Test Problem	MIP soln.	RG %	CPU time (secs)	Iters.	LPR	FRA
1	259637230	0.49	4.1	1108	3	1
2	370792903	17.32	88.2	5305	3	7
3	84327533	10.18	227.5	12340	6	3
4	34704832	10.59	102.5	9435	7	6
5	78278455	1.03	540.7	22516	8	6
6	469914352	33.86	293.1	19006	12	9
7	448381433	14.22	279.4	12883	6	4
8	60272957	6.08	770.8	42031	14	10
9	1023861971	51.94	4003.4	77540	12	11
10	412477641	0.00	1009.4	13447	2	8

Table 4: **Computational Experience Using Heuristic LPH**

Legend: CPU time = Resource usage in CPU secounds on an IBM 3090

Iters = Total number of simplex iterations required to solve the problem

RG = % Relative gap of the MIP solution obtained from the initial linear programming value

LPR = Number of linear programming runs required before discovering the mixed-integer solution.

FRA = Number of units of potential mines having fractional values at the first step.

substantially, and the optimality tolerance will need to be further relaxed to keep these procedures viable. On the other hand, the proposed Heuristic LPH can be expected to remain robust since it only relies on the solution of a limited sequence of linear programming relaxations.

Appendix: Modifications for a Tactical Day-to-Day Operational Model

Sherali and Puri [14] have presented a description of three linear programming tactical models for making day-to-day mining, cleaning, blending, and distribution decisions, given a set of operating mines. The most accurate and detailed of these three models is called "Model 1". In this model, coal is assumed to be shipped out of mines to the silos at the beginning of the time periods, and to be shipped out of the silo units to the customers at the end of the time periods. A maximum of a three period *shipment lag* between coal production at the mines and the final shipment to customers is permitted, based on an estimate of the clearance time at the silos. If t_1 is the time period for a certain mine to silo shipment, and t_2 is that for a continuing silo to customer shipment, the *shipment lag* is given by $t_2 - t_1$. The *transfer lag*, another time lag factor, that indicates the difference between the time of dispatch of a coal shipment from a mine to a silo and its actual arrival time at the silo, is negligible for the problem under study, and is therefore assumed to be zero. (Nonzero transfer lags can, however, be readily accommodated by time-shifting the data.) We also adopt this same structure of the model. However, the following modifications have been made to enhance the problem representation based on the feedback from our case study implementation.

1. A penalty-reward component has been introduced into the objective function, which either penalizes or rewards the quality sent to the customers relative to what is desired. The need for this function arose due to the 1990 Clean Air Act as the customers are now somewhat more liberal about the content of ash in the coal, but are more stringent about its sulfur content. Hence, besides a piecewise linear reward or penalty function imposed on the ash quality of the coal shipped relative to the maximum specified limit, we have also introduced a piecewise linear reward structure in the objective function to reflect the incentive for shipping coal having a better sulfur quality, while restricting a maximum sulfur content in each period as a hard constraint.

2. Due to the 1990 Clean Air Act, coal companies might have to look for different types of cleaning technologies instead of using a single type. "Model 1" can

be given an interpretation so that it can accommodate more than one type of cleaning technology. For example, if there are two types of cleaning technologies used by a coal company, then the silos in J_2 can be divided further into subsets J_{21} and J_{22}, where each J_{21} type of silo constitutes a pair of silos, one of which receives the run-of-mine coal, while the other stores the coal following the cleaning operation, and where each J_{22} type of silo constitutes a similar pair of silos, but representing an alternative cleaning technology. Note that this is principally a data processing, rather than a model oriented modification.

3. The equality constraint that strictly enforces as much amount of coal to be transferred from the Wentz to Bullitt silos as is shipped to stoker customers has been relaxed to an interval constraint, to better reflect the actual practice in this transfer.

Since Sherali and Puri do not provide a mathematical formulation for "Model 1", for the sake of convenience in reference by practitioners, we give below a complete mathematical formulation (including the foregoing modifications), that has been tested and implemented using GAMS. For completeness, we first specify the data requirements along with our notation. Note that the short-term tactical model, in contrast with the long-term strategic model, has to contend with daily storage restrictions at the mines and at the silos, as well as with the dissipation of any initial amount in storage at the silos within a rolling horizon implementation framework.

$t = 1, \ldots, T \equiv$ number of time periods.

$i = 1, \ldots, m \equiv$ number of mines.

$j = 1, \ldots, J \equiv$ number of silo units.

$k = 1, \ldots, K \equiv$ number of customers.

K_{st} = "stoker customers" served by Wentz silos.

p_{it} = production (tons) at mine i, in period t.

a_{it}, s_{it} = ash and sulfur content (%), respectively, in coal produced at mine i, in period t.

SM_i = storage capacity at mine i.

c_i^{sm} = storage cost (per ton) at mine i.

$J_1 \equiv$ ROM silo storage units.

$J_{2W} \equiv$ cleaned silo units at Wentz facility.

$J_{2B} \equiv$ cleaned silo units at Bullitt facility.

$J_2 \equiv$ cleaned silo units $(J_{2W} \cup J_{2B})$.

SS_j = storage capacity of silo unit j. (For $j \in J_2$, this is taken as the sum of the two associated silo storage capacities.)

c_j^{ss} = storage cost per ton of coal at silo j.

q_j^0 = initial amount in storage at silo j.

a_j^0, s_j^0 = ash and sulfur content (%), respectively, in initial storage amount at silo j.

c_{ij}^{sh1} = shipping cost (per ton), from mine i to silo j.

$\alpha_{ij} \in (0,1]$ = total weight attenuation factor (output per ton input) at silo $j \in J_2$ for coal from mine i.

$\beta_{ij} \in (0,1]$ = ash content attenuation factor (output per ton input) at silo $j \in J_2$ for coal from mine i.

$\gamma_{ij} \in (0,1]$ = sulfur content attenuation factor (output per ton input) at silo $j \in J_2$ for coal from mine i.

Note: α_{ij}, β_{ij}, and γ_{ij} are assumed to be 1 for $j \in J_1$.

A_1 = flow arcs (i,j) from mine i to silo j; $F_i^1 = \{ j : (i,j) \in A_1 \}$, $R_j^1 = \{ i : (i,j) \in A_1 \}$.

A_2 = flow arcs (j,k) from silo j to customer k; $F_j^2 = \{ k : (j,k) \in A_2 \}$, $R_k^2 = \{ j : (j,k) \in A_2 \}$.

A_3 = arcs (j,j') representing by-product flow from Wentz silo $j \in J_{2W}$ to Bullitt silo $j' \in J_{2B}$, corresponding to stoker customer shipments; $F_j^3 = \{ j' : (j,j') \in A_3 \}$, $R_j^3 = \{ j : (j,j') \in A_3 \}$.

c_{jk}^{sh2} = shipping costs (per ton), from silo j to customer k.

$c_{jj'}^{sh3}$ = shipping costs (per ton), from Wentz silo $j \in J_{2W}$ to Bullitt silo $j' \in J_{2B}$.

d_{kt} = demand placed by customer k during period t.

ASH_{ukt} = upper limit on ash percentage in coal to be delivered to customer k in period t.

SUL_{ukt} = upper limit on sulfur percentage in coal to be delivered to customer k in period t.

c_{1k}^1 = slope (revenue/ton) of reward function for each % point below the maximum specified limit of ash in coal shipped to customer k.

c^1_{2k} = slope (cost/ton) of penalty function for each % point above the maximum specified limit of ash in coal shipped to customer k.

c^2_{1k} = slope (revenue/ton) of reward function for each % point below the maximum specified limit of sulfur in coal shipped to customer k.

Mathematical Formulation

The decision variables for the tactical model are defined as follows: (i) amount (tons) shipped from mine i to silo j in period t, with continued shipment to customer k in period τ, given by $y^{k\tau}_{ijt}$ (ii) amount (tons) in initial storage at silo j, shipped to customer k in period t, given by y^0_{jkt} (iii) amount (tons) shipped from mine i to silo $j \in J_{2W}$ in time period t_1 which is then shipped to $j' \in J_{2B}$ in period t_2, and finally shipped to customer k in period τ, given by $Y^{k\tau}_{ijt_1j't_2}$. Here, for a given t_1, the permissible values of (t_2, τ) are given by $t_L(t_1)$, subject to a maximum of a three period shipment lag between τ and t_1, and with $t_2 > t_1$, (Note that the initial storage at the silos is also assumed to be dissipated within three periods). Hence, we have,

$$(t_2,\tau) \in t_L(t_1) \equiv \{(t_1+1, t_1+1), (t_1+1, t_1+2), (t_1+2, t_1+2)\}$$

Auxiliary Decision Variables

1. $x^s_{i\delta}$ = slack variable equal to the amount (tons) of coal remaining in storage at mine i during period δ, for $i = 1, \ldots, m$, $\delta = 1, \ldots, T$.

2. $u^s_{j\delta}$ = accumulated storage amount (tons) in silo unit j during period δ, for $j = 1, \ldots, J$, $\delta = 1, \ldots, T$.

3. $ASH_{k\tau}$ = % ash content in blended coal delivered to customer k in period τ, for $k = 1, \ldots, K$, $\tau = 1, \ldots, T$.

   (a) $(ASH_{k\tau})_1$ = % ash content in blended coal delivered to customer k in period τ, that is below the maximum specified limit, for $k = 1, \ldots, K$, $\tau = 1, \ldots, T$.

   (b) $(ASH_{k\tau})_2$ = % ash content in blended coal delivered to customer k in period τ, that exceeds the maximum specified limit, for $k = 1, \ldots, K$, $\tau = 1, \ldots, T$.

4. $SUL_{k\tau}$ = % sulfur content in blended coal delivered to customer k in period τ, for $k = 1, \ldots, K$, $\tau = 1, \ldots, T$.

Objective Function:

$$\text{Minimize} \quad \sum_{j=1}^{J} \sum_{i \in R^1_j} \sum_{k \in F^2_j} \sum_{t=1}^{T} \sum_{\tau=t}^{\min(t+2,T)} (c^{sh1}_{ij} + c^{cl}_{ij} + c^{sh2}_{jk}) y^{k\tau}_{ijt}$$

$$+ \sum_{j\in J_{2W}} \sum_{i\in R_j^1} \sum_{j'\in F_j^3} \sum_{k\in F_{j'}^3} \sum_{t_1=1}^{T} \sum_{(t_2,\tau)\in t_L(t_1)} (c_{ij}^{sh1} + c_{j'k}^{sh2} + c_{jj'}^{sh3} + c_{ij}^{cl}) Y_{ijt_1j't_2}^{k\tau}$$

$$+ \sum_{j=1}^{J} \sum_{k\in F_j^2} \sum_{t=1}^{3} c_{jk}^{sh2} y_{jkt}^0 + \sum_{i=1}^{m} \sum_{t=1}^{T} c_i^{sm} x_{it}^s + \sum_{j=1}^{J} \sum_{t=1}^{T} c_j^{ss} u_{jt}^s$$

$$+ \sum_{k=1}^{K} \sum_{\tau=1}^{T} [-c_{1k}^1 (ASH_{uk\tau} - (ASH_{k\tau})_1) d_{k\tau} + c_{2k}^1 (ASH_{k\tau})_2 d_{k\tau}]$$

$$+ \sum_{k=1}^{K} \sum_{\tau=1}^{T} [-c_{1k}^2 (SUL_{uk\tau} - SUL_{k\tau}) d_{k\tau}]$$

Constraints:

1. *Flow balance and storage at mines:*

$$\sum_{t\le\delta} p_{it} - \sum_{j\in F_i^1} \sum_{t\le\delta} \sum_{k\in F_j^2} \sum_{\tau=t}^{\min(t+2,T)} y_{ijt}^{k\tau} - \sum_{j\in F_i^1\cap J_{2W}} \sum_{t_1\le\delta} \sum_{j'\in F_j^3} \sum_{k\in F_{j'}^2} \sum_{(t_2,\tau)\in t_L(t_1)} Y_{ijt_1j't_2}^{k\tau} = x_{i\delta}^s$$

for $i = 1, \ldots, m$, and $\delta = 1, \ldots, T$

$0 \le x_{i\delta}^s \le SM_i$ for $i = 1, \ldots, m$, and $\delta = 1, \ldots, T$

2. *Storage constraints for silo units:*

$$\sum_{i\in R_j^1} \sum_{k\in F_j^3} \sum_{\delta=t-2}^{t} \sum_{\tau=t}^{\min(\delta+2,T)} \frac{1}{2}(1+\alpha_{ij}) y_{ij\delta}^{k\tau} + \left(q_j^0 - \sum_{\tau<t} \sum_{k\in F_j^2} y_{jk\tau}^0 \right) + \theta_{jt} = u_{jt}^s$$

$0 \le u_{jt}^s \le SS_j$ for $j = 1, \ldots, J$, and $t = 1, \ldots, T$ where,

$$\theta_{jt} = \begin{cases} \sum_{i\in R_j^1} \sum_{j'\in F_j^3} \sum_{k\in F_{j'}^2} \sum_{\delta=t-2}^{t} \sum_{(t_2,\tau)\in t_L(\delta), t_2\ge t} \frac{1}{2}(1+\alpha_{ij})\, Y_{ij\delta j't_2}^{k\tau} & \text{for } j \in J_{2W} \\ \sum_{j'\in R_j^3} \sum_{i\in R_{j'}^1} \sum_{k\in F_j^2} \sum_{(t_2,\tau)\in t_{L_1}(t)} \sum_{t_1=(\tau-2)}^{t_2-1} \frac{1}{2}(1+\alpha_{ij'})\, Y_{ij't_1jt_2}^{k\tau} & \text{for } j \in J_{2B} \\ 0 & \text{otherwise} \end{cases}$$

and where $t_{L_1}(t) = \{(t-1,t), (t,t), (t,t+1)\}$ for defined combinations.

3. *Dissipation of initial storage at silos:*

$$\sum_{k\in F_j^2} \sum_{t=1}^{3} y_{jkt}^0 = q_j^0 \quad \text{for } j = 1, \ldots, J$$

4. *Wentz to Bullitt transfer:*

$$\sum_{j'\in F_j^3}\sum_{k\in F_{j'}^3}\sum_{(t_2,\tau)\in t_L(t)} Y_{ijtj't_2}^{k\tau} \le \sum_{k\in F_j^2\cap K_{st}}\sum_{\tau=t}^{\min(T,t+2)} y_{ijt}^{k\tau} \le 1.1\sum_{j'\in F_j^3}\sum_{k\in F_{j'}^2}\sum_{(t_2,\tau)\in t_L(t)} Y_{ijtj't_2}^{k\tau}$$

for $j \in J_{2W}$, $i \in R_j^1$, and $t = 1,\ldots,T$

5. *Demand constraints for customers:*

$$\sum_{j\in R_k^2}\sum_{i\in R_j^1}\sum_{t=\tau-2}^{\tau} \alpha_{ij}\, y_{ijt}^{k\tau} + \sum_{j\in R_k^2} y_{jk\tau}^0 + \sum_{j'\in R_k^2}\sum_{j\in R_{j'}^3}\sum_{i\in R_j^1}\sum_{t=\tau-2}^{\tau-1}\sum_{t_2=t+1}^{\min(t+2,\tau)} \alpha_{ij}\, Y_{ijtj't_2}^{k\tau} = d_{k\tau}$$

for $k = 1,\ldots,K$ and $\tau = 1,\ldots,T$

6. *Customer product quality constraints:*

$$ASH_{k\tau} = \frac{1}{d_{k\tau}}\left[\sum_{j\in R_k^2}\sum_{i\in R_j^1}\sum_{t=\tau-2}^{\tau} y_{ijt}^{k\tau}\, a_{it}\,\beta_{ij} + \sum_{j\in R_k^2} y_{jk\tau}^0\, a_j^0 + \sum_{j'\in R_k^2}\sum_{j\in R_{j'}^3}\sum_{i\in R_j^1}\sum_{t=\tau-2}^{\tau-1}\sum_{t_2=t+1}^{\min(t+2,\tau)} Y_{ijtj't_2}^{k\tau}\, a_{it}\,\beta_{ij}\right]$$

for $k = 1,\ldots,K$, $\tau = 1,\ldots,T$

$$SUL_{k\tau} = \frac{1}{d_{k\tau}}\left[\sum_{j\in R_k^2}\sum_{i\in R_j^1}\sum_{t=\tau-2}^{\tau} y_{ijt}^{k\tau}\, s_{it}\,\gamma_{ij} + \sum_{j\in R_k^2} y_{jk\tau}^0\, s_j^0 + \sum_{j'\in R_k^2}\sum_{j\in R_{j'}^3}\sum_{i\in R_j^1}\sum_{t=\tau-2}^{\tau-1}\sum_{t_2=t+1}^{\min(t+2,\tau)} Y_{ijtj't_2}^{k\tau}\, s_{it}\,\gamma_{ij}\right]$$

for $k = 1,\ldots,K$, $\tau = 1,\ldots,T$

$$\begin{aligned} ASH_{kt} &= (ASH_{kt})_1 + (ASH_{kt})_2 \\ 0 &\le (ASH_{kt})_1 \le ASH_{ukt} \\ 0 &\le (ASH_{kt})_2 \le \infty \\ 0 &\le SUL_{kt} \le SUL_{ukt} \end{aligned}$$

for $k = 1,\ldots,K$, and $t = 1,\ldots,T$

7. *Nonnegativity constraints:*

$y_{ijt}^{k\tau} \geq 0$, $y_{jkt}^{0} \geq 0$, for $i = 1, \ldots, m$, $j = 1, \ldots, J$, $k = 1, \ldots, K$, $t = 1, \ldots, T$, $\tau = t, \ldots, \min\{T, t+2\}$

$Y_{ijt_1j't_2}^{k\tau} \geq 0$, for $i = 1, \ldots, m$, $j \in J_{2W}$, $j' \in J_{2B}$, $k = 1, \ldots, K$, $t_1 = 1, \ldots, T$, $(t_2, \tau) \in t_L(t_1)$

References

[1] R. Aboudi, A. Hallefjord, C. Helgesen, R. Helming, K. Jornsten, A.S. Pettersen, T. Raum, and P. Spence. A Mathematical Programming Model for the Development of Petroleum Fields and Transport Systems. *Euorpean Journal of Operational Research*, 43:13–25, 1989.

[2] E. Balas and C.H. Martin. Pivot and Complement- A Heuristic for 0-1 Programming. *Management Science*, 26(1):86–96, 1980.

[3] A. Brooke, D. Kendrick, and A. Meeraus. GAMS A User's Guide. The International Bank for Reconstruction and Development, The World Bank, 1988.

[4] W. Candler. Coal Blending – With Acceptance Sampling. *Computers & Operations Research*, 18(7):591–596, 1991.

[5] G.B. Faulkner. Linear Programming Applied to a Mining Smelting Operation. *Canadian Mining & Metallurgical Bulletin*, 60(677):1297–1300, 1967.

[6] M. Gershon. Mine Scheduling Optimization with Mixed Integer Programming. *Mining Engineering*, 35(4):351–354, 1983.

[7] T.B. Johnson. Optimum open-pit mine production scheduling. A Decade of Digital Computation in the Mineral Industry, SME-AIME, New York, USA., 1969.

[8] A.M. Khan. Solid-Waste Disposal with Intermediate Transfer Stations: An Application of the Fixed-Charge Location Problem. *Journal of the Operational Research Society*, 38(1):31–37, 1987.

[9] C.G. Knight and C.B. Manula. The Pennsylvania Coal Model. In *Proceedings of the 14th APCOM Symposium*, SME-AIME, 655–665, New York, 1976.

[10] B.A. Lietaer. A Planning Model for Underground Mines – An Application in a Developing Country. *OMEGA, The International Journal of Management Science*, 5(2):149–159, 1977.

[11] G.L. Nemhauser and L.A. Wolsey. *Integer and Combinatorial Optimization.* John Wiley and Sons Inc., New York, NY, 1988.

[12] J.P. Osleeb and S.J. Ratick. A Mixed Integer and Multiple Objective Programming Model to Analyze Coal Handling in New England. *European Journal of Operational Research*, 12:302–313, 1983.

[13] J.P. Osleeb, S.J. Ratick, P. Buckley, K. Lee, and M. Kuby. Evaluating Dredging and Offshore Loading Locations for U.S. Coal Exports Using the Local Logistics System. *Annals of Operations Research*, 6:163–180, 1986.

[14] H.D. Sherali and R. Puri. Model Development, Testing and Computer Implementation for a Coal Blending and Distribution Problem. *OMEGA, The International Journal of Management Science.* (To appear).

[15] H. Steinmann and R. Schwinn. Computational Experience with a Zero-One Programming Problem. *Operations Research*, 17:917–920, 1969.

[16] R.C. Tomlinson. The Practice of O.R. in Coal Mining. *European Journal of Operational Research*, 1:9–21, 1977.

[17] K.B. Williams and K.B. Haley. A Practical Application of Linear Programming in the Mining Industry. *Operational Research Quarterly*, 10(3):131–138, 1989.

[18] W. Young, J.G. Ferguson, and B. Corbishley. Some Aspects of Planning in Coal Mining. *Operational Research Quarterly*, 14(1):31–45, 1963.

Network Optimization Problems, pp. 263-281
Eds. D.-Z. Du and P.M. Pardalos

Multi-Objective Routing in Stochastic Evacuation Networks

J. MacGregor Smith
Industrial Engineering and Operations Research Department, University of Massachusetts, Amherst, MA 01002, USA

Abstract

A fundamental problem of routing in stochastic queueing networks is the identification of paths which extremize a collection of objective functions. In this paper, an integer set partitioning model with two conflicting objective functions is presented and examined. Also, the properties of the Noninferior set of solutions and the mathematical development of the algorithm for iteratively generating the paths are described. The algorithm is based on a multi-objective k-shortest path algorithm for generating the Noninferior set of paths for which tradeoffs between evacuation time and distance travelled within the network are evaluated. An example of the methodology is also presented.

1 Problem Overview

Fundamentally, the problem of routing customers, occupants, packets, and other transactions in queueing networks is a complex, stochastic, integer, and nonlinear programming problem. These problems are highly transient as well as being multi-objective in nature and there are numerous performance measures inherent within the problem which complicate the different routing strategies. Mathematically, we have a finite queueing network $G(V, E)$, with a finite set of nodes V, and edges(arcs) E, over which multiple classes of customers (occupants) flow from source(s) to sink(s) while a vector of objective functions $\Omega = \{f_1(\bar{x}), f_2(\bar{x}), \ldots, f_p(\bar{x})\}$ is simultaneously extremized subject to a set of constraints on the occupants flowing through the network.

In this paper, some of the mathematical properties, methodology and corresponding algorithms for solving this multi-objective routing problem in queueing networks are presented. One main concern in this paper is to demonstrate how one can model the multi-objective nature of the problem and calculate effective alternative routing strategies that allow the system planner of the queueing network to tradeoff performance between one objective and another.

Much of our past research in this area has considered stochastic evacuation networks [14, 15, 1], and also static and real-time routing in production and manufacturing settings [5, 6, 7, 9]. In these latter studies, *maximum throughput, sojourn time, and average number of customers in the system* have been objectives of interest. In this paper, the focus is on stochastic evacuation networks in which we consider two primary objectives: $f_1(\bar{x})$:= *Total Evacuation Time* and $f_2(\bar{x})$:= *Total Distance travelled.* The methodology and properties found can be generalized for other networks with similar objectives, such as path reliability, see Figure 1.

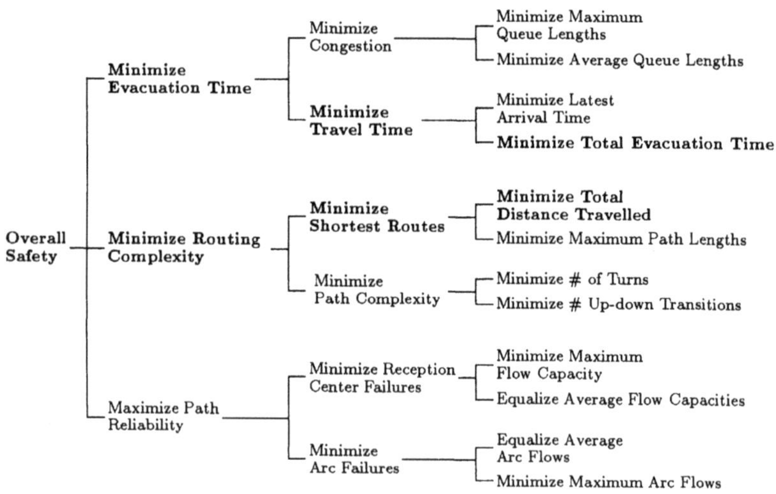

Figure 1: Morphological Diagram of Multi-Objective Approaches

One might think that time and distance are highly correlated and for those situations where no congestion in the network evacuation path exists, this is substantiated. However, when there is congestion in the network and all occupants seek the shortest path, then there is a tradeoff which occurs, where routing some of the occupants on longer paths will reduce the overall evacuation time of all occupants.

The approach that is described is flexible and effective and can be utilized for transient situations where discrete-event simulation models such as **Q-GERT** [12] and their variants, or for steady-state analytical queueing network models such as Mean Value Analysis **QNET-C** [15] are used to evaluate and generate the set of efficient routing paths in an evacuation network.

2 Assumptions and Definitions

By definition, a queueing network (graph) $G(V, E)$ is comprised of a finite set V of nodes (vertices) of size N where $V = \{V_1, V_2, \ldots, V_n\}$ together with a finite set E of arcs $e_k = (v_i, v_j)$ $\forall (i,j)$ nodal pairs. V can further be partitioned into three sets, V_1, which represents the occupant source nodes during the evacuation, V_2 which represents the intermediate nodes during the evacuation, and V_3 which represents the sink or destination nodes of the occupants. The set of arcs represent the different streets, passageways, or routes from V_1 to V_3. Associated with each node $\ell \in V$ and each arc $(v_i, v_j) \in E$ are variables and parameters which represent node and arc processing times, node and arc capacities, arrival times to the network, distances, and occupant population sizes at the source nodes.

Figure 1 illustrates a small evacuation network with many of the parameters and variables of significance to the evacuation planning problem that can be embedded in the network model. Some of the notation most commonly associated with this type of network model is discussed below.

A := There are A chains of occupants labelled $a = 1, 2, \ldots A$. Each occupant chain represents a sequence (vector) of nodes and arcs which the occupant chain population will travel during the evacuation.

$\mathbf{D}$:= A distance matrix where the elements of $\mathbf{D}$, d_{ij} represent the Euclidean or rectilinear distance between nodal pairs $(i, j) \in E$.

E := The network has a finite set E, of arcs(nodal pairs)

$f_j(\bar{x})$:= An objective function evaluating the set of routing alternatives denoted by $(\bar{x})$.

$G(V, E)$; the queueing network (graph)

λ := the arrival rate vector of the occupant classes into the routing alternatives of the network. $\lambda = (\lambda_1, \lambda_2, \ldots, \lambda_k)$ for all occupant sources.

μ := the service rate vector for the nodes and arcs comprising the evacuation network. μ_i represents the service rate of a node while μ_ℓ represents the service

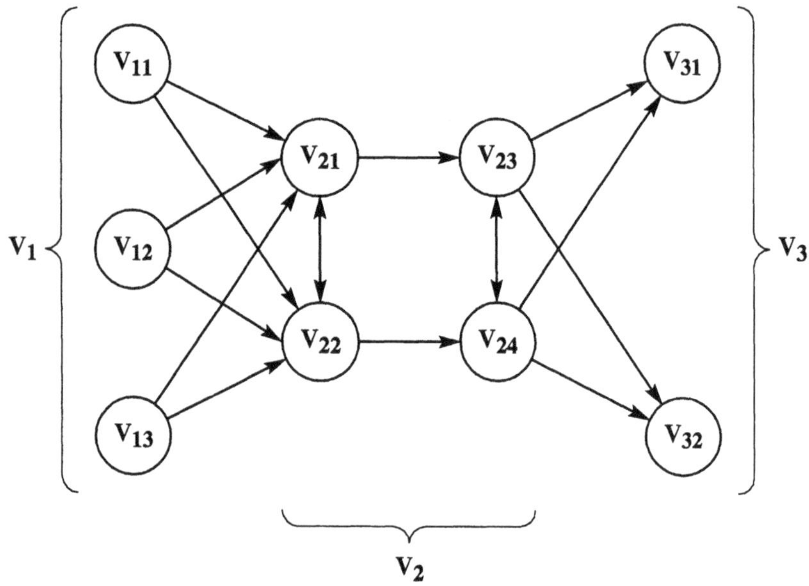

Figure 2: **Example Evacuation Network**

rate of an arc (travel time) between two nodes in G. Each queue is assumed to have infinite waiting room.

NI := a Noninferior evacuation path for an occupant class in an evacuation network.

V := The network model has a finite set V of nodes and further that V is partitioned into three sets: V_1, V_2, V_3 which represent the *source(s), intermediate,& sink(s)* nodes.

Ω := The set of objectives in our routing problem.

Since we are dealing with a multi-objective problem, we need to define the notion of a *Noninferior NI* evacuation path [3].

Definition: $\bar{x}^*$ is said to be a *Noninferior* evacuation path for our evacuation problem defined in §1 if there exists no other feasible evacuation path $\bar{x}$ such that $\bar{f}(\bar{x}) \leq \bar{f}(\bar{x}^*)$ meaning that $\bar{f}_j(\bar{x}) \leq \bar{f}_j(\bar{x}^*)$ for all $j = 1, 2, \ldots, p$ with strict inequality for at least one j.

In other words, if we have a candidate path which we suspect is *NI*, there should be no other *feasible path* which is more minimal in both of the performance measures: *time and distance*. The set of all paths which are *NI* is called the *NI* set.

3 Mathematical Model

There are few mathematical models which have appeared in the literature for generating and evaluating evacuation paths for an occupant population [10, 2, 11].

The model which is presented below is a variation of one model appearing in [11]. It was one of the first to account for the critical features of the stochastic evacuation problem. Another class of models that one might utilize to formulate the problem are those of the class of multi-commodity flow models. Unfortunately, these models will not control the splitting of the occupant population along the different evacuation paths which is problematic since splitting the different source populations will engender confusion and a potential sense of panic among the evacuating occupants. The integer programming model presented below has the desired property to control splitting of the flows.

The multi-objective model of our routing problem is:

$$Minimize\{f_1(\bar{x}); f_2(\bar{x})\}$$

where:

$$(Evacuation Time): \; f_1(\bar{x}) = \sum_i \sum_j \sum_k q_{ijk} \lambda_{ijk} x_{ijk} \tag{1}$$

$$(Distance Travelled): \; f_2(\bar{x}) = \sum_i \sum_j \sum_k d_{ijk} \lambda_{ijk} x_{ijk} \tag{2}$$

subject to:

$$(V_2 \; Arcs): \sum_i \sum_j \sum_k \alpha_{\ell ijk} \lambda_{ijk} x_{ijk} \leq \rho_\ell \quad \forall \ell \tag{3}$$

$$(V_3 \; Sinks): \sum_i \sum_j \sum_k p_{ijk} x_{ijk} \leq C_q \quad \forall q \tag{4}$$

$$(Occupant \; Classes): \sum_k x_{ijk} = 1 \quad \forall ij \tag{5}$$

$$(Routes): x_{ijk} = 0, 1 \quad \forall ijk \tag{6}$$

and where:

x_{ijk}:= 1 if the ith occupant class from the jth source is assigned the kth *NI* route alternative.

$\alpha_{\ell ijk}$: a data coefficient which equals 1 if the ℓth arc is included in the ijk^{th} route assignment and equals 0 otherwise.

ρ_ℓ: maximum allowable traffic along arc ℓ.

C_q: capacity of sink (destination) node q.

p_{ijk}: occupant population of source ij on the kth *NI* route alternative.

q_{ijk}: expected evacuation (sojourn) time of the ijk^{th} occupant class. These values must be calculated from the particular stochastic model used in the evacuation study, see discussion below.

d_{ijk}: average distance travelled for the ijk^{th} occupant class.

Because of the complexity of solving this model directly, an alternative approach which systematically generates feasible routing alternatives to a relaxed version of our mathematical model but at the same time measures the critical objectives of evacuation time and distance travelled is proposed and demonstrated in the next two sections of the paper.

4 Congestion Properties

Some crucial issues guide us in the routing/re-routing process:

- *How should the routing/re-routing process be initialized?*
- *How should alternative NI paths be selected*
- *How should the re-routing process be terminated*

In general, a problem faced with multi-objective programming problems is the often exponential number of *NI* solutions. We would like to limit the exploration of the *NI* solutions to a manageable quantity by decomposing and relaxing the original problem presented in §C so that we systematically treat one objective at a time.

Before we present the relaxed mathematical model along with the mathematical properties, an important definition is needed which concerns the eventual gain in re-routing occupants along longer *NI* paths in the evacuation networks.

Expected Savings: Given a current occupant populations' evacuation route, the potential gain in re-routing an occupant population along some other *NI* route is given as:

$$E_{ij}^{k} = q_{ijk} - [(d_{ij}^{k}/\omega) + q_{ij}^{k}] \quad \forall\ ijk$$

where:

E_{ijk} := is the net increase or decrease in the average egress time per person caused by re-routing occupants to the (k^{th+1}) *NI* route.

q_{ijk} := the sum of the average queue times per person on the original route.

d^k_{ij}:= the increased distance travelled on the (k^{th+1}) *NI* route (e.g. if the kth *NI* route is 100 feet and the k^{th+1} *NI* route is 120 feet, d^k_{ij} is equal to 20 feet i.e. 120 minus 100).

ω := is the average travel speed for d^k_{ij}.

q^k_{ij}:= the sum of the expected queue times per person on the (k^{th+1}) *NI* route.

One important point in what follows is that we assume that we have an accurate estimate of q_{ijk}. For evacuation and networks which are Markovian (Poisson arrivals and Exponential service), this estimation process is exact. For more general networks however, a stochastic approximation technique is required which gives an estimate of the impact of re-routing occupants from their current evacuation path to another *NI* one. Since the alternative evacuation paths are acyclic directed graphs of infinite queues, this estimation problem is difficult, yet not intractable.

Let's present a relaxed version of our original mathematical programming problem which will form the foundation of what is to follow.

$$Minimize\Big\{f_1(\bar{x}); f_2(\bar{x})\}\Big]$$

where:

$$(EvacuationTime): \; f_1(\bar{x}) \;=\; \sum_i\sum_j\sum_k q_{ijk}\lambda_{ijk}x_{ijk} \tag{7}$$

$$(DistanceTravelled): \; f_2(\bar{x}) \;=\; \sum_i\sum_j\sum_k d_{ijk}\lambda_{ijk}x_{ijk} \tag{8}$$

subject to:

$$(Occupant\ Classes): \sum_k x_{ijk} \;=\; 1 \quad \forall ij \tag{9}$$

$$(Routes): x_{ijk} \;=\; 0,1 \quad \forall ijk \tag{10}$$

The relaxed model removes constraint equations (1&2) and concentrates on evacuating the occupant populations along selected routes where *time* is minimized. The relaxation of constraint equations (1&2) is justified in certain evacuation situations where the capacity restrictions are not severe or critical. To find an initial *NI* solution to our relaxed model, we ignore $f_1(\bar{x})$ and solve our network for the first *NI* path for each occupant population using a multi-objective shortest path algorithm such as the one by Climaco and Martins [4]. Thus:

Theorem 1: *The collection of* 1*st shortest paths for each occupant population represents a NI solution for the entire evacuation network.*

Proof: This is a standard result in multi-objective optimization [3], *viz.* selecting one objective, $f_i(\bar{x})$, solving a weighting problem with $w_i = 1$ and $w_j = 0 \forall j \neq i$. Thus, simultaneously minimizing the distance travelled for each occupant population results in the minimum distance travelled for all occupant classes for $f_2(\bar{x})$. ∎

At the same time we generate the first shortest paths, we compute the evacuation (sojourn time) for the occupant population. As we shall see, we approach the solution of our original mathematical model in §**C** oscillating between the set of *NI* paths and the tradeoffs gained in reducing the evacuation time by sending occupant populations along longer paths.

The 1st shortest path solution for each occupant population may result in a unique optimal solution across all populations for both objectives $f_1(\bar{x}), f_2(\bar{x})$, if queueing delays along the routes are not significant. However, we need to ensure this by quantitatively using our notion of *expected savings* to verify this.

Theorem 2: *If the NI solutions at the first stage of the routing process where each occupant source is routed along the* 1*st shortest path an expected savings calculation:*

$$E_{ij}^k = q_{ijk} - [(d_{ij}^k/\omega) + q_{ij}^k] > 0 \quad \textit{for some } ijk$$

then, the 1*st shortest paths are not a unique optimal solution and further rerouting will improve the evacuation time* $f_2(\bar{x})$.

Proof: The proof of this property rests on constructing a Linear/Integer programming problem and its dual which indicates how one can move from one *NI* solution to another on the efficient frontier, if necessary. The Linear/Integer programming problem represents another relaxation of the original model in §**3** where we are only focusing on the expected savings possible by re-routing occupants along longer *NI* paths. This Linear/Integer programming problem can be considered as a way of scalarizing the two objectives into a single objective problem in the *expected savings* of the alternative NI routes.

Let's establish the following *expected savings* problem:

$$Maximize\ Z = \sum_i \sum_j \sum_k E_{ij}^k x_{ijk}$$

$$\text{(Occupant Classes)} \sum_k x_{ijk} \leq 1 \quad \forall ij \tag{11}$$

$$x_{ijk} \leq 0,1 \quad \forall ijk \text{ routes} \tag{12}$$

where

$$E_{ij}^k = [q_{ijk} - [(d_{ij}^k/\omega) + q_{ij}^k] \quad \forall ijk$$

we would like to reroute the occupants perhaps on a longer *NI* path, if the expected savings in evacuation time would be maximized. The unimodularity of the above

Linear/Integer program follows from the 0, 1 properties of the sparse matrix [8]. The unimodularity property allows us to solve for integer solutions using ordinary Linear Programming algorithms.

If we take the dual of the above problem, we gain some insight into whether alternative *NI* paths may result in some savings in $f_1(\bar{x})$. The dual is:

$$Minimize \sum_i \sum_j \pi_{ij}$$

$$\pi_{ij} \geq [q_{ijk} - [(d_{ij}^k/\omega) + q_{ij}^k] \ \forall ijk \tag{13}$$

$$\pi_{ij} \geq 0 \ \forall ij \tag{14}$$

Finally, from the Dual, we obtain the following Complimentary Slackness conditions:

$$\{\pi_{ij} - [q_{ijk} - [(d_{ij}^k/\omega) + q_{ij}^k]\} x_{ijk} = 0 \quad \forall ijk$$

At our current *NI* solution, $x_{ijk} = 1$ for some *NI* alternative for each ij occupant population. By the above complimentary slackness condition, if $x_{ijk} = 1$ then $\Rightarrow$ $\{\pi_{ij} - [q_{ijk} - [(d_{ij}^k/\omega) + q_{ij}^k]\} = 0$.

Therefore, for the other *NI* route alternatives of each occupant population source, the dual feasibility condition remains unsatisfied, i.e.

$$\{\pi_{ij} - [q_{ijk} - [(d_{ij}^k/\omega) + q_{ij}^k]\} < 0$$

if and only if the *Expected Savings* for a route alternative is **positive**, viz. $[q_{ijk} - [(d_{ij}^k/\omega) + q_{ij}^k] > 0$. Thus, the current *NI* solution can be improved if the *Expected Savings* for a *NI* routing alternative is **positive** and a new x_{ijk} will enter the basis.∎

Since in the construction of the Linear/Integer programming problem, more than one occupant population might be re-routed, the *change of basis rule* defined below determines the occupant population that should be chosen. More than one change of basis is not recommended because of the nonlinear interaction effects which might be caused by the multiple re-routings.

Change-of Basis Rule: *At each stage of our re-routing process, we should move to the NI path solution for which:*

$$E^* = \max_{\forall ij \text{ sources}} \{E_{11}, E_{22}, \ldots, E_{IJ}\}$$

Essentially, the rule is a greedy one and has been shown to be effective in practice. It will generate a subset of the *NI* solution space and will not guarantee complete enumeration of the *NI* set. To terminate the re-routing process, we have:

Corollary 1: *If $E^* \leq 0 \forall\ ij$ occupant sources then no more re-routing iterations will result in an improvement in $f_1(\bar{x})$.*

Proof: Again, this result is derived from the Linear/Integer programming problem using the Dual Feasibility and Complimentary Slackness conditions. If the only *NI* route which provides the maximum savings is a current route, for each occupant population source, then no additional *Expected Savings* $E_{ij} > 0 \forall\ ij$ are possible and the process is terminated.

5 Algorithm

The problem we face in our evacuation planning problem is that we do not know *a priori* which paths are NI without assessing the congestion in $G(V, E)$. We must iteratively generate candidate paths, assess the congestion in $G(V, E)$, and then iterate again until the desired tradeoffs between distance travelled and evacuation time is acceptable to the planner. This iterative process leads to the algorithm described below. For product form networks where the estimate of time delays in the *Expected Savings* calculation for re-routing among the alternative *NI* paths can be computed exactly, then the algorithm will guarantee finding a *NI* path for re-routing the occupant classes. For non-product form networks, which are typically the case, we can only approximate these time delays, therefore, the algorithm can only guarantee an approximate *NI* solution. Considering the complexity of the underlying stochastic-integer programming problem, this is a reasonable and practical strategy.

Before presenting the algorithm formally, let us discuss our overall approach to the evacuation planning problem [15]. Our approach for modelling evacuation planning problems has three separate but interrelated steps:

Step 1.0 concerns the *Representation* of the region or facility as a queueing network, while **Step 2.0** concerns the *Analysis* of the queueing network to estimate the critical performance measures of the evacuation. A queueing network model is utilized in order to capture the potential congestion in the network where large populations come together during the evacuation process. Finally, **Step 3.0** concerns the *synthesis* or multi-objective generation of the routing paths for the occupants, where trade offs among the different performance measures estimated in **Step 2.0** of the algorithmic process may be necessary.

The algorithm to facilitate the design methodology can be incorporated into any simulation e.g.**Q-GERT** or analytical model e.g.**QNET-C** to estimate f_1, f_2, and carry out the evacuation planning/routing analysis. To summarize and focus the efforts in this paper, an algorithmic description of **Steps 1.0, 2.0, & 3.0** and it substeps are presented.

Step 1.0: Representation Represent the underlying facility or region as a network $G(V, E)$ where $V :=$ is a finite set of nodes and $E :=$ is a finite set of arcs or nodal pairs.

Step 2.0: Analysis Analyze $G(V, E)$ as a queueing network either with a transient or steady-state model and compute the total evacuation time of the occupant population along with total distance travelled to evacuate given a set of evacuation paths.

Step 3.0 Synthesis Algorithm

Step 3.1: Analyze the queueing output from the evacuation model and compute the set of *NI* evacuation paths which simultaneously minimize time and distance travelled in $G(V, E)$ for each occupant population.

3.1.1 If the set on NI paths are uniquely optimal then

$$E_{ij}^k = q_{ijk} - [(d_{ij}^k/\omega) + q_{ij}^k] \leq 0 \quad \forall ijk$$

go to Step 3.2 otherwise:

3.1.2 Significant queueing (congestion) exists on one or more routes then go to Step 3.3.

Step 3.2: STOP! The NI shortest time/distance routes are optimal and identical and total evacuation time, distance and congestion are minimized.

Step 3.3: Determine the total number of occupants who pass through the queueing area(s) and trace them back to their origins.

Step 3.4: Select the total number of occupants to be re-routed from each source node. The total number of occupants re-routed is correlated to both the size of the queues and the number of occupants on each route. In selecting the population, the analyst should strive to achieve uniformity of occupants and queues on each egress route.

Step 3.5: Re-route the population to the kth route of the NI set of paths where k is selected by employing the following formula:

$$E_{ij}^k = q_{ijk} - [(d_{ij}^k/\omega) + q_{ij}^k] \quad \forall ijk$$

Step 3.6: Select the largest positive E^* for each set of populations to be re-routed, where:

$$E^* = \max_{\forall ij \text{ sources}} \{E_{11}, E_{22}, \ldots, E_{IJ}\}$$

for all possible savings, and then re-run the computer evacuation planning model with the new set of routes, by returning to **Step 2.0** of the *General Algorithm.* If all E_t's are negative, stop! The current set of NI shortest routes used on the previous iteration are selected.

The overall time and space complexity of the algorithm is *exponential* largely because of the time complexity of **Step 2.0.** and the fact that the number of *NI* solution paths for each occupant class may be exponential in the size of the network. Whether one has a product form network or not, the time complexity of **Step 2.0** will be a key bottleneck in the efficiency of the algorithmic re-routing process. Even if an heuristic is utilized in **Step 2.0**, the time and space complexity of the algorithm would then be governed by the exponential number of alternative *NI* paths for each occupant class.

6 Example

In the following section of the paper, an example of the evacuation of three distinct occupant populations is utilized to demonstrate the scope and effectiveness of the previous design methodology to route and re-route the occupants based on the mathematical formulation and subsequent algorithm. The example is taken from [11]. **Q-GERT** is utilized here to estimate the evacuation times and congestion on the NI set of evacuation routes since it is a transient model and represents the most general type of stochastic estimation tool in which the multi-objective routing methodology might be used. The arrivals to the network are from a log-normal distribution and log-normal distributions were used for the arc and nodal service time distributions. More discussion of the parameters are included in [11]. At the end of the example discussion, the use of **QNET-C**, a steady-state analytical queueing network model, is described to estimate the evacuation times.

The example has three occupant groups located at source nodes $\#1, \#2$, and $\#3$ with occupant populations of $40, 10$ and 30 persons respectively. Figure 1 illustrates the sample $G(V, E)$.

As you can see in Table #1, the paths marked with a single * represent the set of NI solutions for each of the occupant populations. The populations are denoted by the large **P** in each of the following tables and the V_3 denotation represents the sink nodes of $G(V, E)$. The other paths were generated by the algorithm, but since we are viewing the evacuation $G(V, E)$ without congestion, the shortest distance paths also correspond to the shortest time paths. The above paths are chosen and the algorithm returns to **Step 2.0** of the *General Algorithm* to assess the congestion in $G(V, E)$. If no significant congestion exists on these routes, the algorithm terminates.

Unfortunately, in our pass through **Step 2.0** significant congestion at nodes #5, #7, and #8 were encountered in the amount of $50.97, 2.217$ and 2.86 time units(t.u.'s.) and a total evacuation time was 158.50 t.u.'s. In iterating again through **Step 3.0** with the above queueing delay estimates along the paths, a different set of NI paths were generated due to congestion, see Table #2. Inspecting Table #2, the paths denoted by the single * are chosen from the set of NI paths, while those denoted by four **** are dominated or else were considered on the previous iteration. We are thus beginning to trade off *distance travelled* in order to seek reductions in *evacuation* *time* for all occupants. The results of the next iteration are displayed in Table #3.

For this iteration #2, the total evacuation time slightly decreased from 158.50 $\rightarrow$ 152.09 t.u.'s. While the delay decreased significantly at node #5 from 50.97 $\rightarrow$ 25.48, the delay at node #8 increased from 2.86 $\rightarrow$ 18.70 which thus resulted in the marginal decrease in total evacuation time. With these new queueing delays, the NI set was again generated and the new NI paths are depicted in Table #3. The new set of paths were chosen and once again the algorithm cycled through **Step 2.0**, but this iteration resulted in a dramatic decrease in evacuation time from 152.09 $\rightarrow$ 114.23 t.u.'s.

It is interesting to note that the program **QNET-C** was run with the same set of paths and the results where similar to the **GERT** runs in that the evacuation times where *165.78, 159.87, and 111.02* for the same iterations. The discrepancy in evacuation times is due to the fact that **QNET-C** utilizes exponential service time distributions for the service times along the evacuation paths so there is a tendency to be pessimistic, while **QGERT** makes no special distribution assumptions.

6.1 Iteration #0

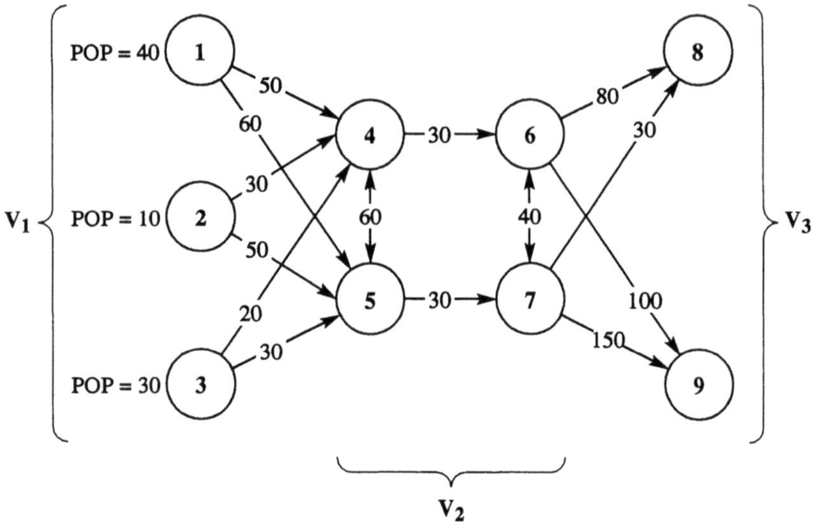

Figure 3: **Initial Evacuation Graph** $G(V, E)$

P=1	**Time**	**Distance**	**Route**
$V_3 = 8$	**35**	**120**	$1 \rightarrow 5 \rightarrow 7 \rightarrow 8$
$V_3 = 9$	**35**	**180**	$1 \rightarrow 4 \rightarrow 6 \rightarrow 9$
P=2	**Time**	**Distance**	**Route**
$V_3 = 8$	**24**	**80**	$2 \rightarrow 5 \rightarrow 7 \rightarrow 8$
$V_3 = 9$	**35**	**160**	$2 \rightarrow 4 \rightarrow 6 \rightarrow 9$
P=3	**Time**	**Distance**	**Route**
$V_3 = 8$	**27**	**90**	$3 \rightarrow 5 \rightarrow 7 \rightarrow 8$
$V_3 = 9$	**50**	**180**	$3 \rightarrow 4 \rightarrow 6 \rightarrow 9$

Table 1: NI set of Evacuation Paths

6.2 Iteration #1: Evacuation Time = 158.50 t.u.

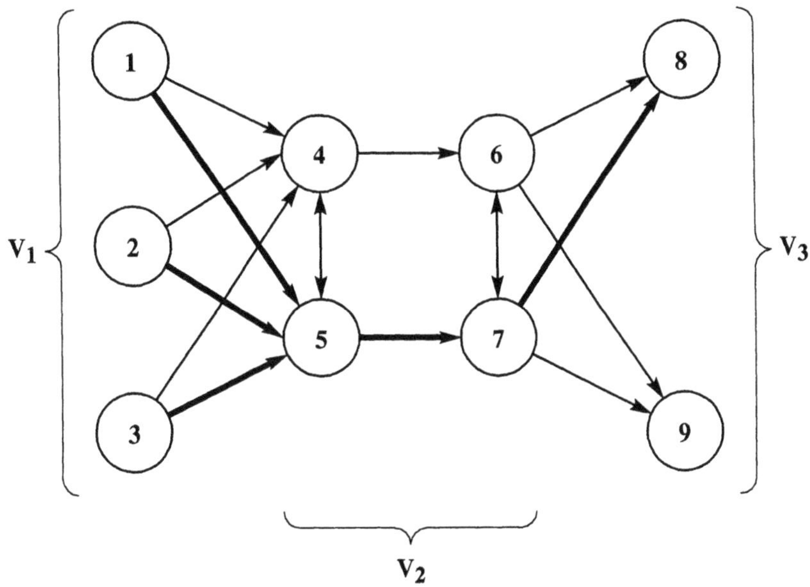

Figure 4: **Evacuation Graph** $G(V, E)$ **w/ 1***st* **paths**

P=1	**Time**	**Distance**	**Route**	**Savings**
$V_3 = 8$	**91**	**120**	$1 \to 5 \to 7 \to 8$	**0**
$V_3 = 8$	**45**	**160**	$1 \to 4 \to 6 \to 8$	**40.61***
$V_3 = 9$	**50**	**180**	$1 \to 4 \to 6 \to 9$	********
P=2	**Time**	**Distance**	**Route**	
$V_3 = 8$	**80**	**80**	$2 \to 5 \to 7 \to 8$	**0***
$V_3 = 8$	**40**	**140**	$2 \to 4 \to 6 \to 8$	**37.65**
$V_3 = 9$	**45**	**120**	$2 \to 4 \to 6 \to 9$	********
P=3	**Time**	**Distance**	**Route**	
$V_3 = 8$	**83**	**90**	$3 \to 5 \to 7 \to 8$	**0***
$V_3 = 8$	**45**	**160**	$3 \to 4 \to 6 \to 8$	**35.05**
$V_3 = 9$	**50**	**180**	$3 \to 4 \to 6 \to 9$	********

Table 2: NI set of Evacuation Paths

6.3 Iteration #2 Evacuation Time = 152.09 t.u.'s

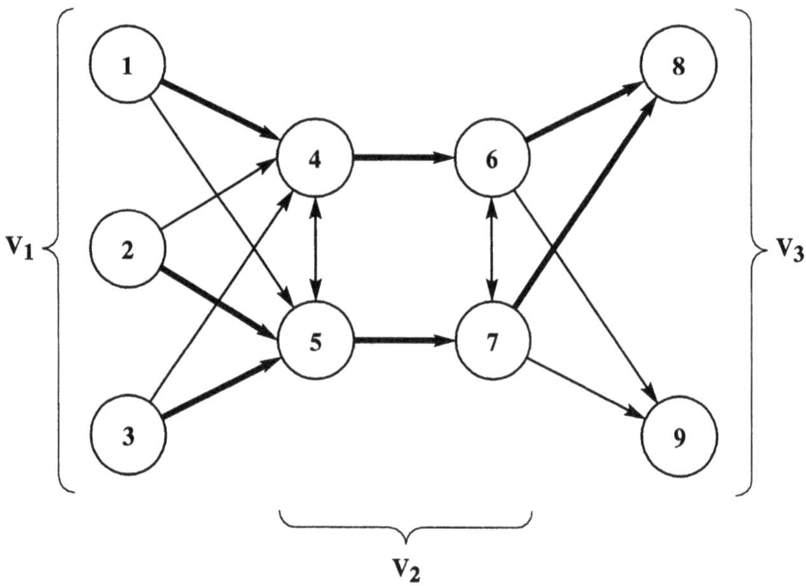

Figure 5: **Evacuation Graph** $G(V, E)$**w/** $2nd$ **set of paths**

P=1	**Time**	**Distance**	**Route**	**Savings**
$V_3 = 8$	**107**	**120**	$1 \to 5 \to 7 \to 8$	********
$V_3 = 8$	**88**	**150**	$1 \to 4 \to 6 \to 7 \to 8$	**0.38**
$V_3 = 9$	**76**	**180**	$1 \to 4 \to 6 \to 9$	**13.50***
P=2	**Time**	**Distance**	**Route**	**Savings**
$V_3 = 8$	**96**	**80**	$2 \to 5 \to 7 \to 8$	**0***
$V_3 = 8$	**61**	**130**	$2 \to 4 \to 6 \to 7 \to 8$	**-12.74**
$V_3 = 9$	**49**	**160**	$2 \to 4 \to 6 \to 9$	**0.39**
P=3	**Time**	**Distance**	**Route**	**Savings**
$V_3 = 8$	**99**	**90**	$3 \to 5 \to 7 \to 8$	**0***
$V_3 = 8$	**66**	**150**	$3 \to 4 \to 6 \to 7 \to 8$	**-15.33**
$V_3 = 9$	**54**	**180**	$3 \to 4 \to 6 \to 9$	**-2.21**

Table 3: NI set of Evacuation Paths

6.4 Iteration #3: Evacuation Time 114.23 t.u.'s

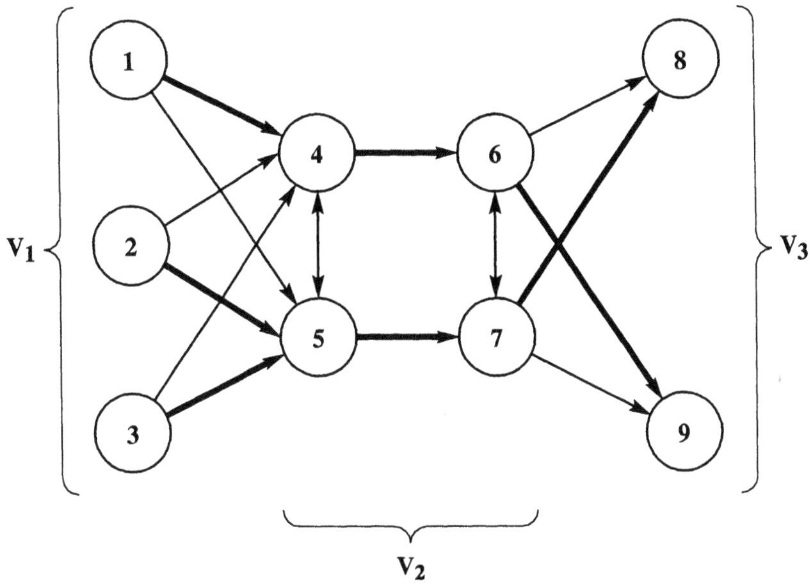

Figure 6: **Evacuation Graph** $G(V, E)$ **3*rd* set of paths**

P=1	Time	Distance	Route	Savings
$V_3 = 8$	107	120	$1 \to 5 \to 7 \to 8$	****
$V_3 = 8$	88	150	$1 \to 4 \to 6 \to 7 \to 8$	****
$V_3 = 9$	78	180	$1 \to 4 \to 6 \to 9$	0*
P=2	**Time**	**Distance**	**Route**	**Savings**
$V_3 = 8$	96	80	$2 \to 5 \to 7 \to 8$	0*
$V_3 = 8$	61	130	$2 \to 4 \to 6 \to 7 \to 8$	-12.72
$V_3 = 9$	51	160	$2 \to 4 \to 6 \to 9$	-15.38
P=3	**Time**	**Distance**	**Route**	**Savings**
$V_3 = 8$	99	90	$3 \to 5 \to 7 \to 8$	0*
$V_3 = 8$	66	150	$3 \to 4 \to 6 \to 7 \to 8$	-15.33
$V_3 = 9$	56	180	$3 \to 4 \to 6 \to 9$	-19.74

Table 4: NI set of Evacuation Paths

Table #4 illustrates the final set of NI paths generated by the algorithm. In cycling through **Step 3.0**, no additional savings are possible by re-routing occupants on longer paths, therefore, the last set of paths yield the most favorable decrease in evacuation time for the occupant population.

7 Summary and Conclusions

In this paper, we have focused on the generation of the *Noninferior NI* set of paths for evacuating occupants in an emergency situation. The complex, multi-objective nature of the problem has been described and mathematical properties and a corresponding algorithm have been developed for generating the set of *NI* evacuation paths which allow for tradeoffs between *evacuation time* and *distance travelled* for the occupant populations. A small example problem also illustrates the iterative process of the algorithm which can be implemented either in a simulation environment *e.g.* **Q-GERT**, or in an analytical model environment *e.g.* **QNET-C**.

Acknowledgement

This material is based upon work supported by the National Science Foundation under grants *#MSM-8715152* and *#MSS-9116666*.

References

[1] J. Ahlberg, Stochastic Queueing Network Program for Evacuation Planning, *Master's Project*. Department of Industrial Engineering and Operations Research, University of Massachusetts, Amherst, MA 01003 (1988).

[2] L.G. Chalmet, R.L. Francis, and P.B. Saunders. Network Models for Building Evacuation. *Management Science* **28**, (1) (1982) 86–105.

[3] V. Chankong and Y.Y. Haimes, *Multiobjective Decision Making: Theory and Methodology* (North-Holland, 1983).

[4] J.C.N. Climaco and E.Q.V. Martins, A Bi-Criterion Shortest Path Algorithm, *European Journal of Operations Research*,**11**(1982) 399-404.

[5] S. Daskalaki and J. MacGregor Smith. The Static Routing Problem in Open Finite Queueing Networks. Presented @ *ORSA/TIMS Meeting, Miami, Florida,* (October 1986.)

[6] S. Daskalaki and J. MacGregor Smith, Optimal Routing and Buffer Space Allocation in Series-Parallel Queueing Networks. Invited presentation to the *EURO/TIMS XXVIII Conference in Paris France,* (July 6–8, 1988).

[7] S. Daskalaki and J. MacGregor Smith, Real Time Routing in Finite Queueing Networks. in *Queueing Networks with Blocking,* eds. H.G. Perros and T. Altiok. (New York: Elsevier Science Publishers B.V., 1989) 313-324.

[8] R Garfinkel and G.L. Nemhauser, *Integer Programming.* (Wiley, 1972).

[9] Hemant Gosavi and J. MacGregor Smith, Heavy Traffic Multi-Commodity Routing in Open Finite Queueing Networks, Paper Presented at the ORSA/TIMS Meeting, Denver, CO. (October 1988.)

[10] R. L. Francis and L.G. Chalmet. Network Models for Building Evacuation: A Prototype Primer. *Unpublished Paper,* Department of Industrial and Systems Engineering, University of Florida, Gainesville, Florida (1980).

[11] C.J. Karbowicz and J. MacGregor Smith, A K-Shortest path Routing Heuristic for Stochastic Evacuation Networks. *Engineering Optimization* **7** (1984) 253–280.

[12] A.Pritsker, *Modelling and Analysis using* **Q-GERT** *Networks.* (Wiley, New York, 1979).

[13] Harvey M. Salkin, *Integer Programming* (Addison-Wesley Publishing Co., Reading Massachusetts, 1975).

[14] J. MacGregor Smith and D. Towsley. The Use of Queueing Networks in the Evaluation of Egress from Buildings. *Environment and Planning* **B-8** (1981) 125–139.

[15] J.MacGregor Smith, **QNET-C**: *An Interactive Graphics Computer Program for Evacuation Planning Proceedings of the Conference on Emergency Planning, SCS Multiconference, 14-16* (January 1987), pp 19–24

Network Optimization Problems, pp. 283-300
Eds. D.-Z. Du and P.M. Pardalos

A Simplex Method for Network Programs with Convex Separable Piecewise Linear Costs and Its Application to Stochastic Transshipment Problems[1]

J. Sun
Department of Industrial Engineering and Management Sciences, Northwestern University, Evanston, IL 60208, USA

K.-H. Tsai
Department of Computer Sciences, National Normal University, Taipei, Taiwan

L. Qi
School of Mathematics, The University of New South Wales, Kensington, NSW2033, Australia

Abstract

This paper is concerned with the pure network program whose objective function is convex, separable, and piecewise linear. We describe a direct simplex algorithm and its implementation for solving such problems. Computational results of applying this algorithm to stochastic transshipment problems are reported. The algorithm keeps the number of variables in the original level by allowing nonbasic variables to take breakpoint values and by using a straightforward pricing strategy similar to the traditional network simplex method. Tree data structure is used to construct efficient implementation. Computational results indicate that the solution time is insensitive to the increase of "piece-number" of the objective function. As a result, we have been able to solve stochastic transshipment problems of more than 30,000 arcs and 3000 nodes, where each node has a discrete random demand of 100 possible values, within one minute on a SUN computer.

[1]The research is supported in part by NSF and Australian Research Counsel.

1 Introduction

This paper is aimed at the following optimization problem

$$\text{(NetPLP)} \begin{cases} \text{minimize} & F(x) = \sum_{j=1}^{n} f_j(x_j) \\ \text{subject to} & Ax = b, \end{cases}$$

where $A \in R^{m\times n}$ is the node-arc incidence matrix of a connected network, $x = (x_1, \cdots, x_n)^T$, $c = (c_1, \cdots, c_n)^T \in R^n$, and $b \in R^m$. Each f_j is a convex piecewise linear function of the single variable x_j and has the following form:

$$f_j(x_j) = \begin{cases} s_{j1}x_j + r_{j1}, & \text{if } c_{j0} \le x_j \le c_{j1}; \\ \vdots & \vdots \\ s_{jk_j}x_j + r_{jk_j}, & \text{if } c_{jk_j-1} \le x_j \le c_{jk_j}; \\ +\infty & \text{otherwise,} \end{cases}$$

where for each j, we call the numbers $c_{j0}, c_{j1}, \ldots, c_{jk_j}$ the *breakpoints* of the variable x_j; they satisfy

$$-\infty \le c_{j0} \le c_{j1} \le \ldots \le c_{jk_j} \le +\infty.$$

Notice that the bound constraints $c_{j0} \le x_j \le c_{jk_j}$, for $j = 1, \cdots, n$, are imposed from the definition of f_j. If all k_j equal to one, this model reduces to the ordinary network linear program.

The problem (NetPLP) arises in at least two areas of operations research. First, it is used to model practical network problems with linear penalties or variable linear costs. Later in this paper, we will show how stochastic transshipment problems with random demands can be formulated as an (NetPLP). Second, it may represent an approximation of a network separable convex program, especially when the exact formula of the cost function is unknown and only experimental data are available.

The third possible application of (NetPLP) is to offer a warm start for nonlinear network algorithms. We admit that a nonlinear network program should be in general approached by nonlinear programming algorithms, but it might be as well important to find a good staring point for those algorithms to be truly efficient. A good algorithm for (NetPLP) could conveniently produce such a warm start by solving a piecewise linearization of the nonlinear problem. In addition, the algorithms for (NetPLP) could also be useful in directly solving some nonlinear problems, as demonstrated by Charnes, Song and Ali [CSA86].

Theoretically speaking, (NetPLP) can be solved by reformulating it as a linear network problem where k_j new arcs are introduced to replace the arc j. However, in practice, we would rather seek algorithms that directly deal with the piecewise linear objective function because reformulation would often increase the number of variables to a prohibited level. Some authors have contributed specific pricing rules for applying the simplex method to certain cases of (NetPLP), for example, see the algorithm of

Ali, Cook and Kress [ACK86] for ordinal ranking problems and the algorithm of Wets [We83] for stochastic programs with simple recourse. The contribution of this paper, however, is to introduce a unified simplex approach for solving (NetPLP) and to elaborate its implementation based on the tree data structure (e.g. Chvatal[C83]). Other than thinking of implicitly dealing with the reformulated linear problem, we develop a direct pricing rule that can be implemented efficiently using the tree data structure.

Although there have been considerable developments in interior point methods for linear programming in recent years, the simplex method remains being very competitive in solving network linear programs. It is our belief that the simplex method would be still efficient in solving the piecewise linear problem. To support our contention, we first test 40 randomly generated (NetPLP) problems similar to that of Klingman et. al [KNS74]. Then we choose four of these problems and observe how the solution time depends on the piece number k_j. We find the increase of solution time is insignificant compared to the increase of $k_1 + \cdots + k_n$.

An important application of our algorithm is to solve stochastic transshipment problems (STP) with discrete random demand. This model include the classical stochastic transportation problem as a special case. Due to the underlying network structure, the stochastic transshipment problem can be efficiently solved by the network piecewise linear programming approach. In our computational test the method has been able to solve a problem of 35,000 arcs and 3000 nodes, with each node having a random demand of 100 possible values, within one minute on a SUN computer.

In the literature, some researchers have proposed direct methods to solve general convex piecewise linear programs (Rockafellar [R84], Fourer [F88], Premoli [P87]). However, to the best of the authors' knowledge, none of those methods has been implemented under the network circumstance. For an extensive literature list of applications of piecewise linear programs, see Fourer [F86].

This paper is organized as follows. In the next section we review the mathematical background. The method and its convergence property are given in Section 3. Section 4 is devoted to implementation techniques. Computational experiment on 40 randomly generated benchmark (NetPLP) is reported in Section 5. The STP is introduced and the corresponding test problems are solved in Section 6. Since only discrete distributions of demand in STP can result in network piecewise linear programs, we discuss how to use the algorithm in STP with continuous distribution and other extensions in Section 7.

2 Background Materials

Nothing much is new in this section. For more detailed introduction of network flow problems and monotropic optimization, we refer the reader to [R84]. Here, we just state some important facts that are to be used in the sequel.

Let $A \in R^{m\times n}$ be the node-arc incidence matrix of a connected network and let $x = (x_1, \cdots, x_n)^T$ be a flow vector. Let $[B, N]$ be a basis-nonbasis partition of the arc index set $J = \{1, \cdots, n\}$. We then have the corresponding partition of the vector $x = (x_B, x_N)$. A variable x_j in the vector x_B is called a basic variable and in the vector x_N a nonbasic variable. The arcs in B form a maximal spanning tree of the network. For simplicity of notation, we will use B to represent both the index set and the tree. It is well-known (e.g. Rockafellar [R84]) that, for each partition $J = B \cup N$, there are a vector $\bar{b}$ and a matrix K such that $x_B = Kx_N + \bar{b}$, is an equivalent system to the system $Ax = b$. Moreover, the matrix K has a combinatorial structure, namely, for each $j \in N$, the j-th column of K is the incidence vector of the unique path in tree B that starts from the head node of arc j and terminates at the tail node of j. Thus the increase (decrease) of the nonbasic variable x_j in the simplex method corresponds to a equal increase (decrease) of the fluxes along the circuit C_j that consists of this path and arc j with j being positive in C_j. We call the incidence vector of the circuit C_j a simplex direction. Let $d^j = (d_1^j, \cdots, d_n^j)$ be this vector. Then we have

$$d_l^j = \begin{cases} 1 & \text{if arc } j \text{ is positive in } C_j, \\ -1 & \text{if arc } j \text{ is negative in } C_j, \\ 0 & \text{if arc } j \text{ is not in } C_j. \end{cases}$$

We now give a new interpretation of the reduced cost in linear programming, which will be instructive later in describing our pricing rule.

Observation. In a network linear program (correspondingly, all $k_j = 1$ in (NetPLP)), if x is a nondegenerate solution, or a degenerate solution with a non-zero ratio, the reduced cost associated with x_j in the simplex method is the directional derivative of F at x along d^j. i. e.

$$F'(x, d^j) = \lim_{t\downarrow 0} \frac{F(x + td^j) - F(x)}{t}.$$

In the case of degenerated solution with zero ratio, $F'(x, d^j) = \infty$.

As a matter of fact, from the theory of monotropic programming the above directional derivative can be computed by

$$F'(x, d^j) = \sum_{l=1}^{n} \max\{d_l^j f_l^-(x_l), d_l^j f_l^+(x_l)\}, \tag{2.1}$$

where f_l^- and f_l^+ are the ordinary left and right derivatives of f_l (could be $\pm\infty$). In the case that all $k_j = 1$,, one has

$$f_l(x_l) = \begin{cases} s_l x_l & \text{if } c_{l0} \le x_l \le c_{l1}, \\ \infty & \text{otherwise.} \end{cases}$$

Therefore, if x is a nondegenerate basic feasible solution in the sense of usual network linear programming, one has $c_{l0} < x_l < c_{l1}$ $\forall\, l \in B$, which implies $f_l^-(x_l) = f_l^+(x_l) =$

$s_l, \forall\, l \in B$. therefore we get

$$F'(x, d^j) = \sum_{l \in C_j^+} s_l - \sum_{l \in C_j^-} s_l = s_j - (p_{jh} - p_{jt}), \tag{2.2}$$

where C_j^+ and C_j^- are the positive and negative arc sets in C_j, respectively, p_{jh} and p_{jt} are the prices at the head and the tail of arc j. The price vector $p = (p_1, \cdots, p_m)^T$ assigns to each node a price such that $p_{lh} - p_{lt} = s_l$ for every $l \in B$.

(2.2) is exactly the formula used to compute the reduced cost in linear network programs. It is also easy to see in the case of degenerate solution with non-zero ratio, all basic arcs with $x_l = c_{l0}$ are in C_j^+, while those with $x_l = c_{lk_l}$ are in C_j^-, so formula (2.2) is still valid. Finally, in the case of degenerate solution with zero ratio, at least one basic arc l with $x_{l0} = c_{l0}$ is in C_j^-, or at least one basic arc l with $x_{l0} = c_{lk_l}$ is in C_j^+, hence the l-th term in (2.1) is $+\infty$, resulting in $F'(x, d^j) = \infty$.

The simplex method looks for a simplex direction d^j, $j \in N$ such that $F'(x, d^j) < 0$. In the piecewise linear case, the simplex direction is defined as either d^j or $-d^j$, and we simply call the derivative (2.1) the *reduced cost* and the right hand side of (2.2) the *nominal cost*. It is the reduced cost, not the nominal cost, determines if d^j is a feasible descend direction. However, since nominal cost can be obtained by relatively small efforts, we use it as an indicator for possible descend simplex directions. In our implementation, we first find a candidate of a descent simplex direction by checking if its nominal cost is negative, then as we go along the circuit C_j to compute the maximal allowable change of flux (ratio test), we successively add the differentials between the reduced cost and the nominal cost to the later one to get the real directional derivative (2.1).

Definition 2.1 A basic solution $x = [x_B, x_N]$ to (NetPLP) is defined as a solution to $Ax = b$, where each nonbasic variable takes one of its breakpoint values. If, in addition, $F(x)$ is finite, then x is called a basic feasible solution.

Using the "tree language", each basic feasible solution of (NetPLP) can be identified as a maximal spanning tree so that $F(x)$ is finite and all arcs not in the tree take breakpoint values.

The following proposition can be derived from a similar property of linear programs.

Proposition 2.2 If (NetPLP) has an optimal solution, then there exists a basic feasible solution which is also optimal.

Suppose that x is a basic feasible solution to problem (NetPLP). One of the fundamental properties of (NetPLP) is the following (see [R84]).

Proposition 2.3 The following statements are equivalent:

(a) x is optimal to (NetPLP).

(b) x is feasible and for all possible partitions $J = B \cup N$, none of the reduced cost of the simplex directions is strictly negative.

(c) x is feasible and there exist a price vector $p \in R^m$ and a differential vector $v = -A^T p$, such that $f_l^-(x_l) \leq v_l \leq f_l^+(x_l)$, for $l = 1, \cdots, n$.

Notice that by the special structure of the node-arc incidence matrix A, one has $v_l = p_{lh} - p_{lt}$. Now we are ready to state the algorithm.

3 The Simplex Algorithm for (NetPLP) and Its Convergence

Algorithm 3.1

Step 0. Find an initial basic feasible solution of (NetPLP) or show that the problem is infeasible. If a basic feasible solution is found, then set this solution as x^0 and $k = 0$. Go to Step 1.

Step 1. Determine by Proposition 2.3(c) whether the current solution is optimal. If not, find a simplex direction associated with a nonbasic variable j such that $F'(x^k, d^j) < 0$. Go to Step 2. By Proposition 2.3(b), such a direction exists.

Step 2. Set $x^{k+1} = x^k + \bar{\alpha} d^j$ where $\bar{\alpha}$ minimizes $F(x^k + \alpha d^j)$ over $\alpha \geq 0$. If such an α does not exist, i.e. $\inf_{\alpha \geq 0} F(x^k + \alpha d^j) = -\infty$, stop; the (NetPLP) has unbounded solution. Otherwise, set $k = k + 1$ and go to Step 1.

It will be shown (Proposition 4.3) that the minimizer $\bar{\alpha}$ in Step 2 can be selected so that x^{k+1} is again a basic feasible solution. If we maintain this machinery in Algorithm 3.1, then it is obvious that in a finite number of iterations, the algorithm will either find an unbounded solution or an optimal basic feasible solution because the total number of basic feasible solution is finite and the strictly decreasing sequence $\{F(x^k)\}$ ensures that no same basic feasible solution can repeat in the sequence $\{x^k\}$.

4 Implementation of the Algorithm

4.1 Data Structure

We use the "tree data structure" that has been long used in many successful implementations of network simplex methods. For a description of it, see [C83]. The arrays used are as follows.

FROM and **TO** arrays. These two arc-length arrays store the initial and ending nodes of arcs.

BREAKPOINT and **SLOPE** arrays. These arrays store the value of breakpoints and slopes for arcs in their natural order. Namely, in the orders of

$$c_{10}, \cdots, c_{1k_1}, \cdots, c_{n0}, \cdots, c_{nk_n} \quad \text{and} \quad s_{11}, \cdots, s_{1k_1}, \cdots, s_{n1}, \cdots, s_{nk_n}.$$

STATE array. This arc-length array records which breakpoint value is being taken by a nonbasic variable.

X array. This node-length array stores the values of basic variables.
P array. This node-length array stores the dual variables, i.e. the price vector.
AU array. This node-length array is used for convenience of adjusting dual variables.

To maneuver operations on a tree, we use four node-length arrays called **PREDECESSOR,DEPTH, THREAD** and **EDGE.** The functions of the first three can be found in [C83]. The last array points to the arc in the tree that is above and incident to a given node. Here by convention, the tree is thought of as upside-down.

4.2 Implementation of Step 0

We use the Gradual Penalty Method (GPM) in Grigoriadis [G86] to solve (NetPLP) in a single phase. It is a so-called big-M method with a self-adaptive mechanism for choosing the number M. It creates an initial base with all-artificial arcs, each with a moderate linear penalty cost, and solve the augmented NetPLP. If the optimal solution contains a positive artificial variable, its penalty is enlarged gradually to certain limit, and the problem is repeatedly solved until the artificial arc has a zero flux, otherwise the problem is claimed infeasible. A empirical formula

$$M = \min\{\mu, 1 + (m-1)\max\{|s_*|, |s^*|\}, s_* + (s^* - s_*)1.5^{[2^{\pi}-2]}\}$$

is used to decide the penalty, where s_* and s^* are the minimum and maximum slopes of the cost functions, respectively; π is a pass number which refers to the times of penalty changes. μ is a threshold value (10^9 in our code).

4.3 Implementation of Step 1

4.3.1 The Setting of the Price Vector

As we mentioned in Section 2, given a tree B, each node i is associated with a price p_i. Thus one has $v_l = p_{lh} - p_{lt}$. In our implementation only p is stored. Initially, the vector p is set so that $v_l = f_l^+(x_l^0)\ \forall l \in B$ (if $f_l^+(x_l^0)$ happen to be $+\infty$, then $v_l = f_l^-(x_l^0)$). Same as in the linear network simplex method, initially p_i's are set to satisfy the system $p_{lh} - p_{lt} = v_l$, $\forall l \in B$. Afterwards p can be updated together with the update of the tree. However, no matter how the vector p is updated, we always keep $f_l^-(x_l) \leq v_l \leq f_l^+(x_l)$ for all $l \in B$, and $p_{lh} - p_{lt} = v_l$, $\forall l \in B$. Especially, after the line search in Step 2, the p_i's related to the simplex direction d^j need to be changed according to the rule

$$p_{lh} - p_{lt} = f_l^+(x_l) \text{ if } d_l^j = 1, \ p_{lh} - p_{lt} = f_l^-(x_l) \text{ if } d_l^j = -1.$$

To save time, these changes are recorded in **AU**, while the minimum ratio $\bar{\alpha}$ is computed. It can be seen that such updating procedure always assigns $v_l = f^+(x_l)$ or $f_l^-(x_l)$, $\forall l \in B$, and the descent property of d^j will ensure $f_j^+(x_j) \neq \infty$ and

$f_j^-(x_j) \neq -\infty$. At the end of an iteration, if there is a pivoting, then p is updated by taking into consideration of both the change of the tree and the value of **AU**.

The above computation procedure for p is completely the same as the linear case, if the objective function happens to be linear.

4.3.2 The Search of a Descent Simplex Direction

The key to find such a direction is the computation of the reduced cost (2.1) in Section 2. We separate the task into two processes. In process 1, we compute the nominal cost as if we were dealing with the linear case. The concrete process is: For each arc $j \in N$, if $p_{jh} - p_{jt} > f_j^+(x_j)$, then d^j, as described in Section 2, is taken as a candidate of the descent direction and the nominal cost is $n_j \equiv f_j^+(x_j) - p_{jh} + p_{jt}$. If $p_{jh} - p_{jt} < f_j^-(x_j)$, then $-d^j$ is taken as a candidate of the descent vector and the nominal cost is $n_j \equiv -f_j^-(x_j) + p_{jh} - p_{jt}$. If none of such j exists, namely for all $j \in N$, $f_j^-(x_j) \leq p_{jh} - p_{jt} \leq f_j^+(x_j)$, then the condition (c) in Proposition 3.3 is satisfied and x is the optimal solution. Otherwise, a sample pricing strategy is adopted to select an entering arc from among a group of candidates. This strategy, with details in [G86], also reduces the danger of cycling when degeneracy arises.

Note that the nominal cost n_j is not necessarily the directional derivative $F'(x, d^j)$ in (2.1). In fact, we always have $n_j \leq F'(x, d^j)$ and the equality is valid only if $v_l = f_l^-(x_l)$ as $d_l^j = -1$ and $v_l = f_l^+(x_l)$ as $d_l^j = 1$, $\forall\, j \in B$. Therefore, if $n_j \geq 0$, then d^j is definitely not a descent direction, while if $n_j < 0$, we still have to compute (2.1) to know whether d^j is really a descent direction. The case of $n_j < 0$ and $F'(x, d^j) \geq 0$ results a degenerate iteration in which we pivot to change the base but the iterative solution remains the same. After that, the process 1 is restarted.

In process 2, the computation of $F'(x, d^j)$ and the line search in Step 2 are carried over simultaneously. Notice that both operations need to travel around the circuit indicated by d^j, we do the two jobs by traveling around the circuit once. We leave the details in Section 4.4.

4.4 The Implementation of Step 2

As mentioned in Section 4.3, suppose that $n_j < 0$, we need to check if d^j is really a descent vector, and if it is, do a line search along d^j. To achieve the two goals together, we first use the **PREDECESSOR, DEPTH,** and **EDGE** arrays to identify the arcs of the circuit associated with d^j. Details may be found in [C83]. Once an arc l in this circuit is identified, compute

$$\text{(a) minimum ratio} : \bar{\alpha} = \min_l \begin{cases} \bar{c}_l - x_l & \text{if } d_l^j = 1 \\ x_l - \underline{c}_l & \text{if } d_l^j = -1 \end{cases},$$

and

$$\text{(b) directional derivative} : \beta = \beta + \begin{cases} f_l^+(x_l) - (p_{lh} - p_{lt}) & \text{if } d_l^j = 1 \\ (p_{lh} - p_{lt}) - f_l^-(x_l) & \text{if } d_l^j = -1 \end{cases},$$

where $\bar{c}_l > x_l$ is the closest breakpoint on the right of x_l and $\underline{c}_l < x_l$ is the closest breakpoint on the left of x_l. If no such breakpoint exists, we regard the difference $\bar{c}_l - x_l$ or $x_l - \underline{c}_l$ as $+\infty$. The initial value of β is $f_j^+(x_j) - p_{jh} + p_{jt} = n_j$ if $d_j^j = 1$, or $-f_j^-(x_j) - p_{jt} + p_{jh}$ if $d_j^j = -1$.

If β become positive or zero at some arc l (note: Initially $\beta < 0$), then d^j is not a descent elementary direction because β increasingly reaches $F'(x, d^j)$ when we travel around the circuit (see Proposition 4.2 below). In this case we adjourn the process and pivot on (j, l). Here arc j is the entering arc and arc l is the leaving one. After pivoting, we go to Step 1 and start the next iteration. If after scanning the entire circuit we still have $\beta < 0$, then d^j is a real descent direction because now $\beta = F'(x, d^j)$ (see Proposition 4.2 below). Set

$$x \leftarrow x + \bar{\alpha} d^j. \tag{4.1}$$

If $\bar{\alpha} = \infty$, the (NetPLP) has unbounded solution. Otherwise, repeat (a)(b) until the pivoting operation happens.

It should be noticed that the values of d_l^j can be obtained through the direction of the arc l in the circuit. Therefore, there is no need to store it. Since d^j could be still a descent direction at the new x obtained in (4.1), we repeat (a) and (b) until pivoting is necessary. Hence the line search can cross several pieces of $F(x)$ instead of just one.

To justify that the implementation really achieves the goal of Algorithm 3.1, we prove the following.

Proposition 4.1 The β computed after scanning the whole circuit is the desired derivative (2.1).

Proof. We only prove the case of $d_j^j = 1$, the case of $d_j^j = -1$ can be discussed similarly. d^j satisfies $Ad^j = 0$ and v satisfies $v = -A^T p$, so we have $v^T d^j = \sum_{l=1}^n d_k^j v_l = 0$. Thus $v_j = -\sum_{l \in B} d_l^j v_l$. At beginning we have $\beta = f_j^+(x_j) - v_j = f_j^+(x_j) + \sum_{l \in B} d_l^j v_l$. In (b), we add $f_l^+(x_l)$ and subtract v_l when $d_l^j - 1$, and add $-f_l^-(x_l)$ and subtract $-v_l$ when $d_l^j = -1$ for each $l \in B$. These operations are equivalent to adding the terms

$$\max\{d_l^j f_l^+(x_l), d_l^j f_l^-(x_l)\} - d_l^j v_l$$

to β. Thus when all $l \in B$ are scanned, we get $\beta = F'(x, d^j)$ according to (2.1).

$(\mathcal{Q.E.D.})$

Proposition 4.2 For any $0 \leq \alpha < \bar{\alpha}$, $F(x + \alpha d^j) = F(x) + \alpha\beta$.

Proof. By the way of determining $\bar{\alpha}$, $F(x)$ is linear on the line segment $[x, x + \alpha d^j]$. Thus $F(x + \alpha d^j) = F(x) + \alpha F'(x, d^j) = F(x) + \alpha\beta$ according to Proposition 4.1.

$(\mathcal{Q.E.D.})$

¿From this proposition, (NetPLP) is unbounded if $\bar{\alpha} = \infty$. We still need to show

Proposition 4.3 After each iteration x remains to be a basic feasible solution.

Proof. A new x is obtained from the current x by a pivoting operation. The criterion for choosing the leaving variable is the change of sign of β. This change is possible only if $f_l^+(x_l) \neq v_l$ or $f_l^-(x_l) \neq v_l$. By the setting of v_l, v_l is either $f_l^+(x_l)$ or $f_l^-(x_l)$. Hence x_l must be a breakpoint and the new x is then a basic solution. The feasibility comes from $Ad^j = 0$ and $F(x + \alpha d^j) < \infty$, which is implied by Proposition 4.2.

$(\mathcal{Q.E.D.})$

In summary, if after the circuit is scanned we have $\beta < 0$, then we do a line search and a real decrease of the objective function is achieved. The new iterative solution is still a basic feasible one. On the other hand during the course if for some arc l in the circuit β becomes nonnegative, then no decrease is made and we merely pivot to another basic feasible solution (a degenerate step). In addition, if there is no "cycling" in consecutive degenerate steps, we will eventually end up with one of the three alternatives: Optimality, unboundedness, or a descent direction.

Although there exist some anti-cycling procedures [F88], for simplicity of coding and efficiency of the algorithm, we did not adopt anyone of them. However, it seems that our policy of choosing the "first arc that makes β negative" as the leaving arc and the sample pricing strategy used in selecting the entering arc have practically prevented the algorithm from cycling. In all tested problems, no cycling is observed.

5 Computational Results

The 40 benchmark problems in Klingman et al. [KNS74] are generated with piecewise linear costs (Although in its original sense the assignment problem should not have piecewise linear cost, we treat it here as a special transportation problem for testing purposes.). Each arc incurs a piecewise linear cost whose breakpoints and slopes are randomly assigned according to uniform distribution. The code is written in integer version of FORTRAN, compiled by F77 with optimization option -O, and executed on a SUN computer. Table 5.1 lists the computational time (in seconds) for these problems which does not include the input-output time.

Table 5.2 presents results for various k_j. Four representative problems are solved repeatedly with respect to different numbers of pieces in their objective functions (without loss of generality, all f_j's are assumed to have the same k_j, because one can introduce redundant breakpoints if necessary). It is interesting to see that the computational time of the method does not change much with respect to the changes of k_j.

Problem Number	Type	m	n	Max. No. of Pieces	CPUTime of NETPLP	Total Number of Iterations
1	Transportation	200	1308	4	0.15	636
2	Transportation	200	1511	4	0.15	566
3	Transportation	200	2000	4	0.20	801
4	Transportation	200	2200	4	0.23	785
5	Transportation	200	2900	4	0.25	881
6	Transportation	300	3174	4	0.34	1158
7	Transportation	300	4519	4	0.50	1531
8	Transportation	300	5169	4	0.62	1708
9	Transportation	300	6075	4	0.53	1550
10	Transportation	300	6320	4	0.59	1746
11	Assignment	400	1500	4	0.30	1252
12	Assignment	400	2250	4	0.37	1533
13	Assignment	400	3000	4	0.48	1936
14	Assignment	400	3750	4	0.57	2235
15	Assignment	400	4500	4	0.59	2294
16	Transshipment	400	1306	4	0.19	1036
17	Transshipment	400	2443	4	0.30	1724
18	Transshipment	400	1306	4	0.17	984
19	Transshipment	400	2443	4	0.33	1718
20	Transshipment	400	1416	4	0.23	1178
21	Transshipment	400	2836	4	0.33	1773
22	Transshipment	400	1416	4	0.21	1056
23	Transshipment	400	2836	4	0.30	1570
24	Transshipment	400	1382	4	0.23	1366
25	Transshipment	400	2676	4	0.47	2648
26	Transshipment	400	1382	4	0.16	906
27	Transshipment	400	2676	4	0.31	2050
28	Transshipment	1000	2900	4	0.83	2258
29	Transshipment	1000	3400	4	1.03	2713
30	Transshipment	1000	4400	4	1.09	3328
31	Transshipment	1000	4800	4	1.09	3316
32	Transshipment	1500	4342	4	1.82	3857
33	Transshipment	1500	4385	4	1.95	4010
34	Transshipment	1500	5107	4	2.05	4391
35	Transshipment	1500	5730	4	2.32	5408
36	Transshipment	8000	15000	4	41.53	15052
37	Transshipment	5000	23000	4	25.35	14768
38	Transshipment	3000	35000	4	17.04	14574
39	Transshipment	5000	15000	4	20.66	11696
40	Transshipment	3000	23000	4	12.86	11983

Table 5.1. Solution Times and Optimal values of 40 Benchmark Problems (on SUN SPARC 2)

No. of pieces	Problem #1	Problem #13	Problem #16	Problem #28
8	0.42	1.60	0.88	2.37
12	0.39	1.35	0.91	2.58
16	0.47	1.74	0.94	2.61
20	0.40	1.90	1.16	2.55
24	0.43	1.25	1.19	2.77
28	0.43	1.30	0.82	1.99
32	0.52	1.41	0.86	2.42
36	0.51	1.22	1.11	2.58
40	0.48	1.29	1.07	2.47
44	0.52	1.33	1.03	2.27
48	0.42	1.17	0.92	2.30
52	0.47	1.19	1.00	2.48
56	0.47	0.96	1.03	2.84
60	0.42	1.23	1.05	2.11
64	0.42	1.02	1.07	2.48
68	0.42	0.91	1.34	2.54
72	0.42	0.88	1.29	2.42
76	0.53	1.00	1.46	2.18
80	0.48	0.99	1.29	2.84
84	0.50	0.88	1.29	2.41
88	0.40	0.68	1.49	2.54
92	0.38	0.84	1.19	2.63
96	0.49	0.88	1.18	2.53
100	0.40	0.91	1.32	2.58

Table 5.2. Solution Times as Number of Pieces Increases (on SUN4/330)

6 The STP and Computational Results

The STP can be described as follows. Suppose that a commodity is manufactured and consumed in m cities and transported between them through a highway network. At city i $(1 \le i \le m)$ there is a random demand $\underset{\sim}{w}_i$ and a fixed supply b_i. The commodity flux x_j along highway link j $(1 \le j \le n)$ has lower and upper limits d_j^- and d_j^+. In accordance with the usual convention, negative flux is understood as positive flux in the opposite direction. The transportation cost along highway j is a convex piecewise linear function $t_j(x_j)$. (e.g. q_j^+ per positive unit and q_j^- per negative unit)

Let E be the $m \times n$ incidence matrix of the network and z_i be the net supply (i.e. the fixed supply minus export) of the commodity at city i. Then in vector notation we have

$$b - Ex = z, \quad d^- \le x \le d^+, \tag{6.1}$$

where $x = (x_1, \cdots, x_n)^T$,, $d^- = (d_1^-, \cdots, d_n^-)^T$, and $d^+ = (d_1^+, \cdots, d_n^+)^T$ belong to R^n, and $b = (b_1, \cdots, b_m)^T$ and $z = (z_1, \cdots, z_m)^T$ belong to R^m. The vector inequalities are understood coordinatewise.

The penalty for surplus or shortage of the commodity at city i is introduced as

$$F_i(z_i, \underset{\sim}{w}_i) = h_i \max\{0, z_i - \underset{\sim}{w}_i\} + l_i \max\{0, \underset{\sim}{w}_i - z_i\},$$

where h_i and l_i are given cost coefficients, $h_i \geq 0$ and $l_i \geq 0$. We want to choose the amount of the commodity transported on each highway (i.e. the vector x), subject to (6.1), so as to minimize the total shipment costs plus the expected total penalty, or in other words to minimize

$$\begin{aligned} \phi(x, z) &= \sum_{j=1}^{n} t_j(x_j) + +\mathcal{E}_w\Big\{\sum_{i=1}^{m} F_i(z_i, \underset{\sim}{w}_i)\Big\} \\ &= \sum_{j=1}^{n} t_j(x_j) + \sum_{i=1}^{m} u_i(z_i), \end{aligned}$$

where $\mathcal{E}_w$ stands for the mathematical expectation with respect to random vector $\underset{\sim}{w} = [\underset{\sim}{w}_1, \cdots, \underset{\sim}{w}_m]^T$ and $u_i(z_i) = \mathcal{E}_w\{F_i(z_i, \underset{\sim}{w}_i)\}$.

It is shown [S86] that u_i is convex and that if $\underset{\sim}{w}_i$ has marginal probability distribution function W_i then the subdifferential of u_i at z_i is given by the formula

$$\partial u_i(z_i) = [(h_i + l_i) \lim_{\xi \uparrow z_i} W_i(\xi) - l_i, (h_i + l_i) \lim_{\xi \downarrow z_i} W_i(\xi) - l_i]. \tag{6.2}$$

Specifically, if $\underset{\sim}{w}_i$ has discrete (marginal) distribution of finite support $\Omega = \{c_{i0}, \cdots, c_{ik_i}\}$, then, from (6.2) the function u_i is convex piecewise linear and it consists of $k_i + 2$ linear pieces with $c_{i0}, \cdots, c_{ik_i}$ being the breakpoints. We refer to this case as the discrete STP.

The works on STP can be looked back upon the paper of Dantzig and Ferguson on stochastic transportation problem [FD56]. Since then, the stochastic transportation problem and other network-related stochastic programs have received considerable attention over the years (see, for instance, [CL77] [B79] [E60] [L80] [P86] [Q84b] [Q85] [Q87] [Wi63] [Co78] [Q84a] [MV88a] [MV88b] [W86] [W89] [WW89], which cover the applications such as traffic control, production planning, and financial investment). Because of the non-bipartite network structure, our STP model is more general than the stochastic transportation problem. On the other hand, due to the explicit form of the recourse function $F_i(z_i, \underset{\sim}{w}_i)$, the model is a special case of stochastic programming with simple recourse [We83]. Existing methods for the stochastic transportation problem do not lend themselves an obvious extension to STP, while general methods for stochastic programming with simple recourse do not take advantage of the network structure and hence tend to be inefficient for STP.

We suggest that the STP problem be solved by explicitly solving its deterministic equivalence. To turn a discrete STP into a piecewise linear network program, we set an artificial node r and m additional arcs, each initiates from a node i of the original network and terminates at r. The flux on arc (i, r) is z_i and its cost is $p_j(z_i)$. Now the STP is equivalent to the following piecewise linear program:

$$\begin{cases} \text{minimize}_{x,z} & \sum_{j=1}^{n} t_j(x_j) + \sum_{i=1}^{m} u_i(z_i) \\ \text{subject to} & \begin{pmatrix} E & I \\ 0 & -e^T \end{pmatrix} \begin{pmatrix} x \\ z \end{pmatrix} = \begin{pmatrix} b \\ -\sum_{i=1}^{m} b_i \end{pmatrix}, \\ & d_j^- \le x_j \le d_j^+ \ \text{ for } j = 1, \cdots, n, \end{cases} \tag{6.3}$$

where I is a unit matrix, $-e^T = (-1, \cdots, -1)$ and $b = (b_1, \cdots, b_m)^T$. The functions t_j and u_i are convex piecewise linear. Thus the STP is a special case of the problem (NetPLP).

In our computational test, the forty benchmark problems developed by Klingman, Napier and Stutz [KNS74] for network linear programming are regenerated under the STP context. Although the random demand does not apply to assignment problems, we still test them and treat them as transportation problems with $b = (1, \cdots, 1, -1, \cdots, -1)^T$. In each problem, we associate to the nodes uniformly distributed demands having 100 possible values. This is probably a typical size of the support for discrete distributions in practice. These problems cover transportation and transshipment networks of various size and have been long used in evaluating and comparing algorithms in network linear programming. Because at this time, no other codes for STP are available to us, we have not been able to compare the efficiency of this method with other methods. However, the solution times in all tested problems are satisfactory — none of them need more than one minute of CPU time.

Table 6.1 listed results for these STPs. The biggest problem in this table is a transshipment problem with 3000 nodes and 35000 arcs. Each arc has a linear shipment cost and each node has a 100-piece expected penalty cost. Overall, we feel that there are at least two advantages to solve STP (including stochastic transportation problems) by using the network piecewise linear algorithm: First, the method is conceptually simpler than other current methods and yet is practically efficient in solving large-scale problems. Second, it is flexible in dealing with different types of problems and complications in practice, for example, it allows the transportation cost also to be piecewise linear.

7 STPC and Other Extensions

Assume that in problem (6.3) the distributions of the random demands at cities are continuous. Thus, $u_i(z_i)$ are convex functions, but not necessarily piecewise linear.

We may first discretize the distributions. This is equivalent to use convex piecewise linear functions $\bar{u}_i(z_i)$ to approximate $u_i(z_i)$. The properties of approximating a convex function by a piecewise linear function were studied by Geoffrion [G77]. In general his result requires a global finer division of the domain in order to obtain higher accuracy. However, for the STP, more grid points may not be necessary. Let us denote the approximation problem by STPD. Suppose that we get an optimal solution $(\bar{x}, \bar{z})$ for STPD by using the (NetPLP) method. We now discuss how to improve this approximation solution for STPC.

Proposition 7.1. STPC has an optimal solution (if it exists) (x^*, z^*) such that x_j^* takes a breakpoint value of the domain of t_j for each j except j belongs to a set of arcs forming a spanning tree in the network.

Proof. Suppose that an optimal solution to STPC is $(\hat{x}, z^*)$. Fix the values of z at z^*. The resulted problem is a piecewise linear network program which has an optimal solution x^* satisfying the requirements.

$(\mathcal{Q.E.D.})$

If $\bar{p}_i$ are good piecewise linear approximations to p_i, we may expect $\bar{x}_j$ to take the same breakpoint value as x_j^* for each j except j belongs to an optimal spanning tree T. We may also assume that $\bar{x}_j$ is in the same linear "piece" of the domain of t_j as x_j^* if $j \in T$. To find x^*, it suffices to find the values of x_j^* for $j \in T$ by fixing the values of x_j for $j \notin T$ at these breakpoint values. Replace x_j $(j \notin T)$ by c_{jk}, where c_{jk} is the corresponding breakpoint values and adjust b_i accordingly. Then it suffices to solve the following problem:

$$(\text{STPT}) \begin{cases} \text{minimize}_{x,z} & \sum_{j\in T} t_j(x_j) + \sum_{i=1}^m u_i(z_i) \\ \text{subject to} & \begin{pmatrix} \hat{E} & I \\ 0 & -e^T \end{pmatrix} \begin{pmatrix} \hat{x} \\ z \end{pmatrix} = \begin{pmatrix} \hat{b} \\ -\sum_{i=1}^m \hat{b}_i \end{pmatrix}, \\ & d_j^- \le x_j \le d_j^+ \quad \text{for } j \in T, \end{cases}$$

where $\hat{E}$ is the incidence matrix of T and $\hat{x} = (x_j)_{j\in T}$. As proved in [Q85], STPT may be solved with the work of solving a one-dimensional monotone equation. This certainly gives us a better approximation solution of STPC.

One of the extensions of STP is the stochastic generalized transshipment problem, i.e., the problem with a magnification or reduction coefficient at each arc. Some versions of such a problem have been discussed in [FD56][E60][Q87]. It is expected that the (NetPLP) method may be generalized to such a problem without big difficulty since the only difference is that the tree structure is replaced by a one-tree structure. See [KH80] and [Q87]. It will be much difficult if we allow the magnification or reduction coefficients are also convex piecewise linear functions. This occurs in the real world. For example, the reduction effect increases in an electricity transmission line as the flow increases. Other extensions of STP includes the multi-commodity case and the stochastic programming problem with network flow recourse, which was first considered by Wallace for fisheries management. See [W86][W89] and [WW89].

Problem Number	Type	m	n	Size of Support	CPU Time (in Sec.)
1	Transportation	200	1311	100	0.36
2	Transportation	200	1500	100	0.33
3	Transportation	200	2007	100	0.48
4	Transportation	200	2205	100	0.44
5	Transportation	200	2900	100	0.50
6	Transportation	300	3150	100	0.63
7	Transportation	300	4500	100	0.93
8	Transportation	300	5170	100	1.07
9	Transportation	300	6095	100	1.14
10	Transportation	300	6311	100	1.21
11	Assignment	400	1500	100	0.38
12	Assignment	400	2250	100	0.49
13	Assignment	400	3000	100	0.56
14	Assignment	400	3750	100	0.52
15	Assignment	400	4500	100	0.69
16	Transshipment	400	1306	100	0.73
17	Transshipment	400	2443	100	1.24
18	Transshipment	400	1306	100	0.77
19	Transshipment	400	2443	100	1.24
20	Transshipment	400	1416	100	0.80
21	Transshipment	400	2836	100	1.44
22	Transshipment	400	1416	100	0.82
23	Transshipment	400	2836	100	1.49
24	Transshipment	400	1382	100	0.81
25	Transshipment	400	2676	100	1.50
26	Transshipment	400	1382	100	0.86
27	Transshipment	400	2676	100	1.55
28	Transshipment	1000	2900	100	2.38
29	Transshipment	1000	3400	100	2.53
30	Transshipment	1000	4400	100	3.35
31	Transshipment	1000	4800	100	3.39
32	Transshipment	1500	4342	100	4.25
33	Transshipment	1500	4385	100	4.14
34	Transshipment	1500	5107	100	4.49
35	Transshipment	1500	5730	100	4.81
36	Transshipment	8000	15000	100	50.48
37	Transshipment	5000	23000	100	28.01
38	Transshipment	3000	35000	100	24.82
39	Transshipment	5000	15000	100	24.91
40	Transshipment	3000	23000	100	18.82

Table 6.1. Solution Times of 40 Benchmark STP (on SUN4/330)

References

[ACK86] I. Ali, W.D. Cook and M. Kress, Ordinal ranking and intensity of preference:

a linear programming approach, *Management Science,* 32(1986)1642-1647.

[C83] V. Chvátal, *Linear Programming,* (Freeman, New York, NY, 1983).

[CSA86] A. Charnes, T. Song and I. Ali, A two-segment approximation algorithm for separable convex programming with linear constraints, *Mathematische Operationsforschung und Statistik-Series Optimization,* 17(1986)147-159.

[Co78] L. Cooper, The stochastic transportation-location problem, *Computer and Mathematics with Applications* 4(1978)265-275.

[CL77] L. Cooper and L.J. LeBlanc, Stochastic transportation problems and other network related convex problems, *Naval Research Logistics Quarterly* 24(1977) 324-336.

[E60] S. Elmaghrabi, Allocation under uncertainty when the demand has a continuous distribution function, *Management Science* 6(1960)270-294.

[F86] R. Fourer, A simplex algorithm for piecewise-linear programming III: Computational Analysis and Applications, *Tech. Report 86-03* Department of IE/MS, Northwestern University, Evanston IL 60208 (1986).

[F88] R. Fourer, A simplex algorithm for piecewise-linear programming: finiteness, feasibility and degeneracy, *Mathematical Programming,* 41(1988)281-316.

[FD56] A. R. Ferguson and G. B. Dantzig, The allocation of aircraft to routes—an example of linear programming under uncertain demand, *Management Science* 3(1956)45-73.

[FF62] L.R. Ford and D.R. Fulkerson, *Flows and Networks*, (Princeton Univ. Press, Princeton, NJ, 1962).

[G77] A.M. Geoffrion, Objective function approximation in mathematical programming, *Mathematical Programming* 13(1977)23-37.

[Gr86] M.D. Grigoriadis, An efficient implementation of the network simplex method, *Mathematical Programming Study* 26(1986)83-111.

[KH80] J.L. Kennington and R.V. Helgason, *Algorithms for Network Programming,* Wiley-Interscience, New York (1980).

[KNS74] D. Klingman, A. Napier and J. Stutz, NETGEN: a program for generating large scale capacitated assignment, transportation, and minimum cost flow network problems, *Management Science* 20(1974)814-821.

[L80] F. Louveaux, A solution method for multi-stage stochastic programs with recourse with applications to an energy investment problem, *Operations Research* 28(1980)889-902.

[MV88a] J. M. Mulvey and H. Vladimirou, Solving multistage stochastic network flows: An application of scenario aggregation, *Tech. Report* SOR-88-1, Dept. of Civil Engineering and Operations Research, Princeton University, Princeton , NJ (1988).

[MV88b] J. M. Mulvey and H. Vladimirou, Stochastic network optimization models for investment planning, *Tech. Report* SOR-88-2, Dept. of Civil Engineering and Operations Research, Princeton University, Princeton , NJ (1988).

[P86] W. B. Powell, A stochastic model of the dynamic vehicle allocation problem, *Transportation Science* 20 (1986) 117-129.

[Q84a] L. Qi, Finitely convergent methods for solving stochastic linear programming and stochastic network flow problems, *Ph. D. Dissertation,* University of Wisconsin, Madison, WI (1984).

[Q84b] L. Qi, The dual forest iteration method for the stochastic transportation problem, *Working Paper* WP-84-59, IIASA, Laxenburg, Austria (1984).

[Q85] L. Qi, Forest iteration method for stochastic transportation problem, *Mathematical Programming Study* 25(1985)142-163.

[Q87] L. Qi, The A-forest iteration method for the stochastic generalized transportation problem, *Mathematics of Operations Research* 12(1987)1-15.

[R84] R.T. Rockafellar, *Network Flows and Monotropic Optimization,* (Wiley-Interscience, New York, 1984).

[S86] J. Sun, On monotropic piecewise quadratic programming, Ph.D Dissertation, University of Washington, Seattle, Washington (1986).

[Sz64] W. Szwarc, The transportation problem with stochastic demands, *Management Science* 11(1964)33-50.

[W86] S. Wallace, Solving stochastic programs with network recourse, *Networks* 16(1986)295-317.

[W89] S. Wallace, Bounding the expected time-cost curve for a stochastic PERT network from below, *Operations Research Letters* 8(1989)89-94.

[We83] R. J-B. Wets, Solving stochastic programs with simple recourse, *Stochastics* 10(1983)219-242.

[Wi63] A.C. Williams, A stochastic transportation problem, *Operations Research* 1(1963)759-770.

[WW89] S. Wallace and R. J-B Wets, Preprocessing in stochastic programming: The case of uncapacitated networks, *ORSA Journal on Computing* 1(1989)252-270.

Network Optimization Problems, pp. 301-331
Eds. D.-Z. Du and P.M. Pardalos

A Bibliography on Network Flow Problems[1]

Marinus Veldhorst
Department of Computer Science, Utrecht University, P.O. Box 80.089, 3508 TB Utrecht, The Netherlands

Abstract

Network flow problems form an important class of research problems in optimization, with many new developments in the last decade. There are many different subclasses of network flow problems, as well as many different techniques to address these problems. In this bibliography we concentrate on combinatorial algorithms for the maximum flow problem and for the minimum cost flow problem with integral capacities and linear cost functions with integral coefficients. Especially results published since 1982 are compiled.

1 Introduction

Network flow problems form an important class of optimization problems on graphs. Basically, we are given a network (directed graph) $G = (V, A)$ and one or more commodities. Some amount of each commodity must be pushed through the network from a number of sources (vertices in G which have a supply of the commodity) to a number of sinks (vertices with a demand of the commodity). At every vertex except the sources and the sinks, the incoming flow of every commodity is pushed on in certain portions over the outgoing arcs. The flow of the commodities may incur some costs, and the amount of flow through the arcs of the network may be restricted by certain capacity constraints and may be subject to losses or gains. Network flow problems usually ask for criteria and efficient algorithms for determining flows of the commodities that satisfy certain conditions.

[1]This work was partially supported by the ESPRIT II Basic Research Actions of the EC under contract no. 7141 (project ALCOM II).

Important subclasses of network flow problems are the maximum flow problems, the minimum cost flow problems, the multicommodity flow problems, and the flows "with losses and gains" (the generalized flow problem). Even within a subclass, problems may vary in such a way that different algorithms are necessary to find the solution for different sorts of problems. For example, the capacity constraints may be nonnegative integral or real numbers, and can be upper or lower bounds on the amount of flow through the arcs; the cost functions of the arcs may be integral or real valued functions and can be linear or nonlinear in the amount of flow through the arc. For special networks (e.g., planar networks) specific algorithms have been designed in order to obtain solutions more efficiently.

Hence, in a bibliography on network flow problems that is not too extensive, one must make a selection from an overwhelming number of publications in this research area. In this bibliography we compiled results which, in our opinion, are interesting from the viewpoint of the design and analysis of algorithms, especially combinatorial algorithms. We concentrated on the maximum flow problems and on the minimum cost flow problems with integral capacities and linear cost functions with integral coefficients. These are, in a way, the most classical network flow problems and were already the subject of most of Ford and Fulkerson's book in 1962. We intend to be complete in these two areas as far as results published after 1982 are concerned. Results that are not published in regular journals or proceedings of conferences, are only included when we consider them important from an historic point of view or when they constitute the current state of the art.
For the more general varieties and subclasses of the network flow problem the bibliography is certainly not complete, but nevertheless we hope to give a rather broad entrance to the scientific literature on these problems.

In the remainder of this introduction we give an overview of the variants of the maximum flow problems and minimum cost flow problems, and mention a number of publications with good introductions, overviews or important contributions to the development of the field of network flow algorithms. After section 1 the bibliography is given[2].

1.1 Maximum Flow Problems

In the (basic) maximum flow problem there is one commodity and a maximum amount of flow of the commodity has to be sent from one source s to one sink t; the flow in each arc $(i,j) \in A$ is not allowed to be more than a given positive integral value (upper bound) $u_{i,j}$. To be precise, in an instance of the maximum flow problem we are given a directed graph $G = (V, A)$, and two specified vertices s and t which are assumed to have no incoming and outgoing arcs, respectively. With each arc

[2]Earlier versions of this bibliography have been published in Algorithms Review, vol. 1 (1990), pp. 97–117, and as Technical Report RUU-CS-91-38, Dept. of Computer Science, University of Utrecht, Utrecht, The Netherlands.

$(i,j) \in A$ is associated a positive integral number $u_{i,j}$, its capacity. A (valid) flow for this instance consists of a real number $f_{i,j}$ for each $(i,j) \in A$ such that the following two conditions are satisfied:

$$0 \leq f_{i,j} \leq u_{i,j} \quad \text{for all } (i,j) \in A$$

$$\sum_{j:(j,i)\in A} f_{j,i} = \sum_{j:(i,j)\in A} f_{i,j} \quad \text{for each } i \neq s, i \neq t$$

The flow has value $\sum_{i:(s,i)\in A} f_{s,i}$. A maximum flow is a valid flow with maximum value among all possible valid flows. The maximum flow problem is the problem to design an efficient algorithm that computes a maximum flow for every given instance.

Pioneering work on the maximum flow problem has been done by Ford and Fulkerson (e.g. [94]). Fundamental improvements have been given by Edmonds and Karp ([76]), Dinic ([72]), Karzanov ([208]), Malhotra et al. ([243]) and Goldberg ([120]). These fundamental improvements often triggered further improvement: faster algorithms could be obtained by incorporation of more sophisticated data structures in the fundamental algorithms (e.g., [111], [333], [130], [125]), or by making proper choices that were left open in the fundamental algorithms (e.g., [344], [130], [52]).
Good introductions or overviews of the algorithmic area of maximum network flow can be found in [5], [61], [94], [177], [194], [231], [287], [344] and [357]. Included in [5] is an excellent historic overview.

For subclasses of maximum flow problems specially designed algorithms may run faster than the general algorithms. For example there are maximum flow algorithms for planar graphs (e.g., [189], [170], [98]), and for bipartite graphs (e.g., [10], [159]). On the other hand one can look for efficient algorithms for the special case that the capacities $u_{i,j}$ are relatively small numbers (e.g., [86], [88]). In case the capacities are bounded by a polynomial in the number of vertices of G, a scaling technique might be useful ([107], [7], [11]).

Algorithms for sequential computers are not necessarily efficient for computers of a different architecture. Hence, several researchers have looked for efficient parallel algorithms (e.g., [328], [130]), and efficient distributed algorithms (e.g., [20], [244], [130], [52]).

Other researchers have considered the maximum flow of random instances of the maximum flow problem (e.g., [150], [335], [268]).

The maximum flow problem can naturally be extended to the so-called multi-terminal flow problem. In this problem one wants to compute the maximum flow values for k source-destination pairs $(s_1,t_1),\ldots,(s_k,t_k)$, simultaneously. Obviously this problem can be solved by separately solving k maximum flow problems (one for each pair (s_i,t_i)), but more efficient algorithms have been designed for the case that G is an undirected graph (e.g., [140], [144], [156]), and for the case G is planar ([252]). Another extension of the maximum flow problem is the parametric flow problem in which the capacities depend on one additional parameter, and one wants to compute some information about how the maximum flow depends on this parameter (e.g., [8],

[115], [158]).
In a third extension there are upper bounds set on the amount of flow streaming through a vertex. Usually problems of this type can be transformed to ordinary maximum flow problems, but by these transformations one may loose several desirable properties of the networks, e.g. planarity ([215]). Other variants of the maximum flow problem can be found in e.g. [262], [272], [279].

As for implementations of the algorithms and their efficiency on existing computers, we refer to e.g. [225], [71], [203], [16] and [14].

1.2 Minimum Cost Flow Problems

In the minimum cost flow problem a nonnegative cost function $c_{i,j}$ is associated with each arc $(i,j) \in A$. Instead of maximizing the flow value, we want to compute the flow of a given value v with minimum cost. The cost of a flow f is defined as $\sum_{(i,j)\in A} c_{i,j}(f_{i,j})$. Arcs may have capacity constraints that consist of nonnegative lower bounds and positive (possibly infinite) upper bounds. The minimum cost flow problem is easily generalized to the minimum cost circulation problem. Here we have a directed graph $G = (V, A)$. With each arc $(i,j) \in A$ is associated a cost function $c_{i,j}$, and a (upper bounding) capacity $u_{i,j}$ (a positive number, possibly infinite). With each vertex $i \in V$ is associated a number b_i; if $b_i < 0$, vertex i has a demand of flow; if $b_i > 0$, i has a surplus of flow. A (valid) circulation consists of a nonnegative real number $f_{i,j}$ for each $(i,j) \in A$ such that the following two conditions hold:

$$0 \leq f_{i,j} \leq u_{i,j} \quad \text{for all } (i,j) \in A$$

$$\sum_{j:(i,j)\in A} f_{i,j} - \sum_{j:(j,i)\in A} f_{j,i} = b_i \quad \text{for all } i \in V$$

and has cost $\sum_{(i,j)\in A} c_{i,j}(f_{i,j})$. The problem is to find a valid circulation of minimum cost. Usually the cost functions are convex or linear.

For the case of linear cost functions with integral coefficients we refer to [5]. It concentrates on combinatorial algorithms, but contains also a historic overview of the different fundamental approaches (e.g., network simplex, primal-dual, out-of-kilter, scaling, relaxation) to the solution of the minimum cost circulation problem. For the case of convex cost functions we refer to [212], [301] and [35].

Other variants of the minimum cost flow problems are mentioned in e.g. [166], [214] and [210]. Minimum cost flow in planar graphs is treated in [183].

As for implementations of the algorithms and their efficiency on existing computers, we refer to e.g. [225], [37], [203], [126] and [122].

References

[1] G. K. Adel'son-Velskii, E. A. Dinic, and A. V. Karzanov. *Flow algorithms.* Science, Moscow, 1975. in Russian.

[2] R. K. Ahuja. Algorithm for the minimax transportation problem. *Naval Res. Log. Quart.*, 33:725–739, 1986.

[3] R. K. Ahuja, J. L. Batra, and S. K. Gupta. A parametric algorithm for the convex cost network flow and related problems. *Europ. J. Oper. Res.*, 16:222–235, 1984.

[4] R. K. Ahuja, A. V. Goldberg, J. B. Orlin, and R. E. Tarjan. Finding minimum-cost flows by double scaling. *Math. Programming*, 53:243–266, 1992.

[5] R. K. Ahuja, T. L. Magnanti, and J. B. Orlin. Network flows. In G. L. Nemhauser, A. H. G. Rinnooy Kan, and M. J. Todd, editors, *Handbooks of Operations Research and Management Science, vol. 1: Optimization*, pages 211–369. North-Holland Publ. Comp., Amsterdam, 1989.

[6] R. K. Ahuja and J. B. Orlin. Improved primal simplex algorithms for the shortest path, assignment and minimum cost flow problems. Technical Report 2090-88, Sloan School of Management, MIT, Cambridge, Mass., 1988.

[7] R. K. Ahuja and J. B. Orlin. A fast and simple algorithm for the maximum flow problem. *Operations Res.*, 37:748–759, 1989.

[8] R. K. Ahuja and J. B. Orlin. Distance directed augmenting path algorithms for maximum flow and parametric maximum flow problems. *Naval Res. Log. Quart.*, 38:413–430, 1991.

[9] R. K. Ahuja and J. B. Orlin. The scaling network simplex algorithm. *Operations Res.*, 40:Supplement S5–S13, 1992.

[10] R. K. Ahuja, J. B. Orlin, C. Stein, and R. E. Tarjan. Improved algorithms for bipartite network flow problems. Technical Report TR-338-91, Department of Computer Science, Princeton University, Princeton, NJ., 1991.

[11] R. K. Ahuja, J. B. Orlin, and R. E. Tarjan. Improved time bounds for the maximum flow problem. *SIAM J. Comput.*, 18:939–954, 1989.

[12] A. I. Ali, R. Padman, and H. Thiagaran. Dual algorithms for pure network problems. *Operations Res.*, 37:159–171, 1989.

[13] I. Ali, D. Barnett, K. Farhangian, J. Kennington, B. Patty, B. Shetty, B. McCarl, and P. Wong. Multicommodity network problems: Applications and computations. *A.I.I.E. Trans.*, 16:127–134, 1984.

[14] F. Alizadeh and A. V. Goldberg. Implementing the push-relabel method for the maximum flow problem on a connection machine. Technical Report STAN-CS-92-1410, Department of Computer Science, Stanford University, Stanford, CA, Feb. 1992.

[15] N. Alon. Generating pseudo-random permutations and maximum flow algorithms. *Inf. Process. Lett.*, 35:201–204, 1990.

[16] R. J. Anderson and J. C. Setubal. On the parallel implementation of Goldberg's maximum flow algorithm. In *Proc. 4th Annual ACM Symp. on Parallel Algorithms and Architectures*, pages 168–177, 1992.

[17] E. M. Arlin and C. H. Papadimitriou. On the complexity of circulations. *J. Algorithms*, 7:134–145, 1986.

[18] J. Aronson and B. Chen. A primary/secondary memory implementation of a forward network simplex algorithm for multiperiod network flow problems. *Comput. Oper. Res.*, 16:379–391, 1989.

[19] A. Assad. Multicommodity network flows - a survey. *Networks*, 8:37–91, 1978.

[20] B. Awerbuch. Reducing complexities of the distributed max-flow and breadth-first-search algorithms by means of network synchronization. *Networks*, 15:425–437, 1985.

[21] F. Barahona and É. Tardos. Note on Weintraub's minimum-cost circulation algorithm. *SIAM J. Comput.*, 18:579–583, 1989.

[22] A. E. Baratz. The complexity of maximum network flow. Technical Report MIT/LCS/TR-230, Lab. for Computer Science, MIT, Cambridge, Mass., 1980.

[23] M. Bazaraa and J. J. Jarvis. *Linear Programming and Network Flows (2nd ed.)*. John Wiley & Sons, New York, 1990.

[24] M. Bellmore and R. R. Vemuganti. On multicommodity maximal dynamic flows. *Operations Res.*, 21:10–21, 1973.

[25] G. E. Bennington. An efficient minimal cost flow algorithm. *Manag. Sci.*, 19:1042–1051, 1973.

[26] C. Berge. *Graphs and Hypergraphs*, chapter 5. North-Holland Publ. Comp., Amsterdam, 1973.

[27] C. Berge and A. Ghouila-Houri. *Programming, Games and Transportation Networks*. John Wiley & Sons, New York, 1962.

[28] D. P. Bertsekas. A unified framework for primal-dual methods in minimum cost network flow problems. *Math. Programming*, 32:125–145, 1985.

[29] D. P. Bertsekas. Distributed asynchronous relaxation methods for linear network flow problems. Technical Report LIDS-P-1986, Lab. for Decision Systems, MIT, Cambridge, Mass., 1986.

[30] D. P. Bertsekas and J. Eckstein. Dual coordinate step methods for linear network flow problems. *Math. Programming*, 42:203–243, 1988.

[31] D. P. Bertsekas and D. El Baz. Distributed asynchronous relaxation methods for convex network flow problems. *SIAM J. Contr. & Optim.*, 25:74–85, 1987.

[32] D. P. Bertsekas, P. A. Hosein, and P. Tseng. Relaxation methods for network flow problems with convex arc costs. *SIAM J. Contr. & Optim.*, 25:1219–1243, 1987.

[33] D. P. Bertsekas and P. Tseng. The relax codes for linear minimum cost network flow problems. In B. Simeone et al., editors, *FORTRAN Codes for Network Optimization*, Annals of Operations Research, vol. 13, pages 125–190, 1988.

[34] D. P. Bertsekas and P. Tseng. Relaxation methods for minimum cost ordinary and generalized network flow problems. *Operations Res.*, 36:93–114, 1988.

[35] D. P. Bertsekas and J. N. Tsitsiklis. *Parallel and Distributed Computation*, chapter 5, 6.5 and 6.6. Prentice-Hall, Inc., Englewood Cliffs, NJ, 1989.

[36] D. Bienstock. Some generalized max-flow min-cut problems in the plane. *Math. Oper. Res.*, 16:310–333, 1991.

[37] R. G. Bland and D. L. Jensen. On the computational behavior of a polynomial-time network flow algorithm. *Math. Programming*, 54:1–40, 1992.

[38] G. Bradley, G. Brown, and G. Graves. Design and implementation of large scale primal transshipment algorithms. *Manag. Sci.*, 24:1–38, 1977.

[39] S. P. Bradley, A. C. Hax, and T. L. Magnanti. *Applied Mathematical Programming*. Addison-Wesley Publ. Comp., New York, 1977.

[40] R. G. Busacker and P. J. Gowen. A procedure for determining a family of minimal-cost network flow patterns. O.R.O. Technical paper 15, Johns Hopkins University, Baltimore, MD, 1961.

[41] R. G. Busacker and T. L. Saaty. *Finite Graphs and Networks: An Introduction with Applications*. McGraw-Hill, New York, 1965.

[42] I. N. Chen. A new parallel algorithm for network flow problems. In T. Y. Feng, editor, *Proc. 1974 Sagamore Computer Conf.*, Lecture Notes in Computer Science, vol. 24, pages 306–307, Springer-Verlag, Berlin, 1975.

[43] I. N. Chen, P. Y. Chen, and T. Y. Feng. Associative processing of network flow problems. *IEEE Trans. Comput.*, C-28:184–190, 1979.

[44] I. N. Chen and T. Y. Feng. A parallel algorithm for maximum flow problem. In *Proc. 1973 Sagamore Computer Conf.*, 1973.

[45] Y. L. Chen and Y. H. Chin. Multicommodity network flows with safety considerations. *Operations Res.*, 40:Supplement S48–S55, 1992.

[46] C. K. Cheng. Ancestor tree for arbitrary multi-terminal-cut functions. *Annals of Operations Res.*, 33:199–213, 1991.

[47] C. K. Cheng and T. C. Hu. Maximum concurrent flow and minimum cuts. *Algorithmica*, 8:233–249, 1992.

[48] J. Cheriyan. Parametrized worst case networks for preflow push algorithms. Technical report, Computer Science Group, Tata Institute of Fundamental Research, Bombay, India, 1988.

[49] J. Cheriyan and T. Hagerup. A randomized maximum-flow algorithm. In *Proc. 30th Annual IEEE Symp. Foundations of Computer Science*, pages 118–123, 1989.

[50] J. Cheriyan, T. Hagerup, and K. Mehlhorn. Can a maximum flow be computed in $o(nm)$ time? In M. Paterson, editor, *Proc. 17th Intern. Coll. on Automata, Languages and Programming*, Lecture Notes in Computer Science, vol. 443, pages 235–248, Springer-Verlag, Berlin, 1990.

[51] J. Cheriyan, T. Hagerup, and K. Mehlhorn. An $o(n^3)$-time maximum-flow algorithm. Technical Report MPI-I-91-120, Max-Planck-institut für Informatik, Saarbrücken, Germany, Nov. 1991.

[52] J. Cheriyan and S. N. Maheshwari. Analysis of preflow push algorithms for maximum network flow. *SIAM J. Comput.*, 18:1057–1086, 1989.

[53] J. Cheriyan and S. N. Maheshwari. The parallel complexity of finding a blocking flow in a 3-layer network. *Inf. Process. Lett.*, 31:157–161, 1989.

[54] R. V. Cherkasky. Algorithm of construction of maximal flow in networks with complexity of $O(V^2\sqrt{E})$ operations. *Math. Methods of Solution of Economical Problems*, 7:112–125, 1977. (In Russian).

[55] T. Cheung. Computational comparison of eight methods for the maximum network flow problem. *ACM Trans. Math. Softw.*, 6:1–16, 1980.

[56] T. Cheung. Graph traversal techniques and the maximum flow problem in distributed computation. *IEEE Trans. Softw. Eng.*, SE-9:504–512, 1983.

[57] N. Christofides. *Graph Theory: An Algorithmic Approach*, chapter 11. Academic Press, New York, 1975.

[58] E. Cohen. Approximate max flow on small depth networks. In *Proc. 33rd Annual IEEE Symp. Foundations of Computer Science*, pages 648–658, 1992.

[59] E. Cohen and N. Megiddo. Algorithms and complexity analysis for some flow problems. In *Proc. 2nd Annual ACM-SIAM Symp. Discrete Algorithms*, pages 120–130, 1991.

[60] E. Cohen and N. Megiddo. New algorithms for generalized network flows. In D. Dolev, Z. Galil, and M. Rodeh, editors, *Theory of Computing and Systems, Proc. ISTCS '92, Haifa, Israel 1992*, Lecture Notes in Computer Science, vol. 601, pages 103–114, Springer-Verlag, Berlin, 1992.

[61] T. H. Cormen, C. E. Leiserson, and R. L. Rivest. *Introduction to Algorithms*, chapter 28. MIT Press, Cambridge, Mass., 1990.

[62] W. Cui. A network simplex method for the maximum balanced flow problem. *J. Oper. Res. Soc. Japan*, 31:551–563, 1988.

[63] W. Cui and S. Fujishige. A primal algorithm for the submodular flow problem with minimum-mean cycle selection. *J. Oper. Res. Soc. Japan*, 31:431–441, 1988.

[64] W. H. Cunningham. A network simplex method. *Math. Programming*, 11:105–116, 1976.

[65] W. H. Cunningham. Theoretical properties of the network simplex method. *Math. Oper. Res.*, 4:196–208, 1979.

[66] W. H. Cunningham and A. Frank. A primal-dual algorithm for submodular flows. *Math. Oper. Res.*, 10:251–262, 1985.

[67] G. B. Dantzig. Application of the simplex method to a transportation problem. In T. C. Koopmans, editor, *Activity Analysis of Production and Allocation*, pages 359–373, John Wiley & Sons, New York, 1951.

[68] G. B. Dantzig. *Linear Programming and Extensions.* Princeton Univ. Press, Princeton, NJ, 1962.

[69] G. B. Dantzig and D. R. Fulkerson. On the max-flow min-cut theorem of networks. In H. W. Kuhn and A. W. Tucker, editors, *Linear Inequalities and Related Systems*, Annals of Mathematics Study, vol. 38, pages 215–221, Princeton Univ. Press, Princeton, NJ, 1956.

[70] U. Derigs. *Programming in Networks and Graphs.* Lecture Notes in Economics and Mathematical Systems, vol. 300. Springer-Verlag, Berlin, 1988.

[71] U. Derigs and W. Meier. Implementing Goldberg's max-flow algorithm, a computational investigation. *Z. Oper. Res.*, 33:383–403, 1989.

[72] E. A. Dinic. Algorithm for solution of a problem of maximum flow in networks with power estimation. *Soviet Math. Dokl.*, 11:1277–1280, 1970.

[73] J. Divoky and M. Hung. Performance of shortest path algorithms in network flow problems. *Manag. Sci.*, 36:661–673, 1990.

[74] J. R. Driscoll, H. N. Gabow, R. Shrairman, and R. E. Tarjan. Relaxed heaps: An alternative to Fibonacci heaps with applications to parallel computations. *Commun. ACM*, 31:1343–1354, 1988.

[75] J. Edmonds and R. Giles. A min-max relation for submodular functions on graphs. *Annals of Discrete Math.*, 1:185–204, 1977.

[76] J. Edmonds and R. M. Karp. Theoretical improvements in algorithmic efficiency for network flow problems. *J. ACM*, 19:248–264, 1972.

[77] J. Elam, F. Glover, and D. Klingman. A strongly convergent primal simplex algorithm for generalized networks. *Math. Oper. Res.*, 4:39–59, 1979.

[78] P. Elias, A. Feinstein, and C. E. Shannon. Note on maximum flow through a network. *IRE Trans. on Inform. Theory*, 2:117–119, 1956.

[79] S. E. Elmaghraby. Sensitivity analysis of multi-terminal network flows. *J. ORSA*, 12:680–688, 1964.

[80] T. R. Ervolina and S. T. McCormick. A strongly polynomial dual cancel and tighten algorithm for minimum cost network flow. Technical Report 90-MSC-010, UBC Faculty of Commerce, 1990.

[81] T. R. Ervolina and S. T. McCormick. A strongly polynomial maximum mean cut cancelling algorithm for minimum cost network flow. Technical Report 90-MSC-009, UBC Faculty of Commerce, 1990.

[82] J. R. Evans. Maximum flow in probabilistic graphs – the discrete case. *Networks*, 6:161–183, 1976.

[83] S. Even. The max-flow algorithm of Dinic and Karzanov. An exposition. Technical Report MIT/LCS/TM-80, Lab. for Computer Science, MIT, Cambridge, Mass., 1976.

[84] S. Even. *Graph Algorithms*, chapter 4, 5, and 10.8. Pitman Publ. Ltd, London, 1979.

[85] S. Even, A. Itai, and A. Shamir. On the complexity of timetable and multicommodity flow problems. *SIAM J. Comput.*, 5:691–703, 1976.

[86] S. Even and R. E. Tarjan. Network flow and testing graph connectivity. *SIAM J. Comput.*, 4:507–518, 1975.

[87] T. E. Feather. *The parallel complextity of some flow and matching problems.* PhD thesis, University of Toronto, Toronto, Canada, 1984.

[88] D. Fernandez-Baca and C. U. Martel. On the efficiency of maximum-flow algorithms on networks with small integer capacities. *Algorithmica*, 4:173–189, 1989.

[89] L. R. Ford and D. R. Fulkerson. Maximal flow through a network. *Canad. J. Math.*, 8:399–404, 1956.

[90] L. R. Ford and D. R. Fulkerson. A simple algorithm for finding maximal network flows and an application to the Hitchcock problem. *Canad. J. Math.*, 9:210–218, 1957.

[91] L. R. Ford and D. R. Fulkerson. Constructing maximal dynamic flows from static flows. *Operations Res.*, 6:419–433, 1958.

[92] L. R. Ford and D. R. Fulkerson. A suggested computation for maximal multicommodity network flow. *Manag. Sci.*, 5:97–101, 1958.

[93] L. R. Ford and D. R. Fulkerson. A network flow feasibility theorem and combinatorial applications. *Canad. J. Math.*, 11:440–450, 1959.

[94] L. R. Ford and D. R. Fulkerson. *Flows in Networks.* Princeton Univ. Press, Princeton, NJ, 1962.

[95] A. Frank. Augmenting graphs to meet edge-connectivity requirements. *SIAM J. Discr. Math.*, 5:25–53, 1992.

[96] A. Frank and É. Tardos. An application of simultaneous Diophantine approximation in combinatorial optimization. *Combinatorica*, 7:49–65, 1987. Preliminary version: An application of simultaneous approximations in combinatorial optimization, Proc. 26th Annual IEEE Symp. Foundations of Computer Science, 459–463, 1985.

[97] H. Frank and I. T. Frisch. *Communication, Transmission, and Transportation networks.* Addison-Wesley Publ. Comp., New York, 1971.

[98] G. N. Frederickson. Fast algorithms for shortest paths in planar graphs, with applications. *SIAM J. Comput.*, 16:1004–1022, 1987.

[99] T. Fujisawa. Maximal flow in a lossy network. In *Proc. Allerton Conf. on Circuit and System Theory*, pages 385–393, 1963.

[100] S. Fujishige. Algorithms for solving the independent-flow problem. *J. Oper. Res. Soc. Japan*, 21:189–204, 1978.

[101] S. Fujishige. A capacity-rounding algorithms for the minimum-cost circulation problem: a dual framework of the Tardos algorithm. *Math. Programming*, 35:298–308, 1986.

[102] S. Fujishige. An out-of-kilter method for submodular flows. *Discrete Applied Math.*, 17:3–16, 1987.

[103] S. Fujishige, A. Nakayama, and W.-T. Cui. On the equivalence of the maximum balanced flow problem amd the weighted minimax flow problem. *Operations Res. Lett.*, 5:207–209, 1986.

[104] S. Fujishige, A. Röck, and U. Zimmermann. A strongly polynomial algorithm for minimum cost submodular flow problems. *Math. Oper. Res.*, 14:60–69, 1989.

[105] D. R. Fulkerson. An out-of-kilter method for minimal cost flow problem. *SIAM J. Appl. Math.*, 9:18–27, 1961.

[106] D. R. Fulkerson and G. B. Dantzig. Computation of maximum flow in networks. *Naval Res. Log. Quart.*, 2:277–283, 1955.

[107] H. N. Gabow. Scaling algorithms for network problems. *J. Comput. Syst. Sci.*, 31:148–168, 1985.

[108] H. N. Gabow and R. E. Tarjan. Faster scaling algorithms for network problems. *SIAM J. Comput.*, 18:1013–1036, 1989.

[109] D. Gale. A theorem on flows in networks. *Pacific J. Math.*, 7:1073–1082, 1957.

[110] D. Gale. Transient flows in networks. *Michigan J. Math.*, 6:59–63, 1959.

[111] Z. Galil. An $O(n^{5/3}m^{2/3})$ algorithm for the maximal flow problem. *Acta Inf.*, 14:221–242, 1980. Preliminary version in Proc. 19th Annual IEEE Symp. Foundations of Computer Science, pages 231–245, 1978.

[112] Z. Galil. On the theoretical efficiency of various network flow algorithms. *Theoretical Comput. Sci.*, 14:103–111, 1981.

[113] Z. Galil and A. Naamad. An $O(EV \log^2 V)$ algorithm for the maximal flow problem. *J. Comput. Syst. Sci.*, 21:203–217, 1980. Preliminary version as "Network flow and generalized path compression" in Proc. 11th Annual ACM Symp. Theory of Computing, pages 13–26, 1979.

[114] Z. Galil and É. Tardos. An $O(n^2(m + n \log n) \log n)$ min-cost flow algorithm. *J. ACM*, pages 374–386, 1988. Preliminary version in Proc. 27th Annual IEEE Symp. Foundations of Computer Science, pages 1–9, 1986.

[115] G. Gallo, M. D. Grigoriadis, and R. E. Tarjan. A fast parametric maximum flow algorithm and applications. *SIAM J. Comput.*, 18:30–55, 1989.

[116] M. R. Garey and D. S. Johnson. *Computers and Intractibility, a Guide to the Theory of NP-Completeness*, chapter A2. W.H. Freeman and Co., San Francisco, 1979.

[117] F. Glover, D. Karney, and D. Klingman. Implementation and computational comparisons of primal, dual and primal-dual computer codes for minimum cost network flow problem. *Networks*, 4:191–212, 1974.

[118] F. Glover, D. Karney, D. Klingman, and A. Napier. A computational study on start procedures, basis change criteria, and solution algorithms for transportation problem. *Manag. Sci.*, 20:793–813, 1974.

[119] A. V. Goldberg. A new max-flow algorithm. Technical Report MIT/LCS/TM-291, Lab. for Computer Science, MIT, Cambridge, Mass., 1985.

[120] A. V. Goldberg. *Efficient graph algorithms for sequential and parallel computers.* PhD thesis, Dept. of Electr. Engin. and Computer Science, MIT, Cambridge, Mass., 1987. Also available als Technical Report TR-374, Lab. for Computer Science, MIT, Cambridge, Mass., 1987.

[121] A. V. Goldberg. Processor-efficient implementation of a maximum flow problem. *Inf. Process. Lett.*, 38:179–185, 1991.

[122] A. V. Goldberg. An efficient implementation of a scaling minimum-cost flow algorithm. Technical Report STAN-CS-92-14139, Department of Computer Science, Stanford University, Stanford, CA, Aug. 1992.

[123] A. V. Goldberg. A natural randomization strategy for multicommodity flow and related problems. *Inf. Process. Lett.*, 42:249–256, 1992.

[124] A. V. Goldberg, M. D. Grigoriadis, and R. E. Tarjan. Efficiency of the network simplex algorithm for the maximum flow problem. Technical Report STAN-CS-89-1248, Department of Computer Science, Stanford University, Stanford, CA, 1989. To appear in *Math. Programming.*

[125] A. V. Goldberg, M. D. Grigoriadis, and R. E. Tarjan. Use of dynamic trees in a network simplex algorithm for the maximum flow problem. *Math. Programming*, 50:277–290, 1991.

[126] A. V. Goldberg and M. Kharitonov. On implementing scaling push-relabel algorithms for the minimum-cost flow problem. Technical Report STAN-CS-92-1418, Department of Computer Science, Stanford University, Stanford, CA, Mar. 1992.

[127] A. V. Goldberg, S. A. Plotkin, and É. Tardos. Combinatorial algorithms for the generalized circulation problem. *Math. Oper. Res.*, 16:351–381, 1991. Preliminary version in Proc. 29th Annual IEEE Symp. Foundations of Computer Science, pages 432–443, 1988.

[128] A. V. Goldberg, S. A. Plotkin, and P. M. Vaidya. Sublinear-time parallel algorithms for matching and related problems. In *Proc. 29th Annual IEEE Symp. Foundations of Computer Science*, pages 174–185, 1988.

[129] A. V. Goldberg, É. Tardos, and R. E. Tarjan. Network flow algorithms. In B. Korte, L. Lovász, H. Prömel, and A. Schrijver, editors, *Flows, paths, and VLSI-layout*, pages 101–164, 1990. Previously published as Tech. Rep. STAN-CS-89-1252, Department of Computer Science, Stanford University, March 1989.

[130] A. V. Goldberg and R. E. Tarjan. A new approach to the maximum flow problem. *J. ACM*, 35:921–940, 1988. Preliminary version in Proc. 18th Annual ACM Symp. Theory of Computing, pages 136–146, 1986.

[131] A. V. Goldberg and R. E. Tarjan. Finding minimum-cost circulations by canceling negative cycles. *J. ACM*, 36:873–886, 1989. Preliminary version in Proc. 20th Annual ACM Symp. Theory of Computing, pages 388–397, 1987.

[132] A. V. Goldberg and R. E. Tarjan. A parallel algorithm for finding a blocking flow in an acyclic network. *Inf. Process. Lett.*, 31:265–271, 1989.

[133] A. V. Goldberg and R. E. Tarjan. Finding minimum-cost circulations by successive approximation. *Math. Oper. Res.*, 15:430–466, 1990. Preliminary version published as MIT/LCS/TM-333, MIT, 1987, and as Solving minimum-cost flow problems by successive approximation, Proc. 19th Annual ACM Symp. Theory of Computing, pages 7–18.

[134] B. Golden and T. L. Magnanti. Deterministic network optimization: A bibliography. *Networks*, 7:149–183, 1977.

[135] D. Goldfarb and M. D. Grigoriadis. A computational comparison of the Dinic and network simplex methods for maximum flow. In B. Simeone, et al., editor, *FORTRAN Codes for Network Optimization*, Annals of Operations Research, vol. 13, pages 83–124, 1988.

[136] D. Goldfarb and J. Hao. A primal simplex algorithm that solves the maximum flow problem in at most $O(nm)$ pivots and $O(n^2m)$ time. *Math. Programming*, 47:353–363, 1990.

[137] D. Goldfarb and J. Hao. On strongly polynomial variants of the network simplex algorithm for the maximum flow problem. *Operations Res. Letters*, 10:383–387, 1991.

[138] D. Goldfarb, J. Hao, and S. Kai. Anti-stalling pivot rules for the network simplex algorithm. *Networks*, 20:79–91, 1990.

[139] L. M. Goldschlager, R. A. Shaw, and J. Staples. The maximum flow problem is log space complete for P. *Theoretical Comput. Sci.*, 21:105–111, 1982.

[140] R. E. Gomory and T. C. Hu. Multi-terminal network flows. *J. SIAM*, 9:551–570, 1961.

[141] R. E. Gomory and T. C. Hu. An application of generalized linear programming to network flows. *J. SIAM*, 10:260–283, 1962.

[142] R. E. Gomory and T. C. Hu. Synthesis of a communication network. *J. SIAM*, 12:348–369, 1964.

[143] M. Gondran and M. Minoux. *Graphs and Algorithms*, chapter 5 and 6. Wiley-Interscience, New York, 1984.

[144] F. Granot and R. Hassin. Multi-terminal maximum flows in node capacitated networks. *Discrete Applied Math.*, 13:157–163, 1986.

[145] F. Granot and M. Penn. On the integral plane two-commodity flow problem. *Operations Res. Lett.*, 11:135–139, 1992.

[146] F. Granot and A. F. Veinott Jr. Substitutes, complements and ripples in network flows. *Math. Oper. Res.*, 10:471–497, 1985.

[147] M. D. Grigoriadis. An efficient implementation of the network simplex method. *Math. Prog. Study*, 26:83–111, 1986.

[148] M. D. Grigoriadis and W. W. White. A partitioning algorithm for the multicommodity network flow problem. *Math. Programming*, 3:157–177, 1972.

[149] G. R. Grimmett and W.-C. S. Suen. The maximal flow through a directed graph with random capacities. *Stochastics*, 8:153–159, 1982.

[150] G. R. Grimmett and D. J. A. Welsh. Flow in networks with random capacities. *Stochastics*, 7:205–229, 1982.

[151] R. C. Grinold. Calculating maximal flows in a network with positive gains. *Operations Res.*, 21:528–541, 1973.

[152] M. Grötschel, L. Lovász, and A. Schrijver. *Geometric Algorithms and Combinatorial Optimization.* Springer-Verlag, Berlin, 1988.

[153] G. Guisewite and P. M. Pardalos. Minimum concave cost network flow problems: applications, complexity, and algorithms. *Annals of Operations Research*, 25:125–190, 1990.

[154] R. P. Gupta. On flows in pseudosymmetric networks. *J. SIAM*, 14:215–225, 1966.

[155] D. Gusfield. Simple constructions for multi-terminal network flow synthesis. *SIAM J. Comput.*, 12:157–165, 1983.

[156] D. Gusfield. Very simple methods for all pairs network flow analysis. *SIAM J. Comput.*, 19:143–155, 1990.

[157] D. Gusfield. Computing the strength of a graph. *SIAM J. Comput.*, 20:639–654, 1991.

[158] D. Gusfield and C. Martel. A fast algorithm for the generalized parametric minimum cut problem and applications. *Algorithmica*, 7:499–519, 1992.

[159] D. Gusfield, C. Martel, and D. Fernandez-Baca. Fast algorithms for bipartite network flow. *SIAM J. Comput.*, 16:237–251, 1987.

[160] D. Gusfield and D. Naor. Efficient algorithms for generalized cut trees. In *Proc. 1st Annual ACM-SIAM Symp. Discrete Algorithms*, pages 422–433, 1990.

[161] H. Hamachar. Numerical investigations on the maximal flow algorithm of Karzanov. *Computing*, 22:17–29, 1979.

[162] H. Hamachar and L. R. Foulds. Algorithms for flows with parametric capacities. *Z. Oper. Res.*, 33:21–37, 1989.

[163] J. Hao and J. B. Orlin. A faster algorithm for finding the minimum cut in a graph. In *Proc. 3rd Annual ACM-SIAM Symp. Discrete Algorithms*, pages 165–174, 1992.

[164] J. K. Hartman and L. S. Lasdon. A generalized upper-bounding algorithm for multicommodity network flow problems. *Networks*, 1:333–354, 1971.

[165] R. Hassin. Maximum flow in (s,t) planar networks. *Inf. Process. Lett.*, 13:107–107, 1981.

[166] R. Hassin. Minimum cost flow in set-constraints. *Networks*, 12:1–21, 1982.

[167] R. Hassin. The minimum cost flow problem: a unifying approach to dual algorithms and a new tree search algorithm. *Math. Programming*, 25:228–239, 1983.

[168] R. Hassin. On multicommodity flow in planar graphs. *Networks*, 14:225–235, 1985.

[169] R. Hassin. Algorithms for the minimum cost circulation problem based on maximizing the mean improvement. *Operations Res. Lett.*, 12:227–233, 1992.

[170] R. Hassin and D. B. Johnson. An $O(n \log^2 n)$ algorithm for maximum flow in undirected planar networks. *SIAM J. Comput.*, 14:612–624, 1985.

[171] R. Hassin and E. Zemel. Probabilistic analysis of the capacitated transportation problem. *Math. Oper. Res.*, 13:80–89, 1988.

[172] R. V. Helgason and J. L. Kennington. An efficient procedure for implementing a dual simplex network flow algorithm. *A.I.I.E. Trans.*, 9:63–68, 1977.

[173] F. L. Hitchcock. The distribution of a product from several sources to numerous facilities. *J. Math. Phys.*, 20:224–230, 1941.

[174] D. S. Hochbaum and A. Segev. Analysis of a flow problem with fixed charges. *Networks*, 19:291–312, 1989.

[175] T. C. Hu. Multicommodity network flows. *Operations Res.*, 11:344–360, 1963.

[176] T. C. Hu. *Integer Programming & Network Flows.* Addison-Wesley Publ. Comp., Reading, Mass., 1969.

[177] T. C. Hu. *Combinatorial Algorithms*, chapter 2.1, 2.2 and 2.3. Addison Wesley Publ. Comp., Reading, Mass., 1982.

[178] T. C. Hu and M. T. Shing. Multiterminal flows in outerplanar graphs. *J. Algorithms*, 4:241–261, 1983.

[179] T. C. Hu and M. T. Shing. A decomposition algorithm for multi-terminal network flows. Technical Report TRCS 84-08, Department of Computer Science, University of California, Santa Barbara, CA, 1984.

[180] C. A. J. Hurkens, A. Schrijver, and É. Tardos. On fractional multicommodity flows and distance functions. *Discrete Math.*, 73:99–109, 1989.

[181] T. Ichimori, H. Ishii, and T. Nishida. Weighted minimax real-valued flow. *J. Oper. Res. Soc. Japan*, 24:52–59, 1981.

[182] H. Imai. On the practical efficiency of various maximum flow algorithms. *J. Oper. Res. Soc. Japan*, 26:61–82, 1983.

[183] H. Imai and K. Iwano. Efficient sequential and parallel algorithms for planar minimum cost flow. In T. Asano, T. Ibaraki, H. Imai, and T. Nishizeki, editors, *Proc. SIGAL International Symposium on Algorithms SIGAL '90*, Lecture Notes in Computer Science, vol. 450, pages 21–30, Springer-Verlag, Berlin, 1990.

[184] M. Iri. A new method of solving transportation-network problems. *J. Oper. Res. Soc. Japan*, 3:27–87, 1960.

[185] M. Iri. *Network Flows, Transportation and Scheduling.* Academic Press, New York, 1969.

[186] A. Itai. Two-commodity flow. *J. ACM*, 25:596–611, 1978.

[187] A. Itai and D. K. Pradhan. Synthesis of directed multicommodity flow networks. *Networks*, 14:213–224, 1984.

[188] A. Itai and M. Rodeh. Scheduling transmissions in a network. *J. Algorithms*, 6:409–429, 1985.

[189] A. Itai and Y. Shiloach. Maximum flows in planar networks. *SIAM J. Comput.*, 8:135–150, 1979.

[190] A. V. Iyer, J. J. Jarvis, and H. D. Ratliff. Hierarchical solution to network flow problems. *Networks*, 20:731–752, 1990.

[191] L. Janiga and V. Koubek. A note on finding cuts in directed planar networks by parallel computation. *Inf. Process. Lett.*, 21:75–78, 1985.

[192] J. J. Jarvis. On the equivalence between node-arc and arc-chain formulations for the multicommodity maximal flow problem. *Naval Res. Log. Quart.*, 16:525–529, 1969.

[193] J. J. Jarvis and A. M. Jezior. Maximal flow with gains through a special network. *Operations Res.*, 20:678–688, 1972.

[194] P. A. Jensen and W. Barnes. *Network Flow Programming.* John Wiley & Sons, New York, 1980.

[195] P. A. Jensen and G. Bhaumik. A flow augmentation approach to the network with gains minimum cost flow problem. *Manag. Sci.*, 23:631–643, 1977.

[196] W. S. Jewell. Optimal flow through networks. Interim Technical Report No. 8, Operations Research Center, MIT, Cambridge, Mass., 1958.

[197] W. S. Jewell. Optimal flow through networks with gains. *Operations Res.*, 10:476–499, 1962.

[198] W. S. Jewell. A primal-dual multicommodity flow algorithm. ORC Report 66-24, Operations Research Center, University of California, Berkeley, CA, 1966.

[199] W. S. Jewell. Multicommodity network solutions. In *Théorie des graphes*. Dunod, Paris, page 183, 1967.

[200] D. B. Johnson. Parallel algorithms for minimum cuts and maximum flows in planar networks. *J. ACM*, 34:950–967, 1987. Preliminary version in Proc. 23rd Annual IEEE Symp. Foundations of Computer Science, pages 244–254, 1982.

[201] D. B. Johnson and S. M. Venkatesan. Using divide and conquer to find flows in directed planar networks in $O(n^{3/2} \log n)$ time. In *Proc. 20th Annual Allerton Conf. on Communication, Control, and Computing*, pages 898–905, Univ. of Illinois, Urbana-Champaign, IL., 1982.

[202] D. B. Johnson and S. M. Venkatesan. Partition of planar flow networks. In *Proc. 24th Annual IEEE Symp. Foundations of Computer Science*, pages 259–264, 1983.

[203] D. S. Johnson and C. C. McGeoch. DIMACS implementation challenge workshop algorithms for network flow and matching. Technical Report 92-4, DIMACS, New Brunswick, NJ, 1992.

[204] E. L. Johnson. Networks and basis solutions. *Operations Res.*, 14:619–624, 1966.

[205] S. Kapoor and P. M. Vaidya. Fast algorithms for convex quadratic programming and multicommodity flows. In *Proc. 18th Annual ACM Symp. Theory of Computing*, pages 147–159, 1986.

[206] R. M. Karp. A characterization of the minimum cycle mean in a digraph. *Discrete Math.*, 23:309–311, 1978.

[207] R. M. Karp, E. Upfal, and A. Wigderson. Constructing a maximum matching is in Random NC. *Combinatorica*, 6:35–48, 1986.

[208] A. V. Karzanov. Determining the maximal flow in a network by the method of preflows. *Soviet Math. Dokl.*, 15:434–437, 1974.

[209] A. V. Karzanov. Half-integral five-terminus flows. *Discrete Applied Math.*, 18:263–278, 1987.

[210] N. Katoh. An efficient algorithm for the bicriteria minimum-cost circulation problem. *J. Oper. Res. Soc. Japan*, 32:420–440, 1989.

[211] J. L. Kennington. Survey of linear cost multicommodity network flows. *Operations Res.*, 26:209–236, 1978.

[212] J. L. Kennington and R. V. Helgason. *Algorithms for Network Programming.* Wiley-Interscience, New York, 1980.

[213] J. L. Kennington and M. Shalaby. An effective subgradient procedure for minimal cost multicommodity flow problems. *Manag. Sci.*, 23:994–1004, 1977.

[214] D. B. Khang and O. Fujiwara. Approximate solutions of capacitated fixed-charge minimum cost network flow problems. *Networks*, 21:689–704, 1991.

[215] S. Khuller and J. Naor. Flow in planar graphs with vertex capacities. Technical Report 90-1089, Computer Science Department, Cornell University, Ithaca, NY, Jan. 1990.

[216] S. Khuller, J. Naor, and P. Klein. The lattice structure of flow in planar graphs. Technical Report UMIACS-TR-2566, Univ. of Maryland Inst. for Advanced Computer Studies, 1990.

[217] S. Khuller and B. Schieber. Efficient parallel algorithms for testing k-connectivity and finding disjoint $s-t$ paths in graphs. *SIAM J. Comput.*, 20:352–375, 1991. Preliminary version in Proc. 30th Annual IEEE Symp. Foundations of Computer Science, pages 288-293, 1989.

[218] A. B. Kinariwala and A. G. Rao. Flow switching approach to the maximum flow problem. *J. ACM*, 24:630–645, 1977.

[219] V. King, S. Rao, and R. E. Tarjan. A faster deterministic maximum flow algorithm. In *Proc. 3rd Annual ACM-SIAM Symp. Discrete Algorithms*, pages 157–165, 1992.

[220] M. Klein. A primal method for minimal cost flows with applications to the assignment and transportation problems. *Manag. Sci.*, 14:205–220, 1967.

[221] P. Klein, A. Agrawal, R. Ravi, and S. Rao. Approximation through multicommodity flow. In *Proc. 31th Annual IEEE Symp. Foundations of Computer Science*, pages 726–737, 1990.

[222] P. Klein, C. Stein, and É. Tardos. Leighton-Rao might be practical: faster approximation algorithms for concurrent flow with uniform capacities. In *Proc. 22th Annual ACM Symp. Theory of Computing*, pages 310–321, 1990.

[223] D. J. Kleitman. An algorithm for certain multicommodity flow problems. *Networks*, 1:75–90, 1971.

[224] J. G. Klincewicz. A Newton method for convex separable network flow problems. *Networks*, 13:427–442, 1983.

[225] D. Klingman, A. Napier, and J. Stutz. NETGEN: A program for generating large scale capacitated assignment, transportation, and minimum cost flow network problems. *Manag. Sci.*, 20:814–821, 1974.

[226] E. Knapp. An exercise in the formal derivation of parallel programs: Maximum flows in graphs. *ACM Trans. Program. Lang. Syst.*, 12:203–223, 1990.

[227] T. C. Koopmans. Optimum utilization of the transportation system. In *Proc. International Statistical Conference*, Washington, D.C., 1947. Also reprinted as supplement to *Econometrica* **17**, 1949.

[228] V. Koubek and A. Riha. The maximum k-flow in a network. In J. Gruska and M. Chytil, editors, *Proc. Mathem. Foundations of Computer Science*, Lecture Notes in Computer Science, vol. 118, pages 389–397, Springer-Verlag, Berlin, 1981.

[229] L. Kucera. Maximum flow in planar networks. In J. Gruska and M. Chytil, editors, *Proc. Mathem. Foundations of Computer Science*, Lecture Notes in Computer Science, vol. 118, pages 418–422, Springer-Verlag, Berlin, 1981.

[230] L. Kucera. Finding a maximum flow in /s,t/-planar network in linear expected time. In M. P. Chytil and V. Koubek, editors, *Proc. Mathem. Foundations of Computer Science*, Lecture Notes in Computer Science, vol. 176, pages 370–377, Springer-Verlag, Berlin, 1984.

[231] E. L. Lawler. *Combinatorial Optimization: Networks and Matroids*, chapter 4, 6.3 and 7.11. Holt, Rinehart and Winston, New York, 1976.

[232] E. L. Lawler. Shortest path and network flow algorithms. *Annuls of Discrete Math.*, 4:251–263, 1979.

[233] E. L. Lawler. An introduction to polymatroidal network flows. In G. Ausiello and M. Lucertini, editors, *Analysis and Design of Algorithms in Combinatorial Optimization*, International Centre for Mechanical Sciences, Courses and Lectures - No. 266, pages 129–146. Springer-Verlag, Vienna, 1981.

[234] E. L. Lawler and C. U. Martel. Computing maximal "polymatroidal" network flow. *Math. Oper. Res.*, 7:334–347, 1982.

[235] T. Leighton, F. Makedon, S. Plotkin, C. Stein, É. Tardos, and S. Tragoudas. Fast approximation algorithms for multicommodity flow problems. In *Proc. 23rd Annual ACM Symp. Theory of Computing*, pages 101–111, 1991.

[236] T. Leighton and S. Rao. An approximate max-flow min-cut theorem for uniform multicommodity flow problems with applications to approximation algorithms. In *Proc. 29th Annual IEEE Symp. Foundations of Computer Science*, pages 422–431, 1988.

[237] T. Lengauer and K. W. Wagner. The binary network flow problem is logspace complete for P. *Theoretical Comput. Sci.*, 75:357–363, 1990. A preliminary version was part of: T. Lengauer and K. W. Wagner, The correlation between the complexities of non-hierarchical and hierarchical versions of graph problems. In: F. J. Brandenburg, G. Vidal-Nacquet and M. Wirsing (eds.), Proc. STACS 87 - 4th Annual Symp. on Theor. Aspects of Computer Science, Lecture Notes in Computer Science, vol. 247, pages 100–113, Springer-Verlag, Berlin, 1987.

[238] R. J. Lipton and R. E. Tarjan. A separator theorem for planar graphs. *SIAM J. Appl. Math.*, 36:177–189, 1979.

[239] M. V. Lomonosov. On the planar integer two-flow problem. *Combinatorica*, 3:207–218, 1983.

[240] M. V. Lomonosov. Combinatorial approaches to multiflow problems. *Discrete Applied Math.*, 11:1–94, 1985.

[241] M. Malek-Zavarei and J. K. Aggarwal. Optimal flow in networks with gains and costs. *Networks*, 1:355–365, 1972.

[242] M. Malek-Zavarei and I. T. Frisch. On the fixed cost flow problem. *Int. J. Control*, 16:897–902, 1972.

[243] V. M. Malhotra, M. P. Kumar, and S. N. Maheshwari. An $O(n^3)$ algorithm for finding maximum flows in networks. *Inf. Process. Lett.*, 7:277–278, 1978.

[244] J. M. Marberg and E. Gafni. An $O(n^2m^{1/2})$ distributed max-flow algorithm. In S. Sahni, editor, *Proc. International Conf. on Parallel Processing*, pages 213–216, 1987.

[245] C. Martel. A comparison of phase and non-phase network flow algorithms. *Networks*, 19:691–705, 1989.

[246] K. Matsumoto, T. Nishizeki, and N. Saito. An efficient algorithm for finding multicommodity flows in planar networks. *SIAM J. Comput.*, 14:289–302, 1985.

[247] K. Matsumoto, T. Nishizeki, and N. Saito. Planar multicommodity flows, maximum matchings and negative cycles. *SIAM J. Comput.*, 15:495–510, 1986.

[248] J. F. Maurras. Optimization of the flow through networks with gains. *Math. Programming*, 3:135–144, 1972.

[249] N. Megiddo. Optimal flows in networks with multiple sources and sinks. *Math. Programming*, 7:97–107, 1974.

[250] N. Megiddo. A good algorithm for lexicographically optimal flows in multi-terminal networks. *Bull. of the AMS*, 83:97–107, 1977.

[251] K. Mehlhorn. *Data structures and Algorithms; vol. 2, Graph Algorithms and NP-completeness*, chapter IV.9. Springer-Verlag, Berlin, 1984.

[252] G. L. Miller and J. Naor. Flow in planar graphs with multiple sources and sinks, extended abstract. In *Proc. 30th Annual IEEE Symp. Foundations of Computer Science*, pages 112–117, 1989.

[253] E. Minieka. Optimal flow in a network with gains. *INFOR*, 10:171–178, 1972.

[254] E. Minieka. Parametric network flows. *Operations Res.*, 20:1162–11678, 1972.

[255] E. Minieka. *Optimization Algorithms for Networks and Graphs.* Marcel Dekker, New York, 1978.

[256] M. Minoux. Résolution des problèmes de multiflots en nombres entier dans les grands résaux. *RAIRO*, 3:21–40, 1975.

[257] M. Minoux. Flots équilibrés et flots avec sécurité. *E.D.F.-Bull. Direction Etudes et Recherches, série C – Mathém., Inform.*, 1:5–16, 1976.

[258] M. Minoux. Multiflots de coût minimal avec fonctions de coût concaves. *Annls Télécommun.*, 31:77–92, 1976.

[259] M. Minoux. A polynomial algorithm for minimum quadratic cost flow problems. *Europ. J. Oper. Res.*, 18:377–387, 1984.

[260] M. Minoux. Network synthesis and optimum network design problems: Models, solution methods and applications. *Networks*, 19:313–360, 1989.

[261] G. J. Minty. Monotone networks. *Proc. Royal Soc. London*, A(257):194–212, 1960.

[262] J. S. B. Mitchell. On maximum flows in polyhedral domains. *J. Comput. Syst. Sci.*, 40:88–123, 1990.

[263] K. Mizuno, S. Mizuno, and M. Mori. A polynomial time interior point algorithm for minimum cost flow problems. *J. Oper. Res. Soc. Japan*, 33:157–167, 1990.

[264] J. Mulvey. Pivot strategies for primal-simplex network codes. *J. ACM*, 25:266–270, 1978.

[265] K. G. Murty. *Linear and Combinatorial Programming.* John Wiley & Sons, New York, 1976.

[266] H. Nagamochi and T. Ibaraki. On max-flow min-cut and integral flow properties for multicommodity flows in directed networks. *Inf. Process. Lett.*, 31:279–285, 1989.

[267] H. Nagamochi and T. Ibaraki. Multicommodity flows in certain planar directed networks. *Discrete Applied Math.*, 27:125–145, 1990.

[268] H. Nagamochi and T. Ibaraki. Maximum flows in probabilistic networks. *Networks*, 21:645–666, 1991.

[269] H. Nagamochi and T. Ibaraki. Computing edge-connectivity in multigraphs and capacitated networks. *SIAM J. Discr. Math.*, 5:54–66, 1992.

[270] A. Nakayama. A polynomial algorithm for the maximum balanced flow problem with a constant balancing rate function. *J. Oper. Res. Soc. Japan*, 29:400–410, 1986.

[271] A. Nakayama. A polynomial-time dual simplex algorithm for the minimum cost flow problem. *J. Oper. Res. Soc. Japan*, 30:265–289, 1987.

[272] A. Nakayama. A polynomial-time binary search algorithm for the maximum balanced flow problem. *J. Oper. Res. Soc. Japan*, 33:1–11, 1990.

[273] A. Nakayama. NP-completeness and approximation algorithm for the maximum integral vertex-balanced flow problem. *J. Oper. Res. Soc. Japan*, 34:13–27, 1991.

[274] T. Nishizeki and N. Chiba. *Planar Graphs: Theory and Algorithms*, chapter 11. Annals of Discrete Mathematics, vol. 32. North-Holland Publ. Comp., Amsterdam, 1988.

[275] H. Okamura. Multicommodity flows in graphs. *Discrete Applied Math.*, 6:55–62, 1983.

[276] H. Okamura and P. D. Seymour. Multicommodity flows in planar graphs. *J. Combin. Theory*, B-31:75–81, 1981.

[277] K. Onaga. Dynamic programming of optimum flows in lossy communication nets. *IEEE Trans. Circuit Th.*, CT-13:282–287, 1966.

[278] K. Onaga. Optimal flows in general communication networks. *J. Franklin Inst.*, 283:308–327, 1967.

[279] J. B. Orlin. Maximum throughput-dynamic networks flows. *Math. Programming*, 27:214–231, 1983.

[280] J. B. Orlin. Genuinely polynomial simplex and non-simplex algorithms for the minimum cost flow problem. Technical Report 1615-84, Sloan School of Management, MIT, Cambridge, Mass., 1984. Also as CWI-OS R8504, Center for Mathematics and Computer Science, Amsterdam, 1985.

[281] J. B. Orlin. Minimum convex cost dynamic network flows. *Math. Oper. Res.*, 9:190–207, 1984.

[282] J. B. Orlin. On the simplex algorithm for networks and generalized networks. *Math. Prog. Stud.*, 24:166–178, 1985.

[283] J. B. Orlin. A faster strongly polynomial minimum cost flow algorithm. In *Proc. 20th Annual ACM Symp. Theory of Computing*, pages 377–387, 1988. To appear in Operations Res.

[284] J. B. Orlin and R. K. Ahuja. New distance-directed algorithms for maximum flow and parametric maximum flow problems. Technical Report 1908-87, Sloan School of Management, MIT, Cambridge, Mass., 1987.

[285] J. B. Orlin and R. K. Ahuja. New scaling algorithms for assignment and minimum cycle mean problems. *Math. Programming*, 54:41–56, 1988.

[286] M. Padberg and G. Rinaldi. An efficient algorithm for the minimum capacity cut problem. *Math. Programming*, 47:19–36, 1990.

[287] C. H. Papadimitriou and K. Steiglitz. *Combinatorial Optimization, Algorithms and Complexity.*, chapter 4.3, 5.6, 6, 7, 9 and 10.3. Prentice-Hall, Inc., Englewood Cliffs, NJ, 1982.

[288] A. B. Philpott. Continuous-time flows in networks. *Math. Oper. Res.*, 15:640–661, 1990.

[289] S. A. Plotkin and É. Tardos. Improved dual network simplex. In *Proc. 1st Annual ACM-SIAM Symp. Discrete Algorithms*, pages 367–376, 1990.

[290] J. Ponstein. On the maximal flow problem with real arc capacities. *Math. Programming*, 3:254–256, 1972.

[291] R. B. Potts and R. M. Oliver. *Flows in Transportation Networks.* Academic Press, New York, 1972.

[292] P. S. Pulat. A decomposition algorithm to determine the maximum flow in a generalized network. *Comput. Oper. Res.*, 16:161–172, 1989.

[293] P. S. Pulat. Maximum outflow in generalized flow networks. *Europ. J. Oper. Res.*, 43:65–77, 1989.

[294] A. P. Punnen. A linear time algorithm for the maximum capacity path. *Europ. J. Oper. Res.*, 53:402–404, 1991.

[295] M. Queyranne. Theoretical efficiency of the algorithm "capacity" for the maximum flow problem. *Math. Oper. Res.*, 5:258–266, 1980.

[296] T. Radzik and A. V. Goldberg. Tight bounds on the number of minimum-mean cycle cancellations and related results. In *Proc. 2nd Annual ACM-SIAM Symp. Discrete Algorithms*, pages 110–119, 1991.

[297] V. Ramachandran. The complexity of minimum cut and maximum flow problems in an acyclic network. *Networks*, 17:387–392, 1987.

[298] K. G. Ramakrishnan. Solving two-commodity transportation problems with coupling constraints. *J. ACM*, 27:736–757, 1980.

[299] J. H. Reif. Minimum s-t cut of a planar undirected network in $O(n \log^2(n))$ time. *SIAM J. Comput.*, 12:71–81, 1983.

[300] H. Röck. Scaling techniques for minimum cost network flows. In U. Pape, editor, *Discrete Structures and Algorithms*, pages 181–191, Carl Hansen Verlag, München, 1980.

[301] R. T. Rockafellar. *Network Flows and Monotropic Optimization.* John Wiley & Sons, New York, 1984.

[302] B. Rothfarb and I. T. Frisch. On the 3-commodity flow problem. *SIAM J. Appl. Math.*, 17:46–58, 1969.

[303] B. Rothfarb, N. P. Shein, and I. T. Frisch. Common terminal multicommodity flow. *Operations Res.*, 16:202–205, 1968.

[304] B. Rothschild and A. Whinston. Feasibility of two commodity network flows. *Operations Res.*, 14:1121–1129, 1966.

[305] B. Rothschild and A. Whinston. On two commodity network flows. *Operations Res.*, 14:377–387, 1966.

[306] G. Ruhe. Parametric maximal flows in generalized networks – complexity and algorithms. *Optimization*, 19:235–251, 1988.

[307] H. M. Safer. Scaling algorithms for distributed max flow. Technical report, Sloan School of Management, MIT, Cambridge, Mass., 1988.

[308] R. Saigal. Multicommodity flows in directed networks. Operations Research Center, University of California, Berkeley, CA, 1968.

[309] M. Sakarovitch. The multicommodity maximum flow problem. ORC Report 66-25, Operations Research Center, University of California, Berkeley, CA, 1968.

[310] M. Sakarovitch. Two commodity network flows and linear programming. *Math. Programming*, 4:1–20, 1973.

[311] B. Schieber and S. Moran. Parallel algorithms for maximum bipartite matchings and maximum 0–1 flows. *J. of Parallel and Distributed Computing*, 6:20–38, 1989.

[312] A. Schrijver. Applications of polyhedral combinatorics to multicommodity flows and compact surfaces. Technical Report CWI-BS-R8921, Center for Mathematics and Computer Science, Amsterdam, 1989.

[313] A. Schrijver. The Klein bottle and multicommodity flows. *Combinatorica*, 9:375–384, 1989.

[314] A. Schrijver. Short proofs on multicommodity flows and cuts. Technical Report CWI-BS-R8922, Center for Mathematics and Computer Science, Amsterdam, 1989.

[315] A. Segall. Decentralized maximum-flow protocols. *Networks*, 12:213–230, 1982.

[316] M. Sengoku, S. Skinoda, and R. Yatsuboshi. On a function for the vulnerability of a directed flow network. *Networks*, 18:73–83, 1988.

[317] M. Serna and P. Spirakis. Tight RNC approximations to maxflow. Technical Report TR 90.01.1, Computer Technology Institute, Patras University, Patras, Greece, 1990.

[318] P. D. Seymour. The matroids with the max-flow min-cut property. *J. Comb. Theory*, B-23:189–222, 1977.

[319] P. D. Seymour. A two-commodity cut theorem. *Discrete Math.*, 23:341–355, 1978.

[320] P. D. Seymour. A short proof of the two-commodity flow theorem. *J. Comb. Theory*, B-26:370–371, 1979.

[321] P. D. Seymour. Four-terminus flows. *Networks*, 10:79–86, 1980.

[322] P. D. Seymour. On odd cuts and planar multicommodity flows. *Proc. London Mathem. Soc.*, 42:178–192, 1981.

[323] F. Shahrokhi. Approximation algorithms for the maximum concurrent flow problem. *ORSA Jrnl. on Computing*, 1:62–69, 1989.

[324] F. Shahrokhi and D. Matula. The maximum concurrent flow problem. *J. ACM*, 37:318–334, 1990.

[325] Y. Shiloach. An $O(nI \log^2 I)$ maximum flow algorithm. Technical Report STAN-78-702, Department of Computer Science, Stanford University, Stanford, CA, 1978.

[326] Y. Shiloach. Multi-terminal $0-1$ flows. *SIAM J. Comput.*, 8:422–430, 1979.

[327] Y. Shiloach. A multi-terminal minimum cut algorithm for planar graphs. *SIAM J. Comput.*, 9:214–219, 1980.

[328] Y. Shiloach and U. Vishkin. An $O(n^2 \log n)$ parallel max-flow algorithm. *J. Algorithms*, 3:128–146, 1982.

[329] M. T. Shing and P. K. Agarwal. Multi-terminal flows in planar networks. Technical Report TRCS 86-07, Department of Computer Science, University of California, Santa Barbara, CA, 1986.

[330] J. F. Sibeyn. A pseudo-polylog time parallel maxflow algorithm. Technical Report RUU-CS-90-17, Department of Computer Science, University of Utrecht, Utrecht, The Netherlands, 1990.

[331] K. Simon. On minimum flow and transitive reduction. In *Proc. 15th Intern. Coll. on Automata, Languages and Programming*, Lecture Notes in Computer Science, vol. 317, pages 535–546, Springer-Verlag, Berlin, 1988.

[332] D. D. Sleator and R. E. Tarjan. An $O(nm \log n)$ algorithm for maximum network flow. Technical Report STAN-CS-80-831, Department of Computer Science, Stanford University, Stanford, CA, 1980.

[333] D. D. Sleator and R. E. Tarjan. A data structure for dynamic trees. *J. Comput. Syst. Sci.*, 26:362–390, 1983.

[334] D. D. Sleator and R. E. Tarjan. Self adjusting binary search trees. *J. ACM*, 32:652–686, 1985.

[335] J. E. Somers. Maximum flow in networks with a small number of random arc capacities. *Networks*, 12:242–253, 1982.

[336] H. Soroush and P. B. Mirchandani. The stochastic multicommodity flow problem. *Networks*, 20:121–155, 1990.

[337] Y. Soun and K. Truemper. Single commodity representation of multicommodity networks. *SIAM J. Algebraic Discrete Methods*, 1:348–358, 1980.

[338] V. Srinivasan and G. L. Thompson. Accelerated algorithms for labeling and relabeling of trees, with applications to distribution problems. *J. ACM*, 19:712–726, 1972.

[339] V. Srinivasan and G. L. Thompson. Benefit-cost analysis of coding techniques for primal transportation problems. *J. ACM*, 20:194–213, 1973.

[340] H. Suzuki, T. Nishizeki, and N. Saito. Algorithms for multicommodity flows in planar graphs. *Algorithmica*, 4:471–501, 1989. Preliminary version in Proc. 17th Annual ACM Symp. Theory of Computing, pages 195–204, 1985.

[341] É. Tardos. A strongly polynomial minimum cost circulation algorithm. *Combinatorica*, 5:247–255, 1985.

[342] É. Tardos. Improved approximation algorithm for concurrent multi-commodity flows. Technical Report 872, School of Operations Research and Industrial Engineering, Cornell University, 1989.

[343] É. Tardos, C. Tovey, and M. Trick. Layered augmented path algorithms. *Math. Oper. Res.*, 11:362–370, 1986.

[344] R. E. Tarjan. *Data Structures and Network Algorithms*, chapter 8. SIAM, Philadelphia, PA, 1983.

[345] R. E. Tarjan. A simple version of Karzanov's blocking flow algorithm. *Operations Res. Lett.*, 2:265–268, 1984.

[346] R. E. Tarjan. Algorithms for maximum network flow. *Math. Prog. Study*, 26:1–11, 1986.

[347] R. E. Tarjan. Efficiency of the primal network simplex algorithm for the minimum-cost circulation problem. *Math. Oper. Res.*, 16:272–291, 1991.

[348] N. Tomizawa. On some techniques useful for solution of transportation network problems. *Networks*, 1:173–194, 1972.

[349] J. A. Tomlin. Minimum-cost multicommodity network flows. *Operations Res.*, 14:45–51, 1966.

[350] L. E. Trotter, Jr. On the generality of multi-terminal flow theory. *Annals of Discrete Math.*, 1:517–525, 1977.

[351] K. Truemper. On max flows with gains and pure minimum cost flows. *SIAM J. Appl. Math.*, 32:450–456, 1977.

[352] K. Truemper. Optimal flows in nonlinear gain networks. *Networks*, 8:17–36, 1978.

[353] K. Truemper. Max-flow min-cut matroids: polynomial testing and polynomial algorithms for maximum flow and shortest routes. *Math. Oper. Res.*, 12:72–96, 1987.

[354] P. Tseng, D. P. Bertsekas, and J. N. Tsitsiklis. Partially asynchronous, parallel algorithms for network flows and other problems. *SIAM J. Control & Optim.*, 28:678–710, 1990.

[355] A. Tucker. A note on the convergence of the Ford-Fulkerson flow algorithm. *Math. Oper. Res.*, 2:143–144, 1977.

[356] P. M. Vaidya. Speeding-up linear programming using fast matrix multiplication, (extended abstract). In *Proc. 30th Annual IEEE Symp. Foundations of Computer Science*, pages 332–337, 1989.

[357] J. van Leeuwen. Graph algorithms. In J. van Leeuwen, editor, *Handbook of Theoretical Computer Science, vol. A: Algorithms and Complexity*, pages 525–631. Elsevier Science Publ., Amsterdam, 1990.

[358] U. Vishkin. A parallel blocking flow algorithm for acyclic networks. *J. Algorithms*, 13:489–501, 1992.

[359] S. W. Wallace. Investing in arcs in a network to maximize the expected max flow. *Networks*, 17:87–103, 1987.

[360] A. Weintraub. A primal algorithm to solve network flow problems with convex costs. *Manag. Sci.*, 21:87–97, 1974.

[361] R. J. Wittrock. Operator assignment and the parametric preflow problem. *Manag. Sci.*, 38:1354–1359, 1992.

[362] R. D. Wollmer. Multicommodity networks with resource constraints: the generalized multicommodity flow problem. *Networks*, 1:245–263, 1972.

[363] M. A. Yakovleva. A problem on minimum transportation cost. In V. S. Nemchinov, editor, *Applications of Mathematics in Economic Research*, pages 390–399, Izdat. Social'no-Ekon. Lit., Moscow, 1959.

[364] N. E. Young, R. E. Tarjan, and J. B. Orlin. Faster parametric shortest path and minimum balance algorithms. *Networks*, 21:205–221, 1991.

[365] N. Zadeh. Theoretical efficiency of the Edmonds-Karp algorithm for computing maximal flows. *J. ACM*, 19:184–192, 1972.

[366] N. Zadeh. A bad network problem for the simplex method and other minimum cost flow algorithms. *Math. Programming*, 5:255–266, 1973.

[367] N. Zadeh. More pathological examples for network flow problems. *Math. Programming*, 5:217–224, 1973.

[368] W. I. Zangwill. Minimum concave cost flows in certain networks. *Manag. Sci.*, 14:429–450, 1968.

[369] C.-Q. Zhang. Minimum cycle coverings and integer flows. *J. Graph Theory*, 14:537–546, 1990.

[370] U. Zimmermann. Minimization on submodular flows. *Discrete Applied Math.*, 4:303–323, 1982.

Network Optimization Problems, pp. 333-353
Eds. D.-Z. Du and P.M. Pardalos

Tabu Search: Applications and Prospects

Stefan Voß
Technische Hochschule Darmstadt, FB 1 / FG Operations Research, Hochschulstraße 1, D - 6100 Darmstadt, Germany

Abstract

Tabu Search is a metastrategy for guiding known heuristics to overcome local optimality. Successful applications of this kind of metaheuristic to a great variety of problems have been reported in the literature. In this paper we consider two applications of tabu search with special emphasis on dynamic tabu list management.

Although still in its infancy, recently some implementations of tabu search on parallel computers have come up. Whereas these implementations are tailored to specific problems we attempt to provide ideas for a more general concept for developing parallel tabu search algorithms.

1 Introduction

Due to the complexity of a great variety of combinatorial optimization problems, heuristic algorithms are especially relevant for dealing with these large scale problems. The main drawback of algorithms such as deterministic exchange procedures is their inability to continue the search upon becoming trapped in a local optimum. This suggests consideration of recent techniques for guiding known heuristics to overcome local optimality. Following this theme, we investigate the application of the *tabu search* metastrategy for solving combinatorial optimization problems.

The first part of this paper considers two specific applications of tabu search to the multiconstraint zero-one knapsack problem and to the quadratic semi-assignment problem. These applications have been performed on a sequential computer. Although usually a very fast method, in real-world applications of e.g. the quadratic semi-assignment problem sometimes very large computation times for tabu search may be observed. Therefore, the idea of parallelization as it recently came up for tabu search, too, may be especially relevant in reducing CPU-times for the algorithms under consideration.

The key issue in designing parallel algorithms is to decompose the execution of the various ingredients of a procedure into processes executable by parallel processors. Improvement procedures like tabu search or simulated annealing at first glance, however, have an intrinsic sequential nature due to the idea of performing the neighbourhood search from one solution to the next. Therefore, there is not yet a common or generally applicable parallelization of tabu search in the literature. In the second part of this paper we attempt to describe some general ideas and a classification scheme for parallel tabu search algorithms.

Following the above framework the paper is organized as follows. In Section 2, we present an outline of tabu search. Section 3 describes our application of tabu search to two combinatorial optimization problems. For the multiconstraint zero-one knapsack problem we report some improvements in comparison to an already known tabu search method. Computational results are performed over a sample of 57 problems known from the literature. With respect to the quadratic semi-assignment problem we review results derived for a real-world application arising in the field of schedule synchronization in public mass transit systems. Before describing some concepts for parallel tabu search algorithms in more detail we briefly discuss the common parallel machine models and algorithms in Section 4. Some examples are given and finally some conclusions are drawn (Section 5). The attempt, of course, is not to give a complete treatment of parallel tabu search but to sketch the potential this area of research carries.

2 Tabu Search

Many solution approaches are characterized by identifying a neighbourhood of a given solution which contains other (*transformed*) solutions that can be reached in a single iteration. A transition from a feasible solution to a transformed feasible solution is referred to as a *move* and may be described by a set of one or more *attributes*. For example, in a zero-one integer programming context these attributes may be the set of all possible value assignments (or changes in such assignments) for the binary variables. Then two attributes e and $\bar{e}$, which denote that a certain binary variable is set to 1 or 0, may be called *complementary* to each other. Following a steepest descent/mildest ascent approach, a move may either result in a best possible improvement or a least deterioration of the objective function value. Without additional control, however, such a process can cause a locally optimal solution to be re-visited immediately after moving to a neighbour.

To prevent the search from endlessly cycling between the same solutions, tabu search may be visualized as follows. Imagine that the attributes of all moves are stored in a list, named a *running list*, representing the trajectory of solutions encountered.

Then, related to a sublist of the running list a so-called *tabu list* may be defined. Based on certain restrictions, it keeps some moves, consisting of attributes complementary to those of the running list, which will be forbidden in at least one subsequent iteration because they might lead back to a previously visited solution. Thus, the tabu list restricts the search to a subset of admissible moves (consisting of admissible attributes or combinations of attributes). This hopefully leads to 'good' moves in each iteration without re-visiting solutions already encountered. A general outline of a tabu search procedure (for solving a minimization problem) may be described as follows:

Tabu Search

Given: A feasible solution x^* with objective function value z^*.
Start: Let $x := x^*$ with $z(x) = z^*$.
Iteration:
while stopping criterion is not fulfilled[1]**do begin**

(1) select best admissible move that transforms x into x' with objective function value $z(x')$ and add its attributes to the running list

(2) perform tabu list management: compute moves to be set tabu, i.e., update the tabu list

(3) perform exchanges: $x := x', z(x) = z(x')$
if $z(x) < z^*$ **then** $z^* := z(x), x* := x$ **endif**

endwhile
Result: x^* is the best of all determined solutions, with objective function value z^*.

* * *

For a background on tabu search and a number of references on successful applications of this metaheuristic see, e.g., Glover (1989, 1990), Domschke et al. (1992), and Glover and Laguna (1992).

Tabu List Management

Tabu list management concerns updating the tabu list, i.e., deciding on how many and which moves have to be set tabu within any iteration of the search. Up to now, usually static methods have been applied in the literature as, e.g., the *tabu navigation method* (TNM).

In TNM, single attributes are set tabu as soon as their complements have been part of a selected move. The attributes stay tabu for a distinct time, i.e. number of iterations, until the probability of causing a solution's re-visit is small. The efficiency of the algorithm depends on the choice of the tabu status duration, i.e. the length tl_size of the underlying tabu list. (In the literature often a 'magic' tl_size=7 is proposed.) For the sake of an improved effectivity, a so-called *aspiration level criterion*

[1] A possible stopping criterion can be, e.g., a prespecified time limit.

is considered, which permits the choice of an attribute even when it is tabu. This can be advantageous when a new best solution may be calculated, or when the tabu status of the attributes prevent any move from feasibility.

The static approach, though successful in some applications, seems to be a rather limited one. Another probably more fruitful idea is to define an attribute as being *potentially tabu* if it belongs to a chosen move and to handle it in a *candidate list* first. Via additional criteria these attributes can be definitely included in the tabu list if necessary, or excluded from the candidate list if possible. Therefore, the candidate list is an intermediate list between a running list and a tabu list. Glover (1990) suggests the use of different candidate list strategies in order to avoid extensive computational effort without sacrificing solution quality. In the sequel, we describe the following dynamic strategies for managing tabu lists: the *cancellation sequence method* (CSM, in a revised version, cf. Dammeyer et al. (1991)), and the *reverse elimination method* (REM).

CSM as well as REM both use additional criteria for setting attributes tabu. The primary goal is to permit the reversion of any attribute but one between two solutions to prevent from re-visiting the older one. To find those *critical* moves, CSM needs a candidate list that contains the complements of attributes being potentially tabu. This *active tabu list* (ATL) is built like the running list where elimination of certain attributes is furthermore permitted. Whenever an attribute of the last performed move finds its complement on ATL this complement will be eliminated from ATL. All attributes between the cancelled one and its recently added complement build a *cancellation sequence* separating the actual solution from the solution that has been left by the move that contains the cancelled attribute. Any attribute but one of a cancellation sequence is allowed to be cancelled by future moves. This condition is sufficient but not necessary, as some aspects have to be taken into account so that CSM works well.

- Making a single attribute tabu prevents many moves which could lead to yet unvisited solutions. (An attribute becomes tabu if its complement is the only attribute of a cancellation sequence. An attribute becomes tabu for one iteration if its complement is the most recent attribute of ATL. Otherwise a cancellation sequence could not be defined between these two attributes.)

- For building a cancellation sequence, the remaining attributes of the older and the current move are not necessarily taken into consideration. This depends on the order in which the move's attributes are added to ATL.

- Those attributes of a move that did not cancel another attribute within a specific cancellation sequence are disregarded when making its last remaining attribute tabu (although they separate two solutions).

Whenever a cancellation sequence includes a smaller one the smaller sequence is said to *dominate* the larger. Then the larger cancellation sequence may be disregarded, because any of its attributes will only become tabu if they are within the smaller sequence, too.

The above mentioned aspects work well for the case that a move consists of exactly one attribute, i.e., when so-called *single-attribute moves* are considered instead of *multi-attribute moves.* In addition, the corresponding parameters have to be chosen appropriately (e.g. the tabu list duration of a tabu attribute, and how to apply the aspiration level criterion). Applying CSM to multi-attribute moves needs additional criteria to prevent errors caused by uncovered special cases. E.g. for *paired-attribute moves* (moves consisting of exactly two attributes) those moves must be prohibited that may cancel a cancellation sequence consisting of exactly two attributes (because none of them is tabu when choosing a move).

The conditions of TNM and CSM need not be necessary to prevent from re-visiting previously encountered solutions. Necessity, however, can be achieved by REM. The idea of REM is that any solution can only be re-visited in the next iteration if it is a neighbour of the current solution. Therefore, in each iteration the running list will be traced back to determine all moves which have to be set tabu (since they would lead to an already explored solution). For this purpose, a *residual cancellation sequence* (RCS) is built up stepwise by tracing back the running list. In each step exactly one attribute is processed, from last to first. After initializing an empty RCS, only those attributes are added whose complements are not in the sequence. Otherwise their complements in the RCS are eliminated (i.e. cancelled). Then at each tracing step it is known which attributes have to be reversed in order to turn the current solution back into one examined at an earlier iteration of the search. If the remaining attributes in the RCS can be reversed by exactly one move then this move is tabu in the next iteration. For single-attribute moves, for instance, the length of an RCS must be one to enforce a tabu move.

Obviously, the execution of REM represents a necessary and sufficient criterion to prevent re-visiting known solutions. Since the computational effort of REM increases if the number of iterations increases, ideas for reducing the number of computations have been developed (cf. Glover (1990) and Dammeyer and Voß(1991a)).

Search Intensification and Search Diversification

A general idea for reducing the computational effort in a tabu search algorithm is that of search intensification using a so-called *short term memory* (cp. the expression *intermediate term memory* in Glover (1989)). Its basic idea is to observe the attributes of all performed moves and to eliminate those from further consideration that have not been part of any solution generated during a given number of iterations. This results in a concentration of the search where the number of neighbourhood solutions in each iteration, and consequently the computational effort, decreases. Obviously the cost of this reduction can be a loss of accuracy.

Correspondingly, a search diversification may be defined as a *long term memory* to penalize often selected assignments. Then the neighbourhood search can be led into not yet explored regions where the tabu list operation is restarted (resulting in an increased computation time). An appealing opportunity for search diversification is created by REM. Let $t > 1$ be integer. If at any tracing step the attributes that have to be reversed to turn the current solution back into an already explored one equal exactly t moves then it is possible to set these moves tabu for the next iteration. Note that for the case of multi-attribute moves, due to various combinations of attributes to moves, even more than t moves may be set tabu in order to avoid different paths through the search space leading to the same solution. Accordingly, search diversification is obvious.

We highlight the case of $t = 2$ as a new stand-alone method called REM2 since in the same number of iterations a larger number of solutions is encountered than with REM. To have $t = 2$ means that all common neighbours of the current solution and of an already explored one are forbidden. These neighbours were implicitly investigated during a former step of the procedure (due to the choice of a best non-tabu neighbour) and need not be looked at again. Therefore, REM2 retains the nice property of being a necessary and a sufficient criterion as mentioned above without having the fear that some solutions would not be encountered, as may be the case for $t > 2$. Nevertheless, from a computational point of view REMt for $t \geq 3$ may be advantageous. However, as also with REM or REM2, *black hole suboptima* may still occur when all moves to neighbourhood solutions are tabu.

For applications and (sequential) comparisons of TNM, CSM, and REM see Dammeyer and Voß(1991b) and Domschke et al. (1992).

3 Applications

In this paragraph we report on two applications of tabu search. The first is the multiconstraint zero-one knapsack problem, which is an example for problems where single-attribute moves are performed. The second example is the quadratic semi-assignment problem, a representative for the use of paired-attribute moves.

Multiconstraint Zero-One Knapsack Problem

The *multiconstraint zero-one knapsack problem* (MCKP) is a special case of general zero-one programming with a great variety of applications in the areas of, e.g., resource allocation and capital budgeting. Given n objects with positive profits c_j and m resources with nonnegative resource consumption values a_{ij} and positive limitations b_i. With binary decision variables x_j the problem may be stated as follows:

$$\text{Maximize} \qquad Z(x) = \sum_{j=1}^{n} c_j \cdot x_j \tag{1}$$

subject to

$$\sum_{j=1}^{n} a_{ij} \cdot x_j \leq b_i \qquad i = 1, \ldots, m \tag{2}$$

$$x_j \in \{0, 1\} \qquad j = 1, \ldots, n \tag{3}$$

Various algorithms for solving this NP-hard problem have been proposed in the literature. Here we refer to Drexl (1988), who developed an efficient simulated annealing algorithm, and to Dammeyer and Voß(1991a) for an implementation of REM and a comparison with simulated annealing. To have a comparative study 57 test problems with known optimal solutions were taken as a reference in both papers. The data for these studies with n varying from 6 to 105 and m from 2 to 30 are fully reproduced in Freville and Plateau (1982, 1990). Dammeyer and Voß(1991a) show that simulated annealing may be outperformed by REM with respect to various criteria. Here we use their implementation of different versions of REM and compare them with the results of REM2 and REM3 as defined in Section 2. Despite the modification caused by this diversification approach all parameter setting is identical. We report some of the details (see Dammeyer and Voß(1991a) for a complete description).

For the purpose of finding a neighbourhood solution within tabu search the following transformation is defined:

Given: A feasible solution $\bar{x} = (x_1, \ldots, x_n)$.
Choose $j^* = \arg\max\{a_{i^*j}/c_j | x_j = 1, j = 1, \ldots, n\}$ with i^* being a bottleneck resource
with $i^* := \arg\max\{\sum_{j=1}^{n} a_{ij} \cdot x_j / b_i | i = 1, \ldots, m\}$ and define $x_j^* := 0$.
feasible := **true**;
while feasible **do**
choose $k^* = \arg\max\{c_k | x_k = 0$ **and** $a_{ik} + \sum_{j=1}^{n} a_{ij} \cdot x_j \leq b_i \quad \forall i = 1, \ldots, m\}$
if k^* exists **then** $x_{k^*} := 1$ **else** feasible := **false endif**;
endwhile;

This so-called DROP/ADD - move may be considered as a multi-attribute move with variable length depending on a specific instance of MCKP. Any such move consists of exactly one *DROP-attribute* j^* and a variable number of *ADD-attributes* according to the choice of elements k^* given in the while-loop. Our implementation considers these moves as *successive multi-attribute moves*, i.e., every move is regarded as a number of separate single-attribute moves. In any iteration the number of traces to be performed, i.e. the number of times an RCS is built, is equal to the number of attributes of the corresponding move.

For each of the 57 test problems 20 initial feasible solutions have been generated. Starting with $x = (0, ..., 0)$ for the first feasible solution we added elements according to the ADD-criterion as applied in the DROP/ADD - move as long as possible. Nine further solutions have been gained from $x = (0, ..., 0)$ by adding randomly chosen elements as long as possible. In the same way we proceeded with an initialization of $x = (1, ..., 1)$. Nine solutions have been obtained by randomly dropping elements until feasibility was achieved, and the tenth by applying a DROP-criterion inverse to the ADD-criterion. Given any of the 20 initial feasible solutions the algorithms terminate whenever there is no improvement of the best feasible solution within a certain number α of iterations. Starting from an initialization $\alpha = n$ this number is increased over time by a factor of 1.1. An additional overall stopping criterion of $10 \cdot n$ iterations did not affect the termination of tabu search. The number of tracing steps for building an RCS was limited to $4 \cdot n$.

Table 1 gives a summarized description of our results. (All programs are implemented in PASCAL and run on an IBM PS 2/70, 386 personal computer.) The first rows show a comparison of REM ($t = 1$), REM2, and REM3 as described above. Then the influence of applying the short and long term memory is analyzed in the next rows. In the short term memory (STM) an element is eliminated from further consideration if it has not been included in any solution examined during α iterations. For a combined long and short term memory (L+STM) the following modification is used. Whenever STM stops a new starting solution is obtained by choosing from those elements that have previously been eliminated according to STM and the algorithm is restarted. This procedure is repeated no more than ten times as long as a new starting solution can be found. Correspondingly in the long term memory (LTM) new starting solutions are obtained from those elements that have not been in any solution for a certain number of iterations. The first two columns of Table 1 show the number of optimal solutions found referring to the 57 test problems and referring to all instances (i.e. from $20 \cdot 57 = 1140$). Correspondingly, the average deviation from optimality is given with respect to the best found solution out of the 20 instances over the 57 test problems and referring to all 1140 instances. The average number of moves gives the average number of neighbourhood exchanges needed with respect to the best found solution over all 57 test problems. Finally, the average CPU-time referring to all 1140 instances is given.

All specified methods behave in nearly the same way. Concerning solution quality, the inclusion of STM into REMt ($t = 1, 2, 3$) gives only slightly worse results, but with a remarkable decrease in CPU-times. The most astonishing entries in Table 1 are the average number of moves needed to find the best solution referring to the most successful out of the sample of 20 trials. REMt with L+STM leads to improved results but with significantly increased CPU-times. If REMt is applied with LTM instead of L+STM the solution quality is slightly affected in some test problems with mostly increased CPU-times.

The modifications of REM proposed in this paper lead to improvements in solution quality with only slightly affected CPU-times. Although the average number of necessary neighbourhood exchanges increases, the CPU-times over all instances even might decrease because the stopping criterion gets active when no further improvements are found in a certain number of iterations. As a recommendation we conclude that REM2 should replace REM whenever this kind of dynamic tabu list management is used.

		number of optimal solutions found		average deviation from optimality (in %)		average no. of moves	average CPU-time
algorithm	t	ref. 57	all instances	ref. 57	all instances	ref. 57	all instances
REM	1	40	283	0.126	3.483	11	4.85
	2	43	302	0.117	2.925	17	5.30
	3	44	223	0.088	2.681	18	4.97
REM	1	39	268	0.130	4.207	9	2.99
with STM	2	43	261	0.119	3.904	12	2.89
	3	41	210	0.113	3.871	17	2.69
REM	1	44	591	0.101	0.558	28	14.59
with L+STM	2	48	617	0.073	0.504	36	19.33
	3	49	605	0.061	0.491	65	19.90
REM	1	45	519	0.095	0.646	42	33.50
with LTM	2	47	677	0.097	0.462	48	32.31
	3	50	633	0.064	0.494	66	30.71

Table 1: Numerical results for MCKP

Further significant improvements for all versions of REM are still possible, e.g., when increasing the factor for modifying α up to 5. The average deviation values for REM2 with LTM decrease to 0.062 and 0.317 with 769 out of 1140 instances solved to optimality, however, with a significant increase in CPU-times.

Quadratic Semi-Assignment Problem

Assigning items to sets such that a quadratic function is minimized may be referred to as the *quadratic semi-assignment problem* (QSAP). This problem indeed is a relaxed version of the well known quadratic assignment problem (QAP) and may be represented in a mathematical model as follows. Given sets $A = \{1, \ldots, m\}$ and $B = \{1, \ldots, n\}$ and a (not necessarily) symmetric cost matrix (c_{ihjk}). With binary variables

$$x_{ih} = \begin{cases} 1 & \text{if } h \in B \text{ is assigned to } i \in A \\ 0 & \text{otherwise} \end{cases}$$

we get the model:

$$\text{Minimize} \qquad Z(x) = \sum_{i=1}^{m} \sum_{h=1}^{n} \sum_{j=1}^{m} \sum_{k=1}^{n} c_{ihjk} \cdot x_{ih} \cdot x_{jk} \tag{4}$$

subject to

$$\sum_{h=1}^{n} x_{ih} = 1 \qquad i = 1, \ldots, m \tag{5}$$

$$x_{ih} \in \{0,1\} \qquad i = 1, \ldots, m, h = 1, \ldots, n \tag{6}$$

QSAP has been formulated in the literature by several authors with respect to different application areas like floor layout planning, certain median problems with mutual communication, and the problem of schedule synchronization in public transit networks (see, e.g., Dutta et al. (1982), Klemt and Stemme (1988), and Chhajed and Lowe (1992)). The problem also arises in certain scheduling problems where the deviation from due dates is penalized by a quadratic function. Here we focus on a real-world application in *schedule synchronization in public mass transit networks* where the objective is to minimize the total transfer waiting times of passengers in a mass transit system which is expressed as the sum of individual waiting times within given operation hours (cf. Domschke (1989), Voß(1990), and Domschke et al. (1992)).

Let there be a number of m lines or routes (Fig. 1 shows an example for a transit network with $m = 3$; note that a line is defined for one direction only). With each route i we join a set $N(i)$. Given a cycle time t_i (in time units, e.g. minutes) for route i then $N(i) = \{1, \ldots, t_i\}$ is a nodeset with each node giving a specific departure time within the cycle time. The departure time is the starting time of i at its first station so that all arrival and departure times furtheron may be easily calculated resulting in a complete time table. All routes have to be scheduled such that the above objective is minimized. Fixing the starting times corresponds to choosing exactly one representative from each set such that the sum of all arc weights of the subgraph induced by these nodes is minimal. Sets A and B denote traffic lines and possible departure times, respectively. The problem size may be referred to as 'number of lines $\times$ cycle time' with identical cycle times for all lines. So, (4) - (6) is a suitable model for schedule synchronization.

In what follows we relate specific issues of tabu search to QSAP. A transition from one feasible solution to another one needs two exchanges within the binary matrix $(x_{ih})_{m \times n}$ such that conditions (5) remain valid. Accordingly moves are *paired-attribute moves* denoting both the assignments and the type of exchange, i.e., selection for or exclusion from the (actual) solution. In more detail each move is described by two attributes of different type belonging to assignments between one element of set A and two different elements of set B. We explain the neighbourhood search by an example (cf. Domschke et al. (1992)):

Consider a QSAP with $A = \{1,2,3\}, B = \{1,2,3\}$, and a matrix of symmetric cost coefficients given in Table 2. Fig. 1 shows an underlying transit network with three routes $r = 1,2,3$ corresponding with the elements of set A. Three possible departure times $t = 1,2,3$ are assumed for each route corresponding with the elements of set B.

Fig. 2 visualizes in three different submatrices the objective function values (multiplied by 0.5 because of the symmetry of the problem) for all 3^3 feasible solutions. In

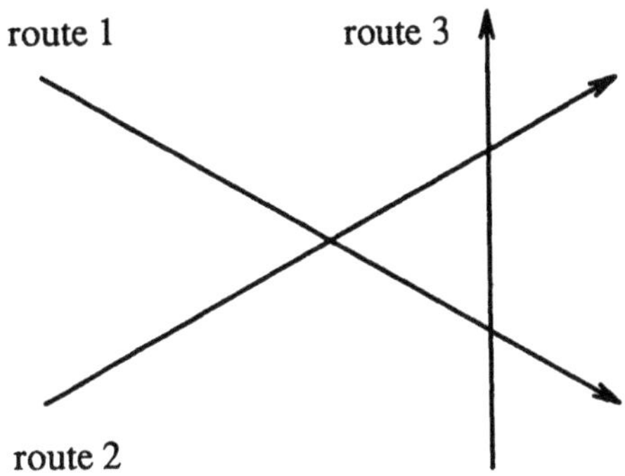

Figure 1: Transit network

	r	1			2			3		
r	t	1	2	3	1	2	3	1	2	3
1	1				8	9	[5]	[1]	6	0
	2				6	6	7	5	7	1
	3				0	3	4	8	7	3
2	1	8	6	0				5	8	2
	2	9	6	3				4	7	8
	3	5	7	4				[0]	8	9
3	1	1	5	8	5	4	0			
	2	6	7	7	8	7	8			
	3	0	1	3	2	8	9			

Table 2: Cost matrix

r1→t1

r		2		
	t	1	2	3
	1	14^{3}	14^{4}	[6]
3	2	22	22	19^{5}
	3	10	17	14^{6}

r1→t2

r		2		
	t	1	2	3
	1	16	15	12^{1}
3	2	21	20	22
	3	9	15	17

r1→t3

r		2		
	t	1	2	3
	1	13	15	12^{2}
3	2	15	17	19
	3	(5)	14	16

Figure 2: Example for TNM

a figurative sense they represent the layers of a threedimensional solution cube each of them with a fixed assignment of route 1 to a specific departure time. As starting solution we have a local optimum solution (1,3,1) by choosing route 1 to start at time 1 (i.e. assignment $r1 \rightarrow t1$), route 2 to start at time 3 ($r2 \rightarrow t3$), and route 3 to start at time 1 ($r3 \rightarrow t1$): $x_{11} = x_{23} = x_{31} = 1$ and $x_{rt} = 0$ otherwise. This solution has an objective function value of 6 (cf. Fig.2) which is calculated by the framed entries of the matrix in Table 2. In addition, Fig. 2 shows all six neighbourhood solutions of (1,3,1) by entries with upper indices. The first and the second neighbourhood solution may be derived by varying times of route 1 with fixed times of routes 2 and 3. The third and fourth may be derived by varying times of route 2 with fixed times of routes 1 and 3 and the fifth and sixth by varying route 3, correspondingly. Changing to (2,3,1) is best possible (in the sense that no better neighbourhood solution can be found). The corresponding move may be described as $(\overline{11}, 12)$ with the attributes $\overline{11}$ indicating that the binary variable x_{11} becomes 0, and 12 indicating that variable x_{12} gets the entry 1.

For applying TNM we assume a tabu list of length tl_size=4. (Note that TNM is very sensitive with respect to tl_size since tl_size=2 would lead to cycling in our example.) As the choice of the first move is not restricted by tabu attributes the move $(\overline{11}, 12)$ is selected resulting in (2,3,1). By this the complements of the corresponding attributes are stored in the tabu list preventing from any exchange of the departure time of route 1 as long as they stay tabu.[2] Thus, in the second iteration only four moves are allowed leading to solutions (2,1,1), (2,2,1), (2,3,2), and (2,3,3), respectively, of which $(\overline{23}, 22)$ is selected increasing the objective function value less than the other allowed moves. The tabu list is updated and now only allows moves according to route 3: $(\overline{31}, 32)$ and $(\overline{31}, 33)$. The better of these two exchanges results in solution (2,2,3). As the new attributes become tabu the oldest tabu attributes are freed again. TNM for this small example continues in a very restricted manner (which is caused by the relation of problem size to tl_size) and finally reaches the optimal solution (3,1,3) in the fifth iteration. In Fig. 2 the trajectory of all performed moves is presented with arrows and Table 3 shows all necessary statistics of the five iterations.

In Domschke et al. (1992) improvement procedures for QSAP are compared. Initial feasible solutions are calculated either randomly or with different versions of a regret heuristic extended by a 2-optimal exchange procedure. To sum up the main results, there is not much difference between the three tested tabu search methods in solution quality if parameters are chosen well. TNM asks for most exact determination of tl_size whereas CSM seems to be more robust according to nonoptimal parameters. CSM proves to be most independent from the starting solution quality, too. REM performs slightly worse but steadily improves a given solution with increasing CPU-time. It is the method that prevents from revisiting solutions best which can easily

[2]Note that for TNM it is not necessary to store the attributes rt when x_{rt} gets the entry 1. The proceeding becomes more transparent, however, when describing the moves in more detail.

iteration	move	solution	tabu list (tabu attributes)
		(1,3,1)	
1	$(\overline{11},12)$	(2,3,1)	$\overline{12},11$
2	$(\overline{23},22)$	(2,2,1)	$\overline{12},11,\overline{22},23$
3	$(\overline{31},33)$	(2,2,3)	$\overline{22},23,\overline{33},31$
4	$(\overline{12},13)$	(3,2,3)	$\overline{33},31,\overline{13},12$
5	$(\overline{22},21)$	(3,1,3)	$\overline{13},12,\overline{21},22$

Table 3: Example

be guaranteed by choosing suitable parameters.

Based on the procedures developed by Domschke et al. (1992) additional computational testing on large scale real-world problems has been performed including REM2 as proposed in Section 2 (as well as sequential testing of the look ahead method proposed in Section 4). The data represent schedule synchronization problems from three German cities with 14, 24, and 27 routes operating in two directions each (i.e. $m \leq 54$). A modification with respect to the above mentioned problem description concerns different cycle times for different routes as well as variable cycle times with $10 \leq t_i \leq 80$ minutes over the day.

The results are quite astonishing in the sense that an analysis of the case studies reveals very specialized data, i.e., the underlying traffic networks have a special shape. In one case it might be characterized as a star network with a central station in the downtown part of the respective city. In the other two cases different lines partly use the same tracks implicating that security distances have to be observed. Motivated by the modelling process above we may still use the QSAP model with slight modifications. This is done as follows (cf. Voß(1990)):

- For any two lines using the same tracks calculate those combinations of departure times which lead to superposition on the commonly used tracks, and define all weights in QSAP corresponding to those combinations to be ∞.

The special structure of the problems leads to the following results. Nearly in all cases the different tabu search methods are able to find only slight improvements over the initialized solutions obtained with the regret and the simple 2-optimal exchange procedures. Even the modifications of REM mentioned above do not give any additional reasonable improvements. Careful analysis of the smallest of the three examples gives us some reasoning (based on the branch and bound approach of Domschke (1989)) that the tabu search methods do not fail but that the initial feasible solutions are close to the optimum or even equal to it. In addition, even local optima close to the optimum have a great number of neighbourhood solutions each with the same objective function value such that additional testing, which is still under way, has to deal with some more sophisticated diversification structure like a combination of REMt for larger t and the long term memory. In addition, another single-attribute approach could be tested with the attributes corresponding to objective function values. With respect to the real-world application, however, the results obtained by our algorithms (including tabu search) are quite satisfactorily and promising.

4 Concepts for Parallel Tabu Search

Parallel Machine Models

Over the years a great variety of architectures have been proposed for parallel computing. The most widely known classification of parallel machine models (although somehow limited) is given by Flynn (1966). He distinguishes four general classes based on the idea of whether single or multiple instruction streams are executed on either one or multiple data set streams:

- **SISD** *(Single Instruction, Single Data)* including the classical sequential computers
- **SIMD** *(Single Instruction, Multiple Data)* including vector computers and array processors
- **MISD** *(Multiple Instructions, Single Data)*
- **MIMD** *(Multiple Instructions, Multiple Data)* with the processors performing each successive set of instructions either simultaneously (synchronous) or independently (asynchronous)

The above classification of parallel machine models may lead to different classes of parallel algorithms. *Vectorized algorithms* operate uniformly on vectors of data sets (SIMD). *Systolic* ones operate rhythmically on streams of data sets (SIMD and synchronous MIMD). *Parallel processing algorithms* operate on a set of synchronously communicating parallel processors (synchronous MIMD). Correspondingly, asynchronous communication leads to *distributed processing algorithms* (asynchronous MIMD and neural networks).

In addition to architectural aspects communication networks are used to classify parallel machine models. For instance, it makes a difference whether processors have simultaneous access to a shared memory, allowing communication between two arbitrary processors in constant time, or whether they communicate through a fixed interconnection network. Less formally, in certain models it is assumed that there is a *master* processor controlling the communication of the network, with the remaining processors of the network called *slaves*. For a comprehensive survey on parallel machines and algorithms see e.g. Akl (1989) and Van Leeuwen (1990).

The quality of parallel algorithms may be judged by a number of quantities, the most important one being the *speedup*, which is the running time of the best sequential implementation of the algorithm divided by the running time of the parallel implementation executed on a number of p processors. Similarly, given a prespecified time limit (cf. footnote 1) a *scaleup* may be defined as the ratio of the average problem sizes solvable with a parallel implementation to a sequential implementation of the algorithm. With heuristics, the solution quality attainable may also be measured. The *processor utilization* or *efficiency* is the speedup divided by p. The best one can achieve is a speedup of p and an efficiency equal to one.

Parallel Tabu Search Algorithms

Due to the success and the underlying simplicity of the main idea of tabu search, recently some implementations on parallel computers have come up tailored to specific problems. Surprisingly, to the best of our knowledge, they are solely devoted to problems using the notion of paired-attribute moves: the travelling salesman problem (see Malek et al. (1989) and Fiechter (1990)), the job shop problem (see Taillard (1989)), and the quadratic assignment problem (see Chakrapani and Skorin-Kapov (1991, 1992), Taillard (1991)).

In a first step we shall describe a *classification of different types of parallelism* that is applicable to most iterative search techniques. Its basis is the idea of having different starting solutions (so-called balls, motivated by the idea of mountains' like solution space where a ball is rolling to find a stable low altitude state) as well as a number of different strategies, e.g. based on various possibilities of the parameter setting or on the tabu list management described in Section 2.

- **SBSS** *(Single Ball, Single Strategy)*

  The algorithm starts from exactly one given feasible solution and performs its moves following exactly one strategy.

- **SBMS** *(Single Ball, Multiple Strategies)*

  The algorithm starts from exactly one given feasible solution by the use of different strategies where each strategy is performed on a different processor.

- **MBSS** *(Multiple Balls, Single Strategy)*

  The algorithm starts from different initial feasible solutions, each on a different processor. The same type of instruction, i.e. strategy, is performed on each processor.

- **MBMS** *(Multiple Balls, Multiple Strategies)*

  The algorithm starts from different initial feasible solutions performing different strategies.

In what follows we discuss the above ideas in more detail with special emphasis on further principles of parallelism within specific strategies. For ease of description we assume the notion of parallel or distributed processing algorithms.

SBSS

The single ball, single strategy idea is the simplest version, and obviously corresponds to the idea of classical sequential computations (cf. the SISD-model). This, however, does not restrict the possibility of parallelization.

Starting from an initial feasible solution, the best move which is not tabu must be performed. The search for this move may be done in parallel by decomposing the set of admissible moves into a number of subsets. E.g. in a master-slave architecture each (slave) processor may evaluate the best move in a specific subset. The best move of each subset is communicated to the master who picks the overall best as the transformed solution and also performs the tabu list management.

To restrict the amount of communication necessary for synchronizing the data each slave could determine the best possible move in its subset without observing any tabu list, while the tabu list in the same time is updated by the master. Then the master picks among all answers the best which is not tabu. If no such move exists, a second trial must be made while each processor has to receive and to observe the tabu list. Otherwise the next iteration is to be performed.

Additional ideas may be developed with respect to the specific strategies. In TNM, the tabu list management may be done by each processor itself by simply providing the most recent move (whose complement will be in the list). In CSM, the master builds the cancellation sequences and partitions them to the slaves, i.e., every slave has to evaluate a certain number of sequences. In subsequent iterations, the attributes of the current moves are communicated. Whenever a cancellation sequence is reduced to 1 it will be re-communicated to the master.

SBMS

In SBMS each processor executes a process which is one of the above tabu search strategies with different tabu conditions and parameters, like e.g. REMt for various t. For TNM this can be different (eventually randomly modified) tabu list lengths; for CSM, different tabu durations may be considered. The (slave) processors are halted after a prespecified time and the results are compared and the best one is calculated. A restart is possible with the best or a good seed solution. Each strategy may take a different path through the search space because of different tabu list management or parameter setting. A restart may be performed either with empty running and tabu lists or with a previously encountered list.

MBSS

The multiple balls approaches start from at most p (the number of processors available) different initial feasible solutions, whose calculation can vary. They may be determined either randomly or by applying different heuristics to the same problem. This may also incorporate ideas involving different diversification and intensification strategies as described above. A third possibility assumes one given feasible solution and starts with a suitable subset of its transformed (neighbourhood) solutions. (Especially with REM2 it may be assured that even in future iterations there is no overlap with the initial feasible solutions of the other processors.) The single strategy approach assumes the application of exactly one tabu search algorithm with the same parameter setting for all processors.

As with SBMS, the processes may be halted after a specific time period to coordinate their results and possibly to initiate a restart with new (hopefully) improved solutions. If the processes are performed synchronously, then the stopping may be initiated after having generated, say, m successive moves. On synchronous MIMD machines the latter approach may be especially relevant. Note that the above-mentioned possibility of parallelization within SBSS is related to a method with $m = 1$ where the best transition is evaluated. With respect to MBSS, this modifies to the evaluation of the p best moves usable for a restart. For $m \geq 2$ this approach may be used as a *look ahead method*, and is especially helpful when evaluating a region of a solution space with almost all neighbours having identical objective function values.

MBMS

The multiple balls, multiple strategies approach subsumes all previous classes, allowing search of the solution space from different starting points with different methods or parameter settings.

Examples

In the sequel we sketch some of the ideas given in the previous sections with respect to well known combinatorial optimization problems. Surprisingly, as mentioned above, we only found some work on problems with the idea of paired-attribute moves to perform the neighbourhood search. Therefore, we start with respect to binary integer programming, exploiting single-attribute moves, as is the case for MCKP described in Section 3.

Consider the SBSS concept. Also consider n decision variables in a binary problem with no (implicit or explicit) restriction on the number of variables set to either 1 or 0. We may define simple ADD- or DROP-moves by complementing the corresponding entries of the binary variables x_i. Assume the existence of $n+2$ processors with $n+2$ being the master processor. The tabu list management is performed by processor $n+1$. In any iteration of the search, each of the synchronously controlled processors

$i \in \{1, \ldots, n\}$ receives the information whose variables' entry has been chosen to be exchanged as the most recent move. This move is performed together with the reversion of x_i. This usually can be done quite efficiently by reconstructing the previous solution stored at i with at most one assignment complemented. Then i offers its objective function value to the master who re-calls all results of processors referring to non-tabu moves (evaluated by processor $n+1$). Obviously this approach may be generalized in various ways to the more general classes described above.

This concept may be applied, for instance, to MCKP, to the well-known warehouse location problem, and to Steiner's problem in graphs. If the number or weighted number of variables with value 1 is limited (as for MCKP) or fixed (as e.g. in the p-median problem) then the same approach may be applied with combined ADD/DROP- or SWAP-moves leading to paired-attribute moves.

Malek et al. (1989) follow the SBMS approach to solve travelling salesman problems (TSP) by TNM with the 2-opt exchange as moves. The tabu attributes follow different strategies in that they are restricted either to one or to the two cities that have been swapped or to the cities and their respective positions in tour. In addition different tabu parameters were used on different processors. For another parallel tabu search algorithm for the TSP see Fiechter (1990).

The quadratic assignment problem (QAP) is treated by Chakrapani and Skorin-Kapov (1991, 1992) by the use of SBSS and TNM with search intensification and search diversification performed sequentially while evaluating the moves in parallel. The set of moves is partitioned into disjoint subsets each one on a different processor as described above. The neighbourhood search is performed by pairwise interchanges such that for $O(n^2)$ processors available all moves can be evaluated in constant time, achieving a speedup of $O(n^2/\log n)$. Battiti and Tecchiolli (1992) use TNM together with a hashing function and compare their algorithm also with a parallel genetic algorithm. Another parallel algorithm for QAP based on TNM (with randomly varying tl_size) has been presented by Taillard (1991). It is an SBSS approach, too. The same idea has also been applied to the job shop as well as to the flow shop problem (see Taillard (1989, 1990)). The latter, in fact, also describes a single-attribute based implementation with attributes corresponding to objective function values. Chakrapani and Skorin-Kapov (1992) is especially relevant since its implementation is based on a connectionist approach related to a Boltzmann machine (cf. Aarts and Korst (1989)).

5 Conclusions

In this paper we have summarized some ideas for developing parallel tabu search algorithms. Motivated by a famous classification scheme for parallel machine models we proposed a classification scheme for parallel tabu search algorithms. While research in this field is still in its infancy we believe that reasonable achievements in the following two aspects will be provided.

- Development of a framework for a general parallel tabu search algorithm that can be applied to a wide range of combinatorial optimization problems.
- Empirical results for parallel tabu search algorithms tailored to specific problems.

Some results known from the literature (cf. Section 4) support this feeling. Despite the emphasis on parallel tabu search, sequential testing is still far from complete. Numerical results for the proposed algorithm REM2 have been reported in this paper, improving the previously known REM. The results are quite encouraging, especially when combining REM2 with the proposed look ahead method for $m = 2$.

In addition, the tabu search metastrategy should be tested on different classes of parallel algorithms and machine models. Especially relevant seems to be a comparison of algorithms tailored to different hardware specifications like vector computers versus synchronous and asynchronous MIMD machines. However, one should take into account identical user specifications with respect to tabu search (e.g. parameter setting, definition of the neighbourhood). Note that our classification scheme is not restriced to parallel tabu search, but may be applied for nearly any iterative search procedure, such as simulated annealing or genetic algorithms.

References

[1] E. Aarts and J. Korst (1989), *Simulated Annealing and Boltzmann Machines* (Wiley, Chichester).

[2] S.G. Akl (1989), *The Design and Analysis of Parallel Algorithms* (Prentice-Hall, Englewood Cliffs).

[3] R. Battiti and G. Tecchiolli (1992), Parallel biased search for combinatorial optimization: genetic algorithms and tabu search. Technical Report 9207-02, IRST, Istituto Trentino di Cultura, Trento.

[4] J. Chakrapani and J. Skorin-Kapov (1991), Massively parallel tabu search for the quadratic assignment problem. Working Paper Harriman School for Management and Policy, State Univ. of New York at Stony Brook.

[5] J. Chakrapani and J. Skorin-Kapov (1992), A connectionist approach to the quadratic assignment problem. *Computers & Operations Research* **19**, 287-295.

[6] D. Chhajed and T.J. Lowe (1992), M-median and M-center problems with mutual communication: solvable special cases. *Operations Research* **40**, S56-S66.

[7] F. Dammeyer, P. Forst and S. Voß(1991), On the cancellation sequence method of tabu search. *ORSA Journal on Computing* **3**, 262-265.

[8] F. Dammeyer and S. Voß(1991a), Dynamic tabu list management using the reverse elimination method. *Annals of Operations Research*, to appear.

[9] F. Dammeyer and S. Voß(1991b), Application of tabu search strategies for solving multiconstraint zero-one knapsack problems. Working paper TH Darmstadt.

[10] W. Domschke (1989), Schedule synchronization for public transit networks. *OR Spektrum* **11**, 17-24.

[11] W. Domschke, P. Forst and S. Voß(1992), Tabu search techniques for the quadratic semi-assignment problem. In: G. Fandel, T. Gulledge and A. Jones (eds.), *New Directions for Operations Research in Manufacturing* (Springer, Berlin), 389-405.

[12] A. Drexl (1988), A simulated annealing approach to the multiconstraint zero-one knapsack problem. *Computing* **40**, 1-8.

[13] A. Dutta, G. Koehler and A. Whinston (1982), On optimal allocation in a distributed processing environment. *Management Science* **28**, 839-853.

[14] C.-N. Fiechter (1990), A parallel tabu search algorithm for large traveling salesman problems. Paper presented at 1st Int. Workshop on Project Management and Scheduling, Compiegne.

[15] M.J. Flynn (1966), Very high-speed computing systems. *Proc. IEEE* **54**, 1901-1909.

[16] A. Freville and G. Plateau (1982), Methodes heuristiques performantes pour les problemes en variables 0-1 a plussieurs constraintes en inegalite. Publication ANO-91, Universite des Sciences et Techniques de Lille.

[17] A. Freville and G. Plateau (1990), Hard 0-1 multiknapsack test problems for size reduction methods. *Investigacion Operativa* **1**, 251-270.

[18] F. Glover (1989), Tabu search - part I. *ORSA Journal on Computing* **1**, 190-206.

[19] F. Glover (1990), Tabu search - part II. *ORSA Journal on Computing* **2**, 4-32.

[20] F. Glover and M. Laguna (1992), Tabu search. Working Paper Univ. of Colorado at Boulder, to appear.

[21] W.D. Klemt and W. Stemme (1988), Schedule synchronization for public transit networks. In: J.R. Daduna and A. Wren (eds.), *Computer-Aided Transit Scheduling*, Lecture Notes in Economics and Mathematical Systems 308 (Springer, Berlin), 327-335.

[22] M. Malek, M. Guruswamy, M. Pandya and H. Owens (1989), Serial and parallel simulated annealing and tabu search algorithms for the traveling salesman problem. *Annals of Operations Research* **21**, 59-84.

[23] E. Taillard (1989), Parallel taboo search technique for the jobshop scheduling problem. Working Paper Ecole Polytechnique Federale de Lausanne.

[24] E. Taillard (1990), Some efficient heuristic methods for the flow shop sequencing problem. *European Journal of Operational Research* **47**, 65-74.

[25] E. Taillard (1991), Robust taboo search for the quadratic assignment problem. *Parallel Computing* **17**, 443-455.

[26] J. Van Leeuwen (1990), *Algorithms and Complexity* (Elsevier, Amsterdam).

[27] S. Voß(1990), Network design formulations in schedule synchronization. Working Paper TH Darmstadt, to appear.

Network Optimization Problems, pp. 355-362
Eds. D.-Z. Du and P.M. Pardalos

The Shortest Path Network and Its Applications in Bicriteria Shortest Path Problems

Guo-Liang Xue
Army High Performance Computing Research Center, University of Minnesota, Suite 101, 1100 South Washington Avenue, Minneapolis, MN 55415, USA

Shang-Zhi Sun
Computer Science Department, University of Minnesota, Minneapolis, MN 55455, USA

Abstract

Let $N = (V, A, l, s)$ be a given network where $G = (V, A)$ is a simple directed graph, V is the set of n vertices, A is the set of e arcs, $l(u, v) \geq 0$ is the *length* of an arc $(u, v) \in A$, and $s \in V$ is the source. The **Shortest Path Network (SPN)** is a subnetwork of N with the property that an s—u path in N is a shortest path in N if and only if it is a path in **SPN**. The **SPN** is a counterpart of the well-known **Shortest Path Tree (SPT)**. Unlike the SPT which may not be unique for a given network, the **SPN** is unique for any given network. Also, the **SPN** provides a unified approach for solving certain kind of bicriteria or multicriteria shortest path problems where one criterion is more important than the others. We present a simple and efficient algorithm for computing the **SPN** with time complexity of $TSP(n, e)$ where $TSP(n, e)$ is the time complexity for solving the one-to-all shortest path problem on a network with n vertices and e arcs. We also present applications of the **SPN**, including a unified approach for solving the maximum capacity shortest path problem, the least risky shortest path problem, and the most reliable shortest path problem.

1 Introduction

Shortest path problems are among the most commonly encountered problems at the interface of Computer Science and Operations Research due to their important applications in communication networks and in road transportation management. In recent years, there have been increased interests in various bicriteria or multicriteria shortest path problems. Examples are the maximum capacity shortest path problem, the least risky shortest path problem, the most reliable shortest path problem, the minimum cost-reliability ratio problem, and the quickest path problem [1, 2, 3, 4, 8, 10, 11, 12].

Since the publication of Dijkstra's famous paper [6] in 1959, there have been many many papers dealing with algorithms for the one-to-all or all-to-all shortest path problems. The fastest algorithm for the one-to-all shortest path problem is the one provided by Fredman and Tarjan [7] which requires $O(e + \log\log n)$ time on a network with n vertices and e arcs by using a data structure called Fibonacci heaps. One important concept associated with the one-to-all shortest path problem is the `Shortest Path Tree (SPT)` [5] or the `Shortest Spanning Tree` [9]. The `SPT` of a network N has a nice property that the unique path from the source to any vertex in `SPT` is guaranteed to be a shortest path in the network N. However, a shortest path in the network N might not be a path in the `SPT`.

In this paper, we introduce a counterpart concept of the `SPT` called the `Shortest Path Network (SPN)` which enables a unified approach for solving certain kind of bicriteria or multicriteria shortest path problems where one criterion is more important than the other criteria and the goal is to find a path which optimizes the other criteria among all the shortest paths with respect to the most important criterion. A simple and efficient algorithm for computing the `SPN` is presented together with examples of its applications.

In section 2, we first observe the deficiency of the ordinary `Shortest Path Tree` in solving bicriteria shortest path problems. We then introduce the concept of `Shortest Path Network` and prove its existence and uniqueness. A simple algorithm is provided which computes the `Shortest Path Network` in time $TSP(n, e)$, where n and e are the number of vertices and number of arcs of the network and $TSP(n, e)$ is the time complexity of solving the one-to-all shortest path problem on that network. In section 3, we show how the `Shortest Path Network` can be used as a useful tool in solving the maximum capacity shortest problem and the least risky shortest path problems. Some conclusions are given in section 4.

2 The Shortest Path Network

Let $N = (V, A, l, s)$ be a given network where $G = (V, A)$ is a simple directed graph, V is the set of n vertices, A is the set of e arcs, $l(u, v) \geq 0$ is the *length* of an arc $(u, v) \in A$, $s \in V$ is the *source*. Applying Dijstra's one-to-all shortest path algorithm [6], we may find, in time $O(n^2)$, the shortest paths from the source s to all the other vertices of N, together with a tree rooted at s with the property that the unique path from s to any vertex u in the tree is also a shortest s—u path in N. Such a subnetwork is usually called a `Shortest Path Tree` [5] which is formally defined as follows.

Definition 2.1. Let $N = (V, A, l, s)$ be a given network with a distinguished source node s. A subnetwork `SPT` of N is be called a `Shortest Path Tree` of N If

(1). the vertex set of `SPT` are all the vertices of N which are reachable from s;
(2). any s—u path in `SPT` is a shortest s—u path in N;
(3). `SPT` is a tree rooted at s.

Given an `SPT` of N and a vertex u in N which is reachable from s, the unique s—u path in `SPT` is a shortest s-u path in N. However, the shortest s—u paths in N may not be unique. Therefore there might be another shortest s—u path in N which is not a path in `SPT`. We are interested in a subnetwork which has the property that an s—u path in N is a shortest s—u path if and only if it is a path in the subnetwork. We will call such a subnetwork a `Shortest Path Network` and it is formally defined below.

Definition 2.2. Let $N = (V, A, l, s)$ be a given network with a distinguished source node s. A subnetwork `SPN` of N will be called a `Shortest Path Network` of N If

(1). the vertex set of `SPN` are all the vertices of N which are reachable from s;
(2). any s—u path in `SPN` is a shortest s—u path in N;
(3). any shortest s—u path in N is a path in `SPN`.

The following theorem establishes the existence and uniqueness of the `SPN` and its characterization.

Theorem 2.1. Let $N = (V, A, l, s)$ be a given network where $G = (V, A)$ is a simple directed graph, V is the set of n vertices, A is the set of e arcs, $l(u, v) \geq 0$ is the *length* of an arc $(u, v) \in A$, $s \in V$ is the *source*. Then the `SPN` for N is unique and is the union of all the shortest s—u paths for $u \in V$ such that there is an s—u path in N.

Proof. Since $l(u, v) \geq 0$ for any arc $(u, v) \in A$, there is a shortest s—u path for a vertex $u \in V$ if and only if u is reachable from s. Let `Union` be the union of all the shortest s—u paths for $u \in V$ such that there is an s—u path in N. We want to show that `Union` is a `Shortest Path Network`. Clearly, the vertex set of `Union` are all the vertices of N which are reachable from s and that any shortest s—u path in N is a path in `Union`. Using the property that *every subpath of a shortest path is itself a shortest path*, one can prove that any s—u path in `Union` is a shortest s—u path in N. This shows that `Union` is a `SPN` for N. Now for any `SPN` of N, property (2) in the definition implies that the `SPN` is a subnetwork of `Union` while property (3) in the definition implies that `Union` is a subnetwork of `SPN`. Therefore `Union` is the unique `SPN` of N. □

Algorithm 2.1.

Step 1. Apply any one-to-all shortest path algorithm on N. Let $d(u)$ be the shortest distance from s to u for any $u \in V$. Let $d(u) = \infty$ when there is no s—u path in N.

Step 2. For each $u \in V$ if $d(u) = \infty$ then delete u from V and delete the arcs from A which are adjacent with u.

Step 3. For each arc $(u, v) \in A$ if $d(u) + l(u, v) > d(v)$ then delete (u, v) from A.

Figure 1: Computing the SPN from a given network.

The above proof also suggests the following algorithm for computing the SPN for a given network.

It is clear that Algorithm 2.1 correctly changes the input network $N = (V, A, l, s)$ to its unique SPN. Since steps 2 and 3 take at most $O(e)$ time, the time complexity of Algorithm 2.1 is $TSP(n, e) + O(e)$, where $TSP(n, e)$ is the time complexity for the one-to-all shortest path problem on a network with n vertices and e arcs. Since $TSP(n, e)$ is always greater than or equal to e, the time complexity of Algorithm 2.1 is $TSP(n, e)$.

In Figure 2 we illustrate a network and its unique `Shortest Path Network`. For clarity, the arcs in the SPN are drawn in thicker lines. It can be easily observed from the figure that the SPN is not a tree and that the given network has more than on SPT's.

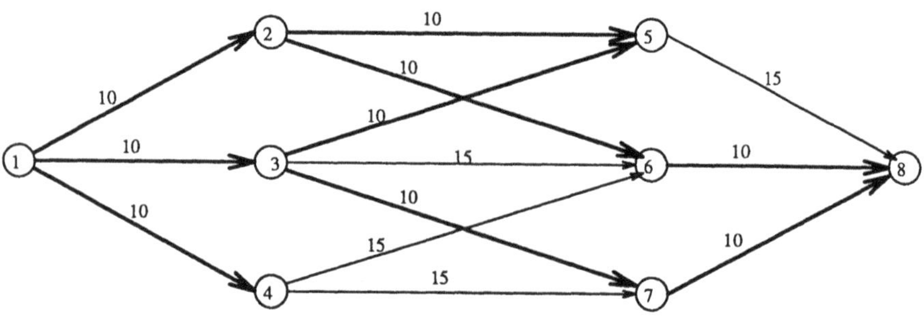

Figure 2: A network and its `Shortest Path Network`.

Once the subnetwork SPN is found, the above mentioned bicriteria shortest path problems all reduce to single criterion problems on SPN. This makes the `Shortest Path Network` a very useful concept in bicriteria shortest path problems. In the next section, we will discuss applications of the SPN.

3 Applications

In the previous section, we have introduced the concept of SPN and presented an algorithm for computing the SPN. In this section, we will show how the SPN can be used to solve various bicriteria shortest path problems in a unified approach. Specifically, we will investigate the maximum capacity shortest path problem, the minimum risky shortest path problem, and the most reliable shortest path problem.

3.1 The Maximum Capacity Shortest Path Problem

Let $N = (V, A, l, s, c)$ be a given network where $G = (V, A)$ is a simple directed graph, V is the set of n vertices, A is the set of e arcs, $l(\bullet)$ and $c(\bullet)$ are weighting functions where $l(u, v) \geq 0$ is the *length* of an arc $(u, v) \in A$ and $c(u, v) \geq 0$ is the *capacity* of that arc, and $s \in V$ is the *source*. We want to find a maximum capacity shortest path from s to u for all $u \in V$, where the length of a path in N is the sum of all the lengths of the arcs on the path while the capacity of a path in N is the minimum of all the capacities of the arcs on that path.

Here we find our first application of the Shortest Path Network. Ignoring the weighting function $c(\bullet)$ for the moment, we may find the SPN of N with respect to $l(\bullet)$. Now, for each $u \in V$, there is an s—u path if and only if u is a vertex of SPN. In addition, an s—u path in N is a maximum capacity shortest s—u path if and only if it is a maximum capacity s—u path in SPN. Therefore, the maximum capacity shortest path problem can be solved by the following algorithm (see Figure 3). For convenience, we will assume that the vertices of the SPN are labeled $1, 2, \cdots, k$ $(\leq n)$ where s is labeled 1.

Algorithm 3.1.
Step 1. Find the SPN with respect to $l(\bullet)$. Delete all the arcs and vertices of N which are not on its SPN.
Step 2. **For** $i = 1$ to k **do begin** $f[i] =$ **false**; $C[i] = c[1, i]$; **end**;
Step 3. $f[1] =$ **true**; $C[1] = \infty$;
  For $i = 1$ **to** $k - 2$ **do**
    choose $u \in argmax\{C[w] | f[w] =$ **false**$\}$;
    $f[u] =$ **true**;
    for $w = 1$ **to** k **do if not** $f[w]$ **and** $\min\{C[u] | c[u, w]\} > C[w]$
    then $C[w] = \min\{C[u], c[u, w]\}$;
  end;

Figure 3: Computing the maximum capacity shortest path.

Step 1 takes $TSP(n, e) + O(e)$ time. Step 2 takes $O(n)$ time. Step 3 takes $O(n^2)$ time. Therefore the time complexity of the algorithm is $O(n^2)$.

3.2 The Least Risky Shortest Path Problem

Let $N = (V, A, l, s, R)$ be a given network where $G = (V, A)$ is a simple directed graph, V is the set of n vertices, A is the set of e arcs, $l(\bullet)$ and $R(\bullet)$ are weighting functions where $l(u,v) \geq 0$ is the *length* of an arc $(u,v) \in A$ and $R(u,v) \geq 0$ is the *risk* of that arc, $s \in V$ is the *source*. We want to find a least risky shortest path from s to u for all $u \in V$, where the length of a path in N is the sum of all the lengths of the arcs on the path while the risk of a path in N is the sum of all the risks of the arcs on that path.

As another application of the shortest path network, we present in Figure 4 an algorithm for solving the least risky shortest path problem.

Algorithm 3.2.
Step 1. Find the `SPN` with respect to $l(\bullet)$. Delete all the arcs and vertices of N which are not in its `SPN`.
Step 2. Solve the one-to-all shortest path problem with respect to $R(\bullet)$.

Figure 4: Computing the least risky shortest path.

Step_1 takes $TSP(n,e) + O(e)$ time. Step 2 takes $TSP(n,e)$ time. Therefore, the time complexity of the algorithm is $TSP(n,e)$.

3.3 The Most Reliable Shortest Path Problem

Let $N = (V, A, l, s, r)$ be a given network where $G = (V, A)$ is a simple directed graph, V is the set of n vertices, A is the set of e arcs, $l(\bullet)$ and $r(\bullet)$ are weighting functions where $l(u,v) \geq 0$ is the *length* of an arc $(u,v) \in A$ and $r(u,v) \in (0,\ 1]$ is the *reliability* of that arc, $s \in V$ is the *source*. We want to find a most reliable shortest path from s to u for all $u \in V$, where the length of a path in N is the sum of all the lengths of the arcs on the path while the reliability of a path in N is the product of all the reliabilities of the arcs on that path.

Now replace the weighting function $r(\bullet)$ by $R(\bullet)$, where $R(u,v) = -\log(r(u,v))$ for each arc $(u,v) \in A$. Note that $R(\bullet) \geq 0$ because $r(\bullet) \in (0,\ 1]$ We will call $R(u,v)$ the *risk* of arc (u,v). Then it is clear that the most reliable shortest path problem on (V, A, s, l, r) is equivalent to the minimum risky shortest path problem on (V, A, s, l, R) which can be solved by Algorithm 3.2.

4 Conclusions

We have introduced the concept of `Shortest Path Network` which is a counterpart of the well-known `Shortest Path Tree`. Unlike the `Shortest Path Tree`, the `SPN` is unique to a given network with a distinguished source. One advantage of the `SPN` over the `SPT` is that any path in the network from the source is a shortest path if and only if it is also a path in the SPN. A simple algorithm for computing the SPN is presented and its time complexity is shown to be the same as that of computing the one-to-all shortest path problem. Examples are given to show the applications of the SPN in some bicriteria shortest path problems.

The SPN greatly simplifies the discussion/solution of certain kind of bicriteria shortest path problems. Therefore, it has both education and practical value. We hope to see more applications of the SPN.

Acknowledgment

The work of the first author was supported in part by the Army Research Office contract number DAAL03-89-C-0038 with the University of Minnesota Army High Performance Computing Research Center. The work of the second author was supported in part by the Computer Science Department of the Univeristy of Minnesota.

References

[1] R.K. Ahuja, Minimum Cost-Reliability Ratio Problem, *Computers and Operations Research*, Vol. 15 (1988), pp. 83-89.

[2] L.D. Bodin, B.L. Golden, A.A. Assad, and M.O. Ball, Routing and Scheduling of Vehicles and Crews: The State of the Art, *Computers and Operations Research*, Vol. 10 (1983), pp. 63-211.

[3] Y.L. Chen and Y.H. Chin, The Quickest Path Problem, *Computers and Operations Research*, Vol. 17 (1990), pp. 153-161.

[4] Y.L. Chen, An Algorithm for Finding the K Quickest Path in a Network, revised in *Computers and Operations Research.*

[5] R. Dial, F. Glover, D. Karney and D. Klingman, A Computational Analysis of Alternative Algorithms and Labeling Techniques for Finding Shortest Path Trees, *Networks*, Vol. 9(1979), pp. 215-248.

[6] E. Dijkstra, A Note on Two Problems in Connection with Graphs, *Numerical Mathematics*, Vol. 1 (1959), pp. 269-271.

[7] M.L. Fredman and R.E. Tarjan, Fibonacci Heaps and Their Uses in Improved Network Optimization Algorithms, *Journal of the Association of Computing Machinery*, Vol. 34 (1987), pp. 596-615.

[8] M. Minoux, Solving Combinatorial Problems with Combined Min-Max-Min-Sum Objective and Applications, *Mathematical Programming*, Vol. 45 (1989), pp. 361-372.

[9] A.R. Pierce, Bibliography on Algorithms for Shortest Path, Shortest Spanning Tree, and Related Circuit Routing Problems, *Networks*, Vol. 5 (1975), pp. 129-149.

[10] J.B. Rosen, S.Z. Sun, and G.L. Xue, Algorithms for the Quickest Path Problem and the Enumeration of Quickest Paths, *Computers and Operations Research*, Vol. 18 (1991), pp. 579-584.

[11] J.B. Rosen and G.L. Xue, Sequential and Distributed Algorithms for the All Pairs Quickest Path Problem, in *Proceedings of the 1991 International Conference on Computing and Information, Ottawa, Canada*, (Springer-Verlag, 1991), pp. 471-473.

[12] G.L. Xue, S.Z. Sun, and J.B. Rosen, Minimum Time Message Transmission in Networks, in *Proceedings of the 1992 International Conference on Computing and Information, May 28-30, 1992, Toronto, Canada*, IEEE Computer Society Press, pp. 22-25.

Network Optimization Problems, pp. 363-386
Eds. D.-Z. Du and P.M. Pardalos

A Network Formalism for Pure Exchange Economic Equilibria

Lan Zhao
Department of Mathematics, SUNY/College at Old Westbury, Old Westbury, NY 11568-0219 USA

Anna Nagurney
School of Management, University of Massachusetts, Amherst, MA 01003 USA

Abstract

In this paper we develop a network formalism for general economic equilibrium problems in the case of pure exchange economies. We first establish that the Walrasian price equilibrium (and its variational inequality formulation) is isomorphic to a network equilibrium problem with special structure. We then propose a general iterative scheme for the computation of the equilibrium prices, which contains, as special cases, the projection method and the relaxation method, and which allows for the full exploitation of the special network structure. Finally, we compare the numerical performance of the projection and the relaxation methods on several economic examples.

1 Introduction

Network equilibrium models have been used to formulate and study competitive phenomena in a wide range of applications in operations research, management science, and, more recently, in economics. Examples include: congested urban transportation systems (see, e.g., [1], [2], [3], [5], [6], [12], [15]), oligopolistic markets ([11]), spatial price equilibrium problems ([8], [10], [14], [17]), disequilibrium problems ([19]), and problems of human migration ([16], [18]).

The aforementioned models, however, have been exclusively partial equilibrium models in that only a subset of agents/activities/commodities has been incorporated within the network equilibrium framework.

In this paper, in contrast, we focus on the general economic equilibrium problem in the case of pure exchange, in which all of the commodities in an economy can be considered. Our approach utilizes the underlying special network structure of the problem, in an abstract setting, in which the nodes do not correspond to locations in space, and in which the links correspond to commodities. This underlying structure, heretofore unidentified and unexplored, motivates the subsequent algorithmic developments in this paper, with the ultimate goal being the computation of large-scale general economic equilibrium problems. As is well-known, the simplicial approximation methods pioneered by Scarf [20] for the computation of economic equilibria, in their present state of development, cannot handle large-scale problems.

In particular, we consider the variational inequality formulation of the problem recently described in Dafermos [9]. Thus far, as discussed therein, the variational inequality approach has been used to obtain only qualitative results, in the form of existence, uniqueness, and stability of pure exchange equilibria, and the computational analogue for this class of problems has not been addressed.

In Section 2 we briefly review the pure exchange or Walrasian price equilibrium model. We then establish that the problem is isomorphic to a particular network equilibrium problem with fixed demand.

In Section 3 we propose a general iterative scheme for the computation of the Walrasian equilibrium price vectors that is based on the general iterative scheme of Dafermos [7], and provide conditions for convergence.

The Walrasian iterative scheme, as is then demonstrated in Section 4, contains, as special cases, both the projection method and the relaxation method. The projection method resolves the Walrasian price equilibrium problem into a series of linear and symmetric network equilibrium problems, each of which, as we also show, can be solved exactly in closed form. The relaxation method, on the other hand, resolves the economic equilibrium problem into a series of nonlinear network flow problems, to which we then apply a network equilibration algorithm for its solution.

In Section 5 we then turn to the empirical performance of the algorithms and compare the efficiency of the relaxation method with that of the projection method on several economic examples.

In Section 6 we summarize and conclude.

The marriage of network theory and variational inequalities has already yielded efficient algorithms for a variety of applications characterized by their large-scale nature. This work brings a class of general economic equilibrium problems under the umbrella of network equilibrium.

2 The Variational Inequality Model of the Pure Exchange Economy and its Isomorphic Network Equilibrium Representation

In this section we first briefly review the pure exchange economic equilibrium model and its variational inequality formulation. We then develop its isomorphic network equilibrium representation. In particular, we consider a pure exchange economy with l commodities, price vector $p = (p_1, p_2, \ldots, p_l)^T$ taking values in the positive orthant R^l_+ and with induced aggregate excess demand function $z(p)$, with components $z_1(p), \ldots, z_l(p)$. As usual, $z(p)$ will be assumed to be homogeneous of degree zero in p, and, therefore, we may normalize prices so that they take values in the simplex:

$$S^l = \{p : p \in R^l_+, \sum_{i=1}^{l} p_i = 1\}. \tag{1}$$

As is standard in general economic equilibrium theory, the aggregate excess demand function must satisfy Walras' law:

$$p^T \cdot z(p) = 0, \quad \forall p \in S^l. \tag{2}$$

We now state the definition of a Walrasian equilibrium.

Definition 1: A price vector $p^* \in S^l$ is called a Walrasian equilibrium if the market is cleared for valuable commodities and is in excess supply for free commodities, that is,

$$\begin{aligned} z_i(p^*) &= 0 \quad \text{if} \quad p_i^* > 0 \\ z_i(p^*) &\leq 0 \quad \text{if} \quad p_i^* = 0. \end{aligned} \tag{3}$$

The following theorem shows us that Walrasian equilibrium price vectors can be characterized as solutions of a variational inequality (see, e.g., Dafermos [9] Theorem 1.1), and is included here for completeness.

Theorem 2.1 *A price vector $p^* \in S^l$ is a Walrasian equilibrium if and only if it satisfies the variational inequality*

$$z(p^*)^T \cdot (p - p^*) \leq 0, \quad \forall p \in S^l \qquad VI(z, S^l).$$

We now establish that the variational inequality model $VI(z, S^l)$ for the Walrasian price equilibrium problem is identical to the variational inequality problem governing a network equilibrium problem with a single origin-destination (O/D) pair and fixed demand.

Consider the following network equilibrium problem: A network is given consisting of a single origin node x, a single destination node y, and with a single origin/destination pair (x, y). There are l links connecting the origin/destination pair

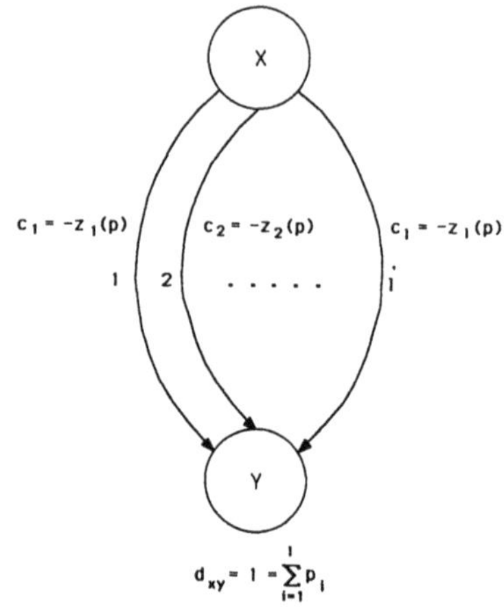

Figure 1: Network equilibrium formulation of the pure exchange economy

(x, y) (cf. Figure 1). A fixed O/D demand d_{xy} is assumed given. Let f_i be the flow passing through link i; $i = 1, \ldots, l$, and let c_i be the user cost associated with link i; $i = 1, \ldots, l$. Group the link loads into a vector $f \in R^l$, and the costs into a vector $c \in R^l$. Assume the general situation that a cost on a link may depend upon the entire link load pattern, that is, $c_i = c_i(f)$. Then f^* is a user equilibrium pattern if and only if no user has any incentive to change his path (which in the model corresponds to a link), that is, mathematically, there exists an ordering of the links $n_i; i = 1, \ldots, l$, such that

$$c_{n_1}(f^*), \ldots, c_{n_s}(f^*) = \lambda \leq c_{n_{s+1}}(f^*) \leq \cdots \leq c_{n_l}(f^*) \tag{4}$$

where

$$f^*_{n_i} \begin{cases} > 0, & i = 1, \ldots, s, \\ = 0, & i = s+1, \ldots, l. \end{cases}$$

As shown in Dafermos ([5], [6]) the above statement is equivalent to the following: A vector $f^* \in K$ is a user equilibrium load pattern if and only if it is a solution to the variational inequality

$$c(f^*)^T \cdot (f - f^*) \geq 0, \quad \forall f \in K, \tag{5}$$

where

$$K \equiv \{f : f \geq 0, \sum_{i=1}^{l} f_i = d_{xy}\}.$$

We now establish the relationship between $VI(z, S^l)$ and the above network equilibrium problem. Consider the demand

$$d_{xy} = 1,$$

the link load pattern

$$f \equiv p,$$

and the user travel cost

$$c(\cdot) \equiv -z(\cdot). \tag{6}$$

The equilibrium condition of the network with the cost vector defined in (6) is:

$$z_i(p^*)\begin{cases} = \lambda, & \text{if} \quad p_i^* > 0 \\ \leq \lambda, & \text{if} \quad p_i^* = 0. \end{cases} \tag{7}$$

Multiplying now the above inequalities by p_i^*; $i = 1, \ldots, l$, summing then the resulting equalities, and using Walras' law, we obtain

$$\lambda = p^{*T} \cdot z(p^*) = 0;$$

thus, the equilibrium condition (7) of the above network with the cost function defined in (6) is identical to the equilibrium condition (3) of the pure exchange economy. Furthermore, variational inequality (5) which governs the traffic network equilibrium problem described above coincides with $VI(z, S^l)$.

Since the variational inequality problem $VI(z, S^l)$ and, hence, the Walrasian equilibrium problem is isomorphic to the above user equilibrium network problem with disjoint paths, we can develop algorithms for the network problem which exploit the disjoint path structure in order to compute the Walrasian price equilibrium.

3 A General Iterative Scheme for the Computation of Walrasian Price Equilibrium

In this section we develop a general iterative scheme for the computation of Walrasian price equilibria, which at each step allows for the exploitation of the special network structure depicted in Figure 1.

In studying algorithms and their convergence, the standard assumption in the economics literature (cf. Scarf [20]) is that the aggregate excess demand function $z(p)$ is well-defined and continuous on all of S^l. In this paper we also make this assumption.

The Iterative Scheme

Construct a smooth function $g(p, q) : S^l \times S^l \mapsto R^l$ with the following properties:

(i) $g(p, p) = -z(p), \quad \forall p \in S^l$,

(ii) for every fixed $p, q \in S^l$, the $l \times l$ matrix $\nabla_p g(p,q)$ is positive definite.

Any smooth function $g(p,q)$ with the above properties generates the following algorithm.

Step 0: Initialization

Start with some $p^0 \in S^l$. Set $k := 1$.

Step 1: Construction and Computation

Compute p^k by solving the variational inequality

$$g(p^k, p^{k-1})^T \cdot (p - p^k) \geq 0, \quad \forall p \in S^l.$$

Step 2: Convergence Verification

If $|p^k - p^{k-1}| \leq \epsilon$, with $\epsilon > 0$, a prespecified tolerance, then stop; otherwise, set $k := k+1$, and go to Step 1.

We denote the above variational inequality by $VI^k(g, S^l)$. Since $\nabla_p g(p,q)$ is positive definite, $VI^k(g, S^l)$ admits a unique solution p^k. Thus, we obtain a well-defined sequence $\{p^k\}$. It is easy to see that if the sequence $\{p^k\}$ is convergent, say $p^k \to p^*$, as $k \to \infty$, then p^* is an equilibrium price vector, that is, it is a solution of variational inequality $VI(z, S^l)$. In fact, on account of the continuity of $g(p,q)$, $VI^k(g, S^l)$ yields

$$-z(p^*)^T \cdot (p - p^*) = g(p^*, p^*)^T \cdot (p - p^*) = \lim_{k \to \infty} g(p^k, p^{k-1})^T \cdot (p - p^k) \geq 0, \, \forall p \in S^l$$

so that p^* is a solution of the original variational inequality $VI(z, S^l)$.

The problem is now to find conditions on $g(p,q)$ which guarantee that the sequence $\{p^k\}$ is convergent. Let $|\cdot|$ denote the usual Euclidean norm in the space R^l and let $\|\cdot\|$ denote the norm of the operator $Q : G^{\frac{1}{2}}V \mapsto R^l$,

$$\|Q\| = \sup_{u \in G^{\frac{1}{2}}V, |u|=1} |Qu| \tag{8}$$

where

$$G(p,q) = \frac{1}{2}(\nabla_p g(p,q) + \nabla_p g^T(p,q)),$$

which, in view of condition (ii), is positive definite,

$$V = \{v : v \in R^l, \sum_{i=1}^{l} v_i = 0\} \tag{9}$$

and

$$G^{\frac{1}{2}}V = \{u : u = G^{\frac{1}{2}}(p,q)v, v \in V\}. \tag{10}$$

We now present conditions for convergence.

Theorem 3.1 *Assume that*

$$\|G^{-\frac{1}{2}}(p^1,q^1)\nabla_q g(p^2,q^2)G^{-\frac{1}{2}}(p^3,q^3)\| < 1, \tag{11}$$

for all $(p^1,q^1),(p^2,q^2),(p^3,q^3) \in S^l$. *Then the sequence* $\{p^k\}$ *obtained by solving* $VI^k(g,S^l)$ *is Cauchy in* S^l.

Proof: Let $p = p^{k+1}$ for $VI^k(g,S^l)$, that is,

$$g(p^k,p^{k-1})^T \cdot (p^{k+1}-p^k) \geq 0, \tag{12}$$

and let $p = p^k$ for $VI^{k+1}(g,S^l)$, that is,

$$g(p^{k+1},p^k)^T \cdot (p^k - p^{k+1}) \geq 0. \tag{13}$$

Adding (12) and (13), we obtain

$$(g(p^k,p^{k-1}) - g(p^{k+1},p^k))^T \cdot (p^{k+1}-p^k) \geq 0, \tag{14}$$

or

$$\begin{aligned}&(g(p^{k+1},p^k) - g(p^k,p^k))^T \cdot (p^{k+1}-p^k)\\ &\leq (g(p^k,p^{k-1}) - g(p^k,p^k))^T \cdot (p^{k+1}-p^k).\end{aligned} \tag{15}$$

By the Mean Value Theorem, there exists a $t \in (0,1)$, such that

$$\begin{aligned}(g(p^{k+1},p^k) - g(p^k,p^k))^T \cdot (p^{k+1}-p^k) &= (p^{k+1}-p^k)^T\\ &\cdot\nabla_p g(tp^k+(1-t)p^{k+1},p^k)\cdot(p^{k+1}-p^k),\end{aligned} \tag{16}$$

or

$$\begin{aligned}&(g(p^{k+1},p^k) - g(p^k,p^k))^T \cdot (p^{k+1}-p^k)\\ &= \frac{1}{2}(p^{k+1}-p^k)^T \cdot (\nabla_p g(tp^k+(1-t)p^{k+1},p^k)\\ &+\nabla_p^T g(tp^k+(1-t)p^{k+1},p^k)\cdot(p^{k+1}-p^k).\end{aligned} \tag{17}$$

Let G_k be defined as

$$G_k = \frac{1}{2}(\nabla_p g(tp^k+(1-t)p^{k+1},p^k) + \nabla_p^T g(tp^k+(1-t)p^{k+1},p^k)). \tag{18}$$

Observe that G_k is symmetric and positive definite. Using now (15), (17), and (18) yields

$$(p^{k+1}-p^k)^T \cdot G_k(p^{k+1}-p^k) \leq (g(p^k,p^{k-1}) - g(p^k,p^k)) \cdot (p^{k+1}-p^k). \tag{19}$$

We define now the inner product on V as

$$(v_1, v_2)_k = v_1^T G_k v_2, \quad \forall v_1, v_2 \in V \tag{20}$$

which induces the norm

$$|v|_k = (v^T G_k v)^{\frac{1}{2}} = |G_k^{\frac{1}{2}} v|, \quad \forall \in V. \tag{21}$$

By applying the Mean Value Theorem, (19) yields

$$|p^{k+1} - p^k|_k^2 \le (p^{k-1} - p^k)^T G_{k-1}^{\frac{1}{2}} G_{k-1}^{-\frac{1}{2}}$$

$$\nabla_q g(p^k, sp^k + (1-s)p^{k-1}) G_k^{-\frac{1}{2}} G_k^{\frac{1}{2}} (p^{k+1} - p^k) \tag{22}$$

for $s \in (0,1)$. Using the Schwarz inequality and condition (11), (22) yields

$$|p^{k+1} - p^k|_k^2 \le |G_{k-1}^{\frac{1}{2}}(p^k - p^{k-1})| \cdot \|G_{k-1}^{-\frac{1}{2}} \nabla_q g(p^k, sp^k + (1-s)p^{k-1}) G_k^{-\frac{1}{2}}\|$$

$$\cdot |G_k^{\frac{1}{2}}(p^{k+1} - p^k)|$$

$$= |p^k - p^{k-1}|_{k-1} \|G_{k-1}^{-\frac{1}{2}} \nabla_q g(p^k, sp^k + (1-s)p^{k-1}) G_k^{-\frac{1}{2}}\| \cdot |p^{k+1} - p^k|_k. \tag{23}$$

Hence,

$$|p^{k+1} - p^k|_k \le \gamma |p^k - p^{k-1}|_{k-1}, \quad k = 1, 2, \ldots, \tag{24}$$

where γ is the maximum over the compact set S^l of the lefthand side of (11), which is less than 1.

¿From (24) we obtain

$$|p^{k+1} - p^k|_k \le \gamma |p^k - p^{k-1}|_{k-1} \le \cdots \le \gamma^k |p^1 - p^0|_0. \tag{25}$$

On the other hand, since $G_k; k = 1, 2, \ldots,$ is nonsingular, for every $(p, q) \in S^l \times S^l$, there is a $\beta > 0$ such that

$$|p^{k+1} - p^k| \le \beta^{-1} |p^{k+1} - p^k|_k, \quad \forall p^{k+1}, p^k;\ k = 0, 1, 2, \ldots. \tag{26}$$

Therefore, (25) yields

$$|p^{k+r} - p^k| \le \sum_{i=k}^{k+r-1} |p^{i+1} - p^i| \le \beta^{-1} \sum_{i=k}^{k+r-1} |p^{i+1} - p^i|_i$$

$$\le \beta^{-1} |p^1 - p^0|_0 \sum_{i=k}^{k+r-1} \gamma^i = \beta^{-1} |p^1 - p^0|_0 \frac{\gamma^k}{1-\gamma} \tag{27}$$

which shows that $\{p^k\}$ is a Cauchy sequence in S^l and the proof is complete.

Remark 1: Naturally, $VI^k(g, S^l)$ should be constructed in such a way so that it is easy to solve. For example, when $\nabla_p g(p,q)$ is also symmetric, $VI^k(g, S^l)$ is equivalent to the convex mathematical programming problem:

Find $p^* \in S^l$ such that

$$F(p^*) = \min_{p \in S^l} F(p), \tag{28}$$

where $F(p)$ is a strictly convex function defined by the line integral

$$F(p) = \int g(p,q)^T dp.$$

Hence, any algorithm suitable for solving (28) can then be used for solving variational inequality $VI^k(g, S^l)$.

Proposition 3.2 *Assume that the Jacobian matrix $\nabla_p g(p,q)$ is also symmetric. Then a necessary condition for (11) to hold is that the Jacobian matrix $\nabla z(p)$ is negative definite over V for any $p \in S^l$, that is,*

$$v^T \nabla z(p) v < 0, \quad \forall v \in V, v \neq 0, \forall p \in S^l. \tag{29}$$

The above condition implies that the function $-z(p)$ is strictly monotone on S^l, that is,

$$(p^1 - p^2)^T \cdot (z(p^1) - z(p^2)) < 0, \quad \forall p^1, p^2 \in S^l, p^1 \neq p^2. \tag{30}$$

Proof: Assume that condition (11) holds and select

$$p^1 = p^2 = p^3 = q^1 = q^2 = q^3.$$

Note that

$$-\nabla_p z(p) = \nabla_p g(p,p) + \nabla_q g(p,p).$$

Therefore, (11) takes the form

$$\|I + G^{-\frac{1}{2}}(p,p)\nabla_p z(p) G^{-\frac{1}{2}}(p,p)\| < 1. \tag{31}$$

Set

$$B(p) = G^{-\frac{1}{2}}(p,p)\nabla_p z(p) G^{-\frac{1}{2}}(p,p). \tag{32}$$

Substituting now (32) into (31) and expanding the lefthand side of (31), we obtain

$$\|I + B\|^2 = \sup_{u \in G^{\frac{1}{2}} V, |u|=1} |(I+B)u|^2$$

$$= \sup_{u \in G^{\frac{1}{2}} V, |u|=1} u^T (I+B)^T (I+B) u = \sup_u (1 + 2u^T B u + u^T B^T B u) < 1 \tag{33}$$

or,

$$2u^T B u < -u^T B^T B u. \tag{34}$$

Since $u = G^{\frac{1}{2}}(p,p)v$, (34) yields

$$2v^T\nabla_p z(p)v < -v^T\nabla_p^T z(p)G^{-\frac{1}{2}}(p,p)G^{-\frac{1}{2}}(p,p)\nabla_p z(p)v$$

$$= -|G^{-\frac{1}{2}}(p,p)\nabla_p z(p)v|^2 \leq 0, \quad \forall v \in V, p \in S^l, v \neq 0.$$

Hence, $\nabla_p z(p)$ is negative definite over V for any $p \in S^l$.

The proof is complete.

We would like to point out that, since $z(p)$ is homogeneous of degree zero, $\nabla z(p)$ cannot be positive definite. Therefore, $z(p)$ is never strictly monotone on a set containing a segment of the ray originating from the origin of the l-dimensional space. However, it can be strictly monotone on the $l-1$ dimensional simplex S^l (see, e.g., [9]).

4 The Projection and Relaxation Methods for the Computation of the Equilibrium Prices

In this section we show that the general iterative scheme induces a projection method and a relaxation method for the computation of the equilibrium prices. We first present the projection method and then the relaxation method. We also propose equilibration algorithms PMN and RMN for the solution of the respective symmetric network equilibrium subproblems with special structure. We note that the network subproblems induced by the projection method are characterized by linear user link cost functions, whereas those induced by the relaxation method are, in general, nonlinear.

a. The Projection Method

The projection method corresponds to the choice

$$g(p,q) = -z(q) + \frac{1}{\rho}G(p-q), \tag{35}$$

where ρ is a positive scalar and G is a fixed, symmetric positive definite matrix. In this case properties (i) and (ii) are satisfied. In fact,

(i) $g(p,q) = -z(p) + \frac{1}{\rho}G(p-p) = -z(p)$,

(ii) $\nabla_p g(p,q) = \rho^{-1}G$, is positive definite and symmetric.

Condition (11) then takes the form

$$\|I + \rho G^{-\frac{1}{2}}\nabla_p z(p)G^{-\frac{1}{2}}\| < 1. \tag{36}$$

The following lemma give conditions under which (36) is satisfied.

Lemma 4.1 *If $-z(p)$ is strongly monotone on S^l, then condition (36) is satisfied.*

Proof: Let $B(p) = G^{-\frac{1}{2}} \nabla_p z(p) G^{-\frac{1}{2}}$. By virtue of the strong monotonicity assumption, the following inequality holds:

$$v^T \nabla_p z(p) v \leq -\alpha |v|^2, \quad \forall v \in V, p \in S^l. \tag{37}$$

Since $z(p)$ is continuously differentiable on S^l, there is a sufficiently large number M bounding $\|\nabla_p^T z(p) G^{-1} \nabla_p z(p)\|$ such that

$$v^T \nabla_p^T z(p) G^{-1} \nabla_p z(p) v \leq M |v|^2, \quad \forall p \in S^l, v \in V.$$

Therefore,

$$\begin{aligned}
\|I + \rho B(p)\|^2 &= \sup_{u \in G^{\frac{1}{2}} V, |u|=1} \{u^T (I + \rho B(p))^T (I + \rho B(p)) u\} \\
&= \sup_{u \in G^{\frac{1}{2}} V, |u|=1} \{1 + 2\rho u^T B u + \rho^2 u^T B^T u\} \\
&= \sup_{v \in V} \{1 + 2\rho v^T \nabla_p z(p) v + \rho^2 v^T \nabla_p z^T(p) G^{-1} \nabla_p z(p) v\} \\
&\leq \sup_{v \in V} \{1 - 2\rho\alpha |v|^2 + \rho^2 M |v|^2\} = \sup_{v \in V} \{1 + \rho |v|^2 (\rho M - 2\alpha)\}.
\end{aligned} \tag{38}$$

The righthand side of (38) is strictly less than 1, whenever $\rho < \frac{2\alpha}{M}$. Thus, condition (11) is satisfied.

The proof is complete.

Remark 2: Define

$$\theta(\rho) = \sup_{v \in V} \{1 - 2\rho\alpha |v|^2 + \rho^2 M |v|^2\}. \tag{39}$$

We observe that it is the value of $\theta(\rho)$ that affects the speed of convergence. In fact, the smaller θ is, the quicker the sequence $\{p^k\}$ converges. From (39) we know that $\theta(\rho)$ is minimized at $\rho = \frac{\alpha}{M}$. Therefore, $\rho = \frac{\alpha}{M}$ is the optimal choice for the projection method.

With such a selected $g(p, q)$, each subproblem $VI^k(g, S^l)$ is isomorphic to the network equilibrium problem with linear link cost functions. In particular, we choose G to be the diagonal positive definite matrix of the form

$$\begin{bmatrix} \alpha_1 & \cdots & 0 \\ \vdots & \ddots & \vdots \\ 0 & \cdots & \alpha_l \end{bmatrix} \tag{40}$$

where α_i; $i = 1, 2, \ldots, l$, is any positive number. A natural choice is to have $\alpha_i = -\frac{\partial z_i}{\partial p_i}|_{p^0}$; $i = 1, 2, \ldots, l$, in which case $VI^k(g, S^l)$ is then isomorphic to the separable network equilibrium problem depicted in Figure 2. We now show that $VI^k(g, S^l)$

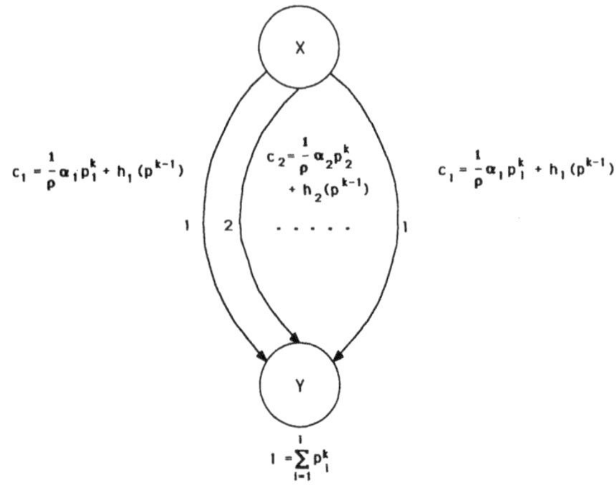

Figure 2: Network equilibrium representation of $VI^k(g, S^l)$ induced by the projection method

can be solved in closed form. We first provide the motivation for the equilibration algorithm which will yield the exact solution to $VI^k(g, S^l)$, and then its statement.

Let the components of $g(p, p^{k-1})$ be given by

$$g_i(p, p^{k-1}) = -z_i(p^{k-1}) + \frac{1}{\rho}\alpha_i(p_i - p_i^{k-1}), \quad i = 1, 2, \ldots, l, \tag{41}$$

and define

$$h_i(p^{k-1}) \equiv -z_i(p^{k-1}) - \frac{1}{\rho}\alpha_i p_i^{k-1}, \quad i = 1, 2, \ldots, l.$$

Then

$$g_i(p, p^{k-1}) = \frac{1}{\rho}\alpha_i p_i + h_i(p^{k-1}), \quad i = 1, 2, \ldots, l. \tag{42}$$

If p^k is a solution of $VI^k(g, S^l)$ (that is, p^k is the corresponding equilibrium of the network depicted in Figure 2), then we have

$$\begin{aligned} g_{n_1}(p^k, p^{k-1}) = g_{n_2}(p^k, p^{k-1}) = \ldots = g_{n_s}(p^k, p^{k-1}) = \lambda \\ \leq g_{n_{s+1}}(p^k, p^{k-1}) \leq \ldots \leq g_{n_l}(p^k, p^{k-1}) \\ p_{n_i}^k > 0, \quad i = 1, \ldots, s \\ p_{n_j}^k = 0, \quad j = s+1, \ldots, l. \end{aligned} \tag{43}$$

Substituting (42) into (43), we obtain

$$\lambda = \frac{1}{\rho}\alpha_{n_i} p_{n_i}^k + h_{n_i}, \quad i = 1, \ldots, s,$$

or

$$p_{n_i}^k = \frac{\rho}{\alpha_{n_i}}(\lambda - h_{n_i}), \quad i = 1, \ldots, s. \tag{44}$$

Summing (44) over i yields

$$\sum_{i=1}^{s} p_{n_i}^k = \lambda\rho \sum_{i=1}^{s} \frac{1}{\alpha_{n_i}} - \sum_{i=1}^{s} \frac{\rho h_{n_i}}{\alpha_{n_i}}. \tag{45}$$

Since $p^k \in S^l$, $p_{n_j} = 0$, for all $j > s$, (45) yields

$$1 = \lambda\rho \sum_{i=1}^{s} \frac{1}{\alpha_{n_i}} - \sum_{i=1}^{s} \frac{\rho h_{n_i}}{\alpha_{n_i}},$$

or

$$\lambda = \frac{1 + \sum_{i=1}^{s} \frac{\rho h_{n_i}}{\alpha_{n_i}}}{\rho \sum_{i=1}^{s} \frac{1}{\alpha_{n_i}}}. \tag{46}$$

Hence, the solution p^k is given by

$$p_{n_i}^k = \frac{\rho}{\alpha_{n_i}}(\lambda - h_{n_i}), \quad i = 1, \ldots, s$$

$$p_{n_j}^k = 0, \quad j = s+1, \ldots, l, \tag{47}$$

where λ is determined through (46).

In order to find λ, we must know the critical index s. Below we describe a procedure for finding the critical index s.

¿From (43), we obtain

$$g_{n_i}(p^k, p^{k-1}) = \lambda, \quad \text{if} \quad p_{n_i}^k > 0$$

which implies

$$h_{n_i}(p^{k-1}) < \lambda, \quad \text{if} \quad p_{n_i}^k > 0, \tag{48}$$

and

$$g_i(p^k, p^{k-1}) \geq \lambda, \quad \text{if} \quad p_{n_i}^k = 0$$

which implies

$$h_{n_i}(p^{k-1}) \geq \lambda, \quad \text{if} \quad p_{n_i}^k = 0. \tag{49}$$

Hence, s is an index such that

$$\lambda = \frac{1 + \rho \sum_{i=1}^{s} \frac{h_{n_i}}{\alpha_{n_i}}}{\rho \sum_{i=1}^{s} \frac{1}{\alpha_{n_i}}} \leq h_{n_{s+1}}. \tag{50}$$

We are now ready to state the following algorithm for solving subproblem $VI^k(g, S^l)$ where $g(\cdot)$ is specified by (35) and (40).

Algorithm PMN (equilibration algorithm for Projection Method Network subproblems)

Step 0: Sort

Sort the numbers $h_{n_i}; i = 1, 2, \ldots, l$, in nondescending order, and relabel them accordingly. Assume, henceforth, that they are relabeled. Also, define $h_{l+1} \equiv \infty$.
Set $L := 1$.

Step 1: Computation

Compute

$$\lambda^L = \frac{1 + \rho \sum_{i=1}^{L} \frac{h_i}{\alpha_i}}{\rho \sum_{i=1}^{L} \frac{1}{\alpha_i}}.$$

Step 2: Evaluation

If $h_L < \lambda^L \leq h_{L+1}$, let $s = L$, $\lambda = \lambda^L$, and go to Step 3; otherwise, set $L := L + 1$, and go to Step 1.

Step 3: Update

Set

$$p_i^k = \frac{\rho}{\alpha_i}(\lambda - h_i), \quad i = 1, 2, \ldots, s$$

$$p_j^k = 0, \quad j = s+1, s+2, \ldots, l.$$

The algorithm converges in a finite number of steps (cf. Dafermos and Sparrow [12]).

b. The Relaxation Method

The relaxation method corresponds to the choice

$$g_i(p, q) = -z_i(q_1, \ldots, q_{i-1}, p_i, q_{i+1}, \ldots, q_l), \quad \forall i = 1, 2, \ldots, l. \tag{51}$$

In this case properties (i) and (ii) are also satisfied. In fact,
(i) $g(p, p) = -z(p)$,

(ii) $\nabla_p g(p, q) = \begin{bmatrix} -\frac{\partial z_1}{\partial p_1} & \cdots & 0 \\ \vdots & \ddots & \vdots \\ 0 & \cdots & -\frac{\partial z_l}{\partial p_l} \end{bmatrix}$ is a diagonal matrix.

By recalling the properties of the aggregate excess demand function $z(p)$, one deduces that it is reasonable to assume that

$$\frac{\partial z_i}{\partial p_i} < 0, \quad \forall i = 1, 2, \ldots, l. \tag{52}$$

Hence, $\nabla_p g(p,q)$ is positive definite. Furthermore,

$$\nabla_q g(p,q) = \begin{bmatrix} 0 & \frac{\partial z_1}{\partial p_2} & \cdots & \frac{\partial z_1}{\partial p_l} \\ \frac{\partial z_2}{\partial p_1} & 0 & \cdots & \frac{\partial z_2}{\partial p_l} \\ \vdots & \vdots & \ddots & \vdots \\ \frac{\partial z_l}{\partial p_1} & \cdots & \frac{\partial z_l}{\partial p_{l-1}} & 0 \end{bmatrix},$$

and

$$\nabla_p g^{-\frac{1}{2}}(p_1,q_1)\nabla_q g(p_2,q_2)\nabla_p g^{-\frac{1}{2}}(p_3,q_3)$$

$$= \begin{bmatrix} 0 & \frac{\partial z_1}{\partial p_2}(-\frac{\partial z_1}{\partial p_1})^{-\frac{1}{2}}(-\frac{\partial z_2}{\partial p_2})^{-\frac{1}{2}} & \cdots \\ \frac{\partial z_2}{\partial p_1}(-\frac{\partial z_2}{\partial p_2})^{-\frac{1}{2}}(-\frac{\partial z_2}{\partial p_2})^{-\frac{1}{2}} & 0 & \cdots \\ \vdots & \vdots & \ddots \\ \frac{\partial z_l}{\partial p_1}(-\frac{\partial z_1}{\partial p_1})^{-\frac{1}{2}}(-\frac{\partial z_l}{\partial p_l})^{-\frac{1}{2}} & \cdots & \cdots \end{bmatrix}. \quad (53)$$

We now state:

Theorem 4.2 *Let*

$$-\frac{\partial z_T}{\partial p_T} = \min_i\{-\frac{\partial z_i}{\partial p_i}\} \quad (54)$$

and assume that

$$-\frac{\partial z_T}{\partial p_T} > \sum_{k\neq i}\frac{\partial z_i}{\partial p_k}(-\frac{\partial z_k}{\partial p_k})^{-\frac{1}{2}}(-\frac{\partial z_T}{\partial p_T})^{\frac{1}{2}}, \quad \forall i = 1,2,\ldots,l. \quad (55)$$

Then condition (11) of Theorem 3.1 holds.

Proof: Introduce the norm $|x|_\infty = \max_i\{|x_i|\}$ in the Euclidean space, which leads to the norm $\|\cdot\|_\infty$ for any operator Q:

$$\|Q\|_\infty = \sup_x |Qx|_\infty. \quad (56)$$

We use the norm

$$|v|_k^\infty = |G_k^{\frac{1}{2}} v|_\infty \quad (57)$$

in the proof of Theorem 3.1. Then condition (11) in Theorem 3.1 becomes

$$\|G^{-\frac{1}{2}}(p^1,q^1)\nabla_q g(p^2,q^2)G^{-\frac{1}{2}}(p^3,q^3)\|_\infty < 1. \quad (58)$$

Substituting (53) into (11) yields

$$\|G^{-\frac{1}{2}}(p^1,q^1)\nabla_q g(p^2,q^2)G^{-\frac{1}{2}}(p^3,q^3)\|_\infty$$

$$\leq \max_i \{ \sum_{k \neq i} | \frac{\partial z_i}{\partial p_k} | (-\frac{\partial z_i}{\partial p_i})^{-\frac{1}{2}} (-\frac{\partial z_k}{\partial p_k})^{-\frac{1}{2}} \}$$

$$\leq (-\frac{\partial z_T}{\partial p_T})^{-1} \max_i \{ \sum_{k \neq i} | \frac{\partial z_i}{\partial p_k} | (-\frac{\partial z_k}{\partial p_k})^{-\frac{1}{2}} (-\frac{\partial z_T}{\partial p_T})^{-\frac{1}{2}} \}. \tag{59}$$

By virtue of assumption (55), the righthand side of (59) is strictly less than 1, that is, condition (11) holds.

Remark 3: Note that $(-\frac{\partial z_k}{\partial p_k})^{-\frac{1}{2}}(-\frac{\partial z_T}{\partial p_T})^{\frac{1}{2}} \leq 1$, and, hence, (55) is a diagonal dominance condition which has been imposed in the literature to ensure the global stability of the tatonnement process (see, e.g., Cornwell [4]).

Recalling that $\nabla_p g(p,q)$ is diagonal and positive definite, and observing that the diagonal elements $-\frac{\partial z_i}{\partial p_i}$ depend only on p_i, we see that $VI^k(g, S^l)$ is equivalent to the separable strictly convex mathematical programming problem

$$\min_{p \in S^l} F(p) = \min_{p \in S^l} \{ \int_0^p g(p, p^{k-1})^T dp \}$$

$$= \min_{p \in S^l} \{ - \sum_{i=1}^{l} \int_0^{p_i} z_i(p_1^{k-1}, \ldots, p_{i-1}^{k-1}, p_i, q_{i+1}^{k-1}, \ldots, q_l^{k-1} dp_i \} \tag{60}$$

which can be solved, in general, by any efficient mathematical programming algorithm.

Next we design an algorithm for solving $VI^k(g, S^l)$, where $g(\cdot)$ is now specified by (51). The algorithm exploits the special network structure of the problem. For a graphical depiction, see Figure 3.

Algorithm RMN (equilibration algorithm for Relaxation Method Network subproblems)

Step 0: Initialization

Start with the feasible point $p^{k-1} \in S^l$ obtained by solving $VI^{k-1}(g, S^l)$ and let $n = k-1$.

Step 1: Selection

Select m and s such that

$$g_m(p^n, p^{k-1}) = \max_{\{i, p_i^n > 0\}} \{ g_i(p^n, p^{k-1}) \},$$

or

$$-z_m(p_1^{k-1}, \ldots, p_{m-1}^{k-1}, p_m^n, p_{m+1}^{k-1}, \ldots, p_l^{k-1})$$

$$= \max_{\{i, p_i^n > 0\}} \{ -z_i(p_1^{k-1}, \ldots, p_{i-1}^{k-1}, p_i^n, p_{i+1}^{k-1}, \ldots, p_l^{k-1}) \}, \tag{61}$$

$$g_s(p^n, p^{k-1}) = \min_i \{ g_i(p^n, p^{k-1}) \},$$

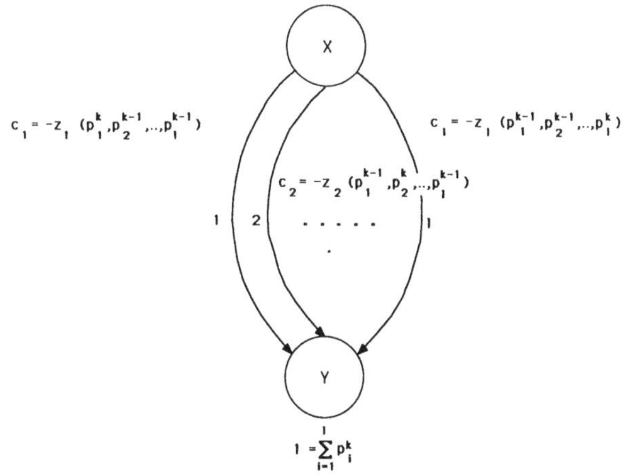

Figure 3: Network equilibrium representation of $VI^k(g, S^l)$ induced by the relaxation method

or

$$\begin{aligned} &-z_s(p_1^{k-1}, \ldots, p_{s-1}^{k-1}, p_s^n, p_{s+1}^{k-1}, \ldots, p_l^{k-1}) \\ &= \min_i\{-z_i(p_1^{k-1}, \ldots, p_{i-1}^{k-1}, p_i^n, p_{i+1}^{k-1}, \ldots, p_l^{k-1})\}. \end{aligned} \tag{62}$$

If $|g_m(p^n, p^{k-1}) - g_s(p^n, p^{k-1})| \leq \epsilon$, for $\epsilon > 0$ a preset convergence tolerance, then stop. The current p^n is a solution of $VI^k(g, S^l)$. Otherwise, go to Step 2.

Step 2: Equilibration

Equilibrate g_m and g_s by solving the following one-dimensional mathematical programming problem for δ:

$$\begin{aligned} \min\ &|z_m(p_1^{k-1}, \ldots, p_{m-1}^{k-1}, p_m^n - \delta, p_{m+1}^{k-1}, \ldots, p_l^{k-1})| \\ &-z_s(p_1^{k-1}, \ldots, p_{s-1}^{k-1}, p_s^n + \delta, p_{s+1}^{k-1}, \ldots, p_l^{k-1}), \end{aligned} \tag{63}$$

subject to $0 \leq \delta \leq p_m^n$.

Suppose that δ^n is the solution of the above minimization problem. Let

$$p_i^{n+1} = p_i^n, \quad \forall i \neq m, s,$$

$$\begin{aligned} p_m^{n+1} &= p_m^n - \delta^n \\ p_s^{n+1} &= p_s + \delta^n \end{aligned} \tag{64}$$

Table 1: Parameters for a 4 commodity, 2 consumer economy

a^i_j, w^i_j	$j=1$	$j=2$	$j=3$	$j=4$
$i=1$	0.1, 20.0	0.5, 30.0	0.1, 6.0	0.3, 2.0
$i=2$	0.2, 10.0	0.4, 30.0	0.2, 16.0	0.2, 2.0

and go back to Step 1 with $n = n + 1$.

The sequence $\{p^n\}$ thus obtained converges to the solution of $VI^k(g, S^l)$, which can be seen by the fact that

$$F(p^{n+1}) < F(p^n) \tag{65}$$

where $F(\cdot)$ is the objective function of (60).

We conclude this section by pointing out the economic meaning of the convergence condition (36) of the projection method and convergence condition (55) of the relaxation method: If the price of a commodity is a decreasing function of the demand for this commodity and is affected principally by the demand for this commodity, then conditions (36) and (55) can be expected to hold.

5 Numerical Examples

In this section, we illustrate the performance of the projection method and the relaxation method on several numerical examples. The aggregate excess demand functions in the economies are derived from Cobb-Douglas utility functions and are of the form:

$$z_j(p) = \sum_{i=1}^{m} \frac{p^T W^i}{\sum_{j=1}^{l} a^i_j} \left(\frac{a^i_j}{p_j}\right) - \sum_{i=1}^{m} w^i_j, \quad j = 1, \ldots, l, \tag{66}$$

where W^i is the vector with components $\{w^i_1, \ldots, w^i_l\}$. For completeness and ease of reproducibility, we give the data for the examples below.

Example 1:

There are 4 commodities and 2 consumers in this economy. The coefficients a^i_j and w^i_j are given in Table 1.

Example 2:

The second example is taken from Eaves [13] and the data are given in Table 2. In this economy there are eight commodities and five consumers.

Example 3:

The data for this example, consisting of 15 commodities and 4 consumers in the economy, are reported in Table 3.

Example 4:

Table 2: Parameters for an 8 commodity, 5 consumer economy

a^i_j, w^i_j	$i = 1$	$i = 2$	$i = 3$	$i = 4$	$i = 5$
$j = 1$	0.3,3.0	0.0,0.0	0.0,0.0	0.0,0.0	0.0,4.0
$j = 2$	0.0,0.0	0.0,15.	1.0,0.0	0.0,0.0	0.0,0.0
$j = 3$	.13,0.0	0.0,0.0	0.0,0.0	0.0,5.0	0.0,0.0
$j = 4$	0.0,3.0	0.0,0.0	0.0,0.0	.73,4.0	.47,13.0
$j = 5$	0.0,3.0	1.0,2.0	0.0,3.0	0.0,0.0	0.0,0.0
$j = 6$	0.0,5.0	1.0,0.0	0.0,0.0	0.0,0.0	.11,0.0
$j = 7$	.38,2.0	1.0,0.0	0.0,0.0	0.0,4.0	.05,6.0
$j = 8$	.19,0.0	1.0,0.0	0.0,0.0	.27,4.0	.37,6.0

Table 3: Parameters for a 15 commodity, 4 consumer economy

a^i_j, w^i_j	$i = 1$	$i = 2$	$i = 3$	$i = 4$
$j = 1$	.05, 2.0	.02, 10.0	.06, 12.0	.20, 10.0
$j = 2$	.02, 4.0	.30, 1.0	.03, 14.0	.00, 5.0
$j = 3$	.03, 6.0	.02, 4.0	.01, 16.0	.01, 7.0
$j = 4$	.04, 8.0	.04, 2.0	.05, 14.0	.02, 9.0
$j = 5$	.06, 10.0	.04, 2.0	.01, 3.0	0.0, 1.0
$j = 6$	.06, 12.0	.06, 12.0	0.0, 5.0	.05, 2.0
$j = 7$	.03, 14.0	.03, 14.0	.01, 7.0	.02, 4.0
$j = 8$	.01, 16.0	.01, 16.0	.02, 9.0	.03, 6.0
$j = 9$	.05, 14.0	.05, 14.0	.00, 1.0	.04, 8.0
$j = 10$	.05, 12.0	.05, 12.0	.05, 12.0	.02, 5.0
$j = 11$	.20, 10.0	.05, 2.0	.20, 10.0	.02, 7.0
$j = 12$	.30, 8.0	.02, 4.0	.30, 8.0	.02, 7.0
$j = 13$	.02, 4.0	.02, 5.0	.02, 4.0	.02, 6.0
$j = 14$	.04, 2.0	.02, 6.0	.04, 2.0	.04, 5.0
$j = 15$	.04, 2.0	.02, 7.0	.04, 2.0	.06, 10.0

Table 4: Parameters for a 20 commodity, 4 consumer economy

a_j^i, w_j^i	$i=1$	$i=2$	$i=3$	$i=4$
$j=1$	.01, 3.0	.20, 10.0	.05, 2.0	.20, 10.0
$j=2$	.00, 5.0	.30, 8.0	.02, 4.0	.30, 8.0
$j=3$	.01, 7.0	.01, 3.0	.03, 6.0	.02, 4.0
$j=4$	.02, 9.0	.00, 5.0	.04, 8.0	.04, 2.0
$j=5$	.00, 1.0	.01, 7.0	.06, 10.0	.04, 2.0
$j=6$	.05, 2.0	.02, 9.0	.06, 12.0	.06, 12.0
$j=7$	.02, 4.0	.00, 1.0	.03, 14.0	.03, 14.0
$j=8$	.03, 6.0	.02, 4.0	.01, 16.0	.01, 16.0
$j=9$	.04, 8.0	.04, 2.0	.01, 14.0	.05, 14.0
$j=10$	.06, 12.0	.04, 2.0	.01, 3.0	.05, 12.0
$j=11$	.06, 12.0	.06, 12.0	.00, 5.0	.01, 3.0
$j=12$	.03, 14.0	.03, 14.0	.01, 7.0	.00, 5.0
$j=13$	.01, 16.0	.01, 16.0	.02, 9.0	.01, 7.0
$j=14$	.05, 14.0	.05, 14.0	.00, 1.0	.02, 9.0
$j=15$	.05, 12.0	.05, 12.0	.05, 12.0	.00, 1.0
$j=16$	.20, 10.0	.05, 2.0	.20, 12.0	.05, 2.0
$j=17$	.30, 8.0	.02, 8.0	.30, 4.0	.02, 4.0
$j=18$	.02, 4.0	.03, 6.0	.02, 4.0	.03, 6.0
$j=19$	.04, 2.0	.04, 8.0	.04, 2.0	.04, 8.0
$j=20$	.04, 2.0	.06, 10.0	.04, 2.0	.06, 10.0

The fourth example consists of 20 commodities and 4 consumers with the coefficients given in Table 4.

Example 5:

The fifth example consists of 25 commodities and 4 consumers in the economy and the coefficients are given in Table 5.

Both the projection method and the relaxation method were coded in FORTRAN. The projection method was embedded with PMN and the relaxation method with RMN. The golden section method was used to solve the one variable minimization problem encountered in RMN. In the projection method we chose the matrix $G = \{\frac{\partial z_i}{\partial p_i}, i = 1, \dots, l\}|p^0$ and $\rho = .8$ for $l = 4, m = 2$, $\rho = .5$ for $l = 8, m = 5$, $\rho = .5$ for $l = 15, m = 4$, $\rho = .1$ for $l = 20, m = 4$, and $\rho = .5$ for $l = 25, m = 4$. The codes were implemented on an IBM 3090 at Brown University, and the FORTVS compiler was used for compilation. Both algorithms were initialized with $p^0 = (\frac{1}{l}, \dots, \frac{1}{l})$, and the termination criterion was $|p^k - p^{k-1}| \leq 10^{-6}$. As we can see from Table 6, the projection method converged faster than the relaxation method, even though the

Table 5: Parameters for a 25 commodity, 4 consumer economy

a^i_j, w^i_j	$i = 1$	$i = 2$	$i = 3$	$i = 4$
$j = 1$	.01, 3.0	.20, 10.0	.02, 6.0	.20, 10.0
$j = 2$	.02, 5.0	.30, 8.0	.02, 5.0	.30, 8.0
$j = 3$	.02, 6.0	.01, 3.0	.03, 6.0	.02, 4.0
$j = 4$	.02, 7.0	.02, 5.0	.04, 8.0	.04, 2.0
$j = 5$	.02, 6.0	.02, 6.0	.06, 10.0	.04, 2.0
$j = 6$	.02, 5.0	.02, 7.0	.05, 2.0	.06, 12.0
$j = 7$	.00, 5.0	.02, 6.0	.02, 4.0	.03, 14.0
$j = 8$	.01, 7.0	.02, 5.0	.03, 6.0	.01, 16.0
$j = 9$	.02, 9.0	.00, 5.0	.04, 8.0	.05, 14.0
$j = 10$	.00, 1.0	.01, 7.0	.06, 10.0	.05, 12.0
$j = 11$	.05, 2.0	.02, 9.0	.06, 12.0	.01, 3.0
$j = 12$	.02, 4.0	.00, 1.0	.03, 14.0	.00, 5.0
$j = 13$	.03, 6.0	.02, 4.0	.01, 16.0	.01, 7.0
$j = 14$	.04, 8.0	.04, 2.0	.05, 14.0	.02, 9.0
$j = 15$	.06, 10.0	.04, 2.0	.01, 3.0	.00, 1.0
$j = 16$	.06, 12.0	.06, 12.0	.00, 5.0	.05, 2.0
$j = 17$	.30, 14.0	.03, 14.0	.01, 7.0	.02, 4.0
$j = 18$	.01, 16.0	.01, 16.0	.02, 9.0	.03, 6.0
$j = 19$	.05, 14.0	.05, 14.0	.00, 1.0	.04, 8.0
$j = 20$	.05, 12.0	.05, 12.0	.05, 12.0	.02, 5.0
$j = 21$	.20, 10.0	.05, 2.0	.20, 10.0	.02, 6.0
$j = 22$	.30, 8.0	.02, 4.0	.30, 8.0	.02, 7.0
$j = 23$	.02, 4.0	.02, 5.0	.02, 4.0	.02, 6.0
$j = 24$	.04, 2.0	.04, 6.0	.04, 2.0	.04, 5.0
$j = 25$	.04, 2.0	.02, 7.0	.04, 2.0	.06, 10.0

Table 6: Numerical Results for the Projection and Relaxation Methods

Example Number	Number of Iterations		CPU Time (seconds)	
	Relaxation	Projection	Relaxation	Projection
1	11	18	.20	.05
2	28	117	1.64	.20
3	5	80	1.02	.11
4	4	91	1.97	.36
5	4	64	3.49	.38

number of iterations in the projection method is larger than the number of iterations in the relaxation method, for the same degree of accuracy. This is most likely due to the fact that the projection method solves the network subproblems $VI^k(g, S^l)$ in closed form. As the scale of the economy becomes larger, the advantage of the projection method becomes more significant.

6 Summary and Conclusions

In this paper we have developed a network formalism for the study of a class of general economic equilibrium problems - that of pure exchange or Walrasian price equilibrium problems. We first established that the Walrasiàn price equilibrium problem is isomorphic to a network equilibrium problem with special structure and, hence, the corresponding variational inequality formulations are one and the same.

We then turned to the computation of the equilibrium patterns. We proposed a general iterative scheme for the computation of the Walrasian price equilibrium, which was then shown to induce the projection method and the relaxation method, as special cases. In particular, the projection method resolves the network equilibrium problem into linear symmetric network equilibrium problems, for which an equilibration algorithm termed PMN was then proposed. The relaxation method, on the other hand, resolves the problem into separable nonlinear problems, for which an equilibration algorithm named RMN was developed.

Finally, we presented numerical results for five examples which demonstrated that the projection method in combination with PMN consistently outperformed the relaxation method in combination with RMN. This is due, in part, to the simplicity of the network equilibrium subproblems which were then solved in closed form. As the scale of the problems increased, the relative efficiency of the projection method vis a vis the relaxation method also increased, suggesting that the full exploitation of the underlying network structure of this class of general economic equilibrium problems will enable the computation of large-scale problems in practice.

References

[1] H. Z. Aashtiani and T. L. Magnanti, Equilibria on a congested transportation network *SIAM Journal on Algebraic and Discrete Methods* **2** (1981) 213-226.

[2] M. Beckmann, C. B. McGuire, and C. B. Winsten, *Studies in the Economics of Transportation* (Yale University Press, New Haven, Connecticut, 1956).

[3] D. P. Bertsekas and E. Gafni, Projection methods for variational inequalities and application to the traffic assignment problem, *Mathematical Programming Study* **17** (1982) 139-159.

[4] R. Cornwell, *Introduction to the Use of General Equilibrium Analysis* (North-Holland, Amsterdam, The Netherlands, 1984).

[5] S. Dafermos, Traffic equilibrium and variational inequalities, *Transportation Science* **14** (1980) 42-54.

[6] S. Dafermos, The general multimodal traffic equilibrium problem, *Networks* **12** (1982) 57-72.

[7] S. Dafermos, An iterative scheme for variational inequalities, *Mathematical Programming* **15** (1983) 40-47.

[8] S. Dafermos, Isomorphic multiclass spatial price and multimodal traffic network equilibrium models, *Regional Science and Urban Economics* **16** (1986) 197-209.

[9] S. Dafermos, Exchange price equilibria and variational inequalities, *Mathematical Programming* **46** (1990) 391-402.

[10] S. Dafermos and A. Nagurney, Sensitivity analysis for the general spatial economic equilibrium problem, *Operations Research* **32** (1984) 1069-1086.

[11] S. Dafermos and A. Nagurney, Oligopolistic and competitive behavior of spatially separated markets, *Regional Science and Urban Economics* **17** (1987) 245-254.

[12] S. C. Dafermos and F. T. Sparrow, The traffic assignment problem for a general network, *Journal of Research of the National Bureau of Standards* **73B** (1969) 91-118.

[13] B. C. Eaves, Where solving for stationary points by LCPs is mixing Newton iterates, in *Homotopy Methods and Global Convergence*, B. C. Eaves, F. J. Gould, H. O. Peitgen, and M. J. Todd, editors (Plenum Press, New York, 1983) pp. 63-78.

[14] M. Florian and M. Los, A new look at static spatial price equilibrium models, *Regional Science and Urban Economics* **12** (1982) 579-597.

[15] M. Florian and H. Spiess, The convergence of diagonalization algorithms for asymmetric network equilibrium problems, *Transportation Research* **16B** (1982) 477-483.

[16] A. Nagurney, Migration equilibrium and variational inequalities, *Economics Letters* **31** (1989) 109-112.

[17] A. Nagurney and D. S. Kim, Parallel computation of large-scale dynamic market network equilibria via time period decomposition, *Mathematical and Computer Modelling* **15** (1991) 55- 67.

[18] A. Nagurney, J. Pan, and L. Zhao, Human migration networks, *European Journal of Operational Research* **59** (1992) 262-274.

[19] A. Nagurney and L. Zhao, A network equilibrium formulation of market disequilibrium and variational inequalities, *Networks* **21** (1991) 102-132.

[20] H. Scarf (with T. Hansen), *Computation of Economic Equilibria* (Yale University Press, New Haven, Connecticut, 1973).

Network Optimization Problems, pp. 387-401
Eds. D.-Z. Du and P.M. Pardalos

Steiner Problem in Multistage Computer Networks

Sourav Bhattacharya
Bhaskar Dasgupta[1]
Computer Science Department, University of Minnesota, Minneapolis, MN 55455

Abstract

Multistage computer networks are popular in parallel architectures and communication applications. We consider the message communication problem for the two types of multistage networks: one popular for parallel architectures and the other popular for communication networks. A subset of the problem can be equated to the *Steiner tree* problem for multistage graphs. Inherent complexities of the problem is shown and polynomial-time heuristics are developed. Performance of these heuristics is evaluated using analytical as well as simulation results.

1 Introduction

Multistage interconnection networks (MINs) are popular among parallel architecture and/or communication network topologies. An $N \times log_2N$ element MIN consists of log_2N stages of N elements each. A common pictorial view of an $N \times log_2N$ MIN is to collect N elements in a stage (vertically) and arrange $log_2N + 1$ such stages horizontally one after the other. MINs offer a good balance between network cost and performance. They are often characterized as intermediate $\{O(N \times log_2N)\}$ cost networks falling within the two extreme cases: fully connected $\{O(N^2)$ cost$\}$ and bus connected $\{O(N)$ cost$\}$. Architectural and other topological properties of MIN may be found in [8].

[1]Supported in part by NSF grant CCR-9208913

1.1 Two Versions of MINs

Let $S_{i,j}$ denote the i-th stage j-th row element in an $N \times log_2N$ MIN, $0 \leq i \leq log_2N, 0 \leq j \leq N-1$. We consider source-to-source wrap-around MINs only, i.e., when $\forall j : S_{0,j} = S_{log_2N,j}$. These networks can allow multiple passes using the wrap-around connections. Depending on the role of intermediate stage elements, two types of MINs are possible as outlined below:

- *Intermediate stages as switches only:* This type is popular in parallel architecture applications. Here the source end (leftmost) and the destination end (rightmost stage) constitute of processors, while the intermediate elements are bare switches which interconnect various sources and destinations. Such MINs are of commercial usage in parallel processors, e.g., the BBN Butterfly machine. We refer to MINs of this type as *type-1 MIN*.

- *Intermediate stages as processors:* This type is common in communication network applications. Here the intermediate stage elements are identical to the source or destination stage processors, i.e., they can have their own message traffic. Example of such MINs can be found in [10]. We refer to MINs of this type as *type-2 MIN*.

1.2 Communication in MINs

Depending on the number of destinations involved in a communication in MIN, three types can be classified: *one-to-one*, *one-to-many* and *one-to-all*. These are commonly known as *routing*, *multicast* and *broadcast*. In this article we focus ourselves to the multicast problem for MINs. Note that routing & broadcast are two special instances of multicast and do not offer any opportunity for traffic reduction.

The multicast problem specifies a source node and a set of k destination nodes. Without loss of generality we assume the source node to be $S_{0,0}$. Destination nodes are spread over the MIN, $1 \leq k \leq N$ ($k = 1 \equiv$ routing, $k = N \equiv$ broadcast). Objective of the multicast problem is to transmit the message from the source node to the destination nodes.

Flow-control Mechanism

For multihop networks, various form of switching and flow-control mechanisms have evolved. *Store and forward* is a traditional approach to message communication. *Virtual cut-through, wormhole, deflection routing* etc. have been subsequently proposed. A survey can be found in [11, 4]. We assume packetized message communication, where packets are independently flown through the network. Our focus is to estimate (and possibly reduce) the overall traffic overhead in message communications.

1.3 Optimality Criteria in MIN Multicast

Two possible criteria to measure the optimality of MIN multicast communication are to minimize one of the following two objective functions:

- the total traffic generated in the network (each occupied link of the network counts as one unit of traffic.
- the hops-distance between the source node and any destination node.

The *traffic* metric makes the problem equivalent to the *Steiner problem for MIN*, while the *time* metric is a different dimension altogether. These two metrics work in the dual sense. Reducing one increases the other and vice versa. Thus, we focus on the *traffic* metric only. Considerations along the *time* metric is an open problem.

2 Multistage Interconnection Networks

We consider *type 1* MINs with the cube network topology. These class of networks (e.g., baseline, delta, generalized cube, indirect binary-cube, omega, banyan [8]) have been proposed as fixed-degree alternative to hypercube architecture. They are popular in switching and communication applications. They can also emulate the performance of hypercube in most applications (e.g., the CCC architecture [12]). Let MIN_d denote a d dimensional generalized MIN.

2.1 Formulation of the Traffic Reduction Problem.

We consider multicasting on MIN_d which are *unique path networks.* Given a set of k multicast destinations (D_i, $1 \leq i \leq k$) and a source node S in MIN_d, the path from S to any particular D_i is fixed. However, it is clear that for a given set of multicast destinations, the total traffic generated in MIN_d depends on the relative order in which d different dimensions are arranged. This leads to our problem formulation as (see Section 2.1.1 for practical applicability):

> *Given a set of destination nodes, traffic optimum multicasting in MIN_d is to find a permutation of the d dimensions (each stage of MIN_d is allocated to one particular dimension value) so that the total traffic is minimized.*

Unfortunately, this problem is NP-complete as shown by the next theorem. Hence, we need to investigate the possibility of designing efficient heuristics for this problem.

Theorem 2.1 *The traffic optimum multicasting problem is NP-complete.*

Proof sketch: The problem is obviously in NP. To show NP-hardness one can reduce the *space minimized full trie* problem, which is shown to be NP-complete in [3, 6], to this problem. Details are available in [1]. □

2.1.1 Design Issues

Any hardware implementation of a MIN_d would assume an ordering among the d dimensions. In such cases, online dimension ordering (as required by the traffic reduction criterion in this paper) in a MIN_d may be argued from the practical viewpoint. We identify the following situations as practical applications.

(1) *Communication networks* often use MINs. Traditional hardware implementation of switches at every intermediate stages have been replaced using Wave-Time Division Multiplexors (WTDM) over passive stars [5]. The actual interconnection is formed by wavelength (frequency) or time-slot assignment of different nodes, i.e., by *firmware* control. A *firmware* controlled design can be changed without changing the underlying hardware. Thus, it is possible to re-order the dimensions in a MIN_d dynamically. Every stage may have to configure to at most d possible dimensions, for which the wave/time assignments can be pre-computed and stored.

(2) If the traffic pattern is known and repetitive (as may happen in periodically occurring similar message communications) then from the above optimum dimensional ordering for each multicasting instance one can derive the most common pattern and design the MIN_d using the corresponding optimum dimensional ordering. The idea here is to achieve traffic optimality for *most* multicasting instances which leads to an overall traffic reduction.

(3) *Hierarchical hypercubes* are designed for several practical reasons [9]. Such hierarchical designs limit the availability of different dimensions at any node. Only a certain set of dimensions can be availed at each node. This imposes a hierarchy among dimensions in a routing/ multicasting operation. In some other cases, even with complete hypercubes routing/ multicasting is done in hierarchical fashion, imposing a (arbitrary) desired ordering among dimensions [4]. With these applications our results and optimality ordering among dimensions can be used as a measure whether or not a particular multicast operation is generating optimal traffic. Note that a hypercube with hierarchically ordered dimensions can be treated as a MIN_d for analysis purpose and results from the latter can be used for the former.

2.2 Greedy Heuristic

Let $Reach_p$ denote the number of nodes which received a copy of the message at stage p. Let k_d be the dimension between stage p and stage $p+1$. We define an *expansion ratio* $F_{k_d} = \frac{Reach_{p+1}}{Reach_p}$. Intuitively, this fraction F_{k_d} indicate how much the size of the multicast destination is increasing at every stage. This expansion ratio depends of the dimension, stage position and on the set of all prior dimensions served already. For the sake of brevity we treat this *all previous* information as part of the stage information and denote compactly using the stage number position. Now, the total traffic equals

$$\sum_{p=1}^{p=d} Reach_p = F_{k_1} \times [1 + F_{k_2} \times [1 + \ldots + F_{k_{d-1}} \times [1 + F_{k_d}] \ldots]]$$

Our objective is to arrive at values of $k_1, k_2, \ldots, k_d$, such that the above expression is minimized. Note that F_{k_i} $(1 \leq i \leq d)$ varies between 1 and 2 and have real values.

We propose the following greedy algorithm which works stage by stage and selects one dimension in each stage. After d iterations of the algorithm the complete permutation of d dimensions in MIN_d are generated.

Greedy heuristic:

> *At each stage select dimension k_i, where $\forall j$, $F_{k_j} \geq F_{k_i}$. In case of a tie, anyone of the smallest F_{k_i} dimension may be chosen. A particular dimension k_i used in a preceding stage may not be repeated in a subsequent stage.*

Regarding time and space complexities of the above heuristic, it is easy to prove the following theorem.

Theorem 2.2 *The greedy heuristic runs in $O(k.d^3)$ time, performs $O(k.d^4)$ bitwise arithmetic operations and uses $O(k.d)$ space.*

The following theorem, whose proof can be found in [1], shows local optimality of dimension ordering of the greedy heuristic.

Theorem 2.3 *The greedy heuristic leads to a locally optimal ordering of the dimensions, i.e., orders the dimensions so that no* **adjacent pairwise interchange** *of the dimensions can minimize the total traffic.*

Table 1 compares the time and space complexities of the greedy heuristic with two other known exponential time optimum strategies[1].

Algorithm	Time (in bit operations)	Space (in bits)
Direct Permutation	$O(k \cdot d^3 \cdot d!)$	$O(k \cdot d)$
Dynamic Programming	$O(k \cdot d^2 \cdot 2^d)$	$O(k \cdot d + 2^d \cdot d)$
Heuristic	$O(k \cdot d^4)$	$O(k \cdot d)$

Table 1: Multicast in MIN_d with k destinations: Summary of time and space complexity of different solutions.

2.3 Performance

This section analytically compares the greedy heuristic with "randomly ordered dimensions" approach as well as the optimal algorithm. Detailed proofs of all the results are available in [1]. For our analysis purpose, we characterize the *destination node set* into two classes: the "complete subcube multicast" and the "incomplete subcube multicast". Each one of these cases are explained below and worst (average) case performance of the greedy heuristic is compared with that of optimal algorithm as well as random dimension ordering. Let traffic "overhead" comparison between two approaches be denoted as the absolute difference in the traffic generated by those two individual approaches[2]. For example, Overhead(greedy, optimum) = greedy traffic - optimum traffic, Overhead(random, greedy) = traffic in "random ordering" - greedy traffic.

2.3.1 Complete Subcube Multicast

In this case the set of multicast destinations, D_i, $1 \leq i \leq k$, forms a complete subcube. Thus, $i = 2^r$, for some integer r and the set $[D_i]$ can form a r-dimensional subcube. Let $CSM(r)$ denote this situation.

Theorem 2.4 *The greedy heuristic produces optimum traffic for the complete subcube multicast case. Also, in the worst CSM(r) case,*

$$Overhead(random, optimum) = Overhead(random, greedy) = (d-r) \times (2^r - 1).$$

The next theorem gives the probabilistic traffic overhead of "random dimension ordering" as opposed to the worst case performance as stated above.

Theorem 2.5 *In the average $CSM(r)$ case, the random dimensions ordering approach incurs a traffic overhead Overhead(random, optimum) = Overhead(random, greedy) =* $\sum_{p=1}^{p=r} Q_r(p) \times (N_p - N_0)$, *where*

$$N_p = \sum_{i=1}^{i=p} 2^i + 2^p \times (d-r) + 2^p \times \sum_{i=1}^{i=p} 2^i$$

$$Q_r(p) = \frac{\binom{r}{p} \cdot (d-r) \cdot (d+p-r-1)! \cdot (r-p)!}{d!}$$

[2] A random dimension ordering occurs when d dimensions of MIN_d are randomly ordered.

2.3.2 Incomplete Subcube Multicast

In this case not all the 2^r destination nodes are present in the set of multicast destinations. Thus, the set of multicast destinations (D_i, $1 \leq i \leq k$) forms an incomplete r-dimensional subcube, where r is the minimum dimension value to include all those destination nodes. Let $ISM(r)$ denote this situation. The following theorems give worst case and average case performance ratios of various strategies. First we consider the case when k is a power of 2.

Theorem 2.6 *In the case of incomplete r-subcube multicast with $k = 2^j$ destinations,*

$$\begin{aligned} \textit{Overhead(greedy, optimum)} &\leq ((r-j) \times (k-1)) \\ \textit{Overhead(random, greedy)} &\leq [(d - 2r + log_2 k) \times (k-1)] \end{aligned}$$

Since $j = log_2 k$, Overhead(greedy,optimum)$\approx (r - log_2 k) \times k$. For a given r, this value is maximized when $k = 2^{r-1}$. Thus, the worst case performance degradation suffered by the greedy heuristic equals 2^{r-1}.

Next we generalize the value of k to a non-power of 2.

Lemma 2.7 *In $ISM(r)$, with $k = 2^j + l$ nodes ($0 \leq l \leq 2^j - 1$), Overhead(greedy, optimum) = A(d,k,r,j), where*

$$\begin{aligned} A(d,k,r,j) &= [(d-r) + \textstyle\sum_{i=0}^{i=j+1} 2^i + (r-j-1) \times k] - \\ &\quad [(d-r) + (r-j) + \textstyle\sum_{i=0}^{i=j} 2^i] \\ &\approx 2^{j+1} + (r-j-1) \times (k-1) \end{aligned}$$

and, Overhead(random,greedy) = B(d,k,r,j), where

$$\begin{aligned} B(d,k,r,j) &= [(r-j) + \textstyle\sum_{i=0}^{i=j} 2^i + (d-r) \times k] - \\ &\quad [(d-r) + \textstyle\sum_{i=0}^{i=j+1} 2^i + (r-j-1) \times k] \\ &\approx (d + j + 1 - 2r) \times (k-1) - 2^{j+1} \end{aligned}$$

The following lemma gives estimates for the average performance. Let

$$P_k(r) = \frac{\binom{d}{r} \cdot \sum_{j=0}^{r} (-1)^{r-j} \binom{r}{j} 2^{d-j} \binom{2^j}{k}}{\binom{2^d}{k}}$$

We reuse notations of the previous lemma for brevity.

Lemma 2.8 *In $ISM(r)$, with $k = 2^j + l$ nodes ($0 \leq l \leq 2^j$ -1), on an average case:*

$$\begin{aligned} \textit{Overhead(greedy, optimum)} &= \textstyle\sum_{p=0}^{p=r-1} P_{k=2^p}(r) \times A(d,k,r,p) \\ \textit{Overhead(random, greedy)} &= \textstyle\sum_{p=0}^{p=r-1} P_{k=2^p}(r) \times B(d,k,r,p) \end{aligned}$$

The above result for the average case is based on equal distribution of destination nodes among the given nodes.

2.4 Simulation Performance

Often a MIN_d design implicitly assumes ascending or descending order among its d dimensions. For example, in a MIN_3 an increasing order would be [0,1,2], while a decreasing order would be [2,1,0]. Such is also the usual practice in hierarchical routing in hypercube, where dimensions are treated one after another alike distinct stages of MIN_d. Clearly, these *linearly ordered dimensions* approaches cannot generate traffic optimal multicasting in MIN.

We compare performance of the greedy heuristic with the *linearly ordered dimensions* heuristic approach and demonstrate the advantages. We show for randomly generated M multicast destinations how much traffic can be reduced (on an average) if the locally optimal greedy dimension ordering approach proposed in this paper is followed. We present simulation results towards this.

Our simulation implemented four situations: the exhaustive optimum traffic generation approach, greedy approach, linearly increasing and linearly decreasing (hereafter referred as 'increasing' and 'decreasing' respectively). Performance of the "increasing" and "decreasing" cases are found similar, and hence we report only the "increasing" case. We consider three different dimension values 4, 5 and 6 (i.e., MIN_4, MIN_5 and MIN_6). Number of multicast destinations are varied as 1%, 2%, 5%, 10%, 20%, 50%, 80%, 90%, 95% and 99% of the total number of nodes in the cube. For each multicast set size 30 random distributions were generated and averages taken to introduce the effect of *large numbers*. Thus, each algorithm is run 300 times with every dimension - leading to a total of 3600 experiments.

The proposed greedy algorithm produces optimal multicast traffic in most ($\approx$ 90%) of the cases. Thus the greedy heuristic is "almost always" optimum. We present simulation results showing the *miss-rate*, i.e., the percentage of test runs in which the proposed greedy algorithm does not coincide with the optimum multicast algorithm. Even in cases where the greedy algorithm deviates from optimum solution, the deviation is found to be small. This section also shows the relationship of *optimal* traffic cost (T) to the number of multicast destinations (M). We present simulation results to show variation of T for different values of M.

Let T_O, T_G and T_I denote the traffic generated using the *optimum* (exhaustively generated), *greedy* and *linearly ordered increasing* approaches respectively. Thus, the traffic overhead using *greedy* and *increasing* approaches equal $(T_G - T_O)$ and $(T_I - T_O)$ respectively. Fig.1a shows the average value of these parameters for MIN_d with d=4, 5, 6.

Let "miss-rate" be defined as the percentage of simulation runs in which a particular heuristic approach (e.g., the greedy approach or the increasing approach) differs from the optimum approach. This percentage shows the frequency by which the heuristic deviates from the optimum solutions. A low value of miss-rate indicates that the corresponding heuristic is 'almost always' optimum. Fig.1b shows the miss-rate *greedy*, *increasing* heuristics for different dimensions.

Observation 1: The greedy approach misses the optimum solution rarely (e.g.,

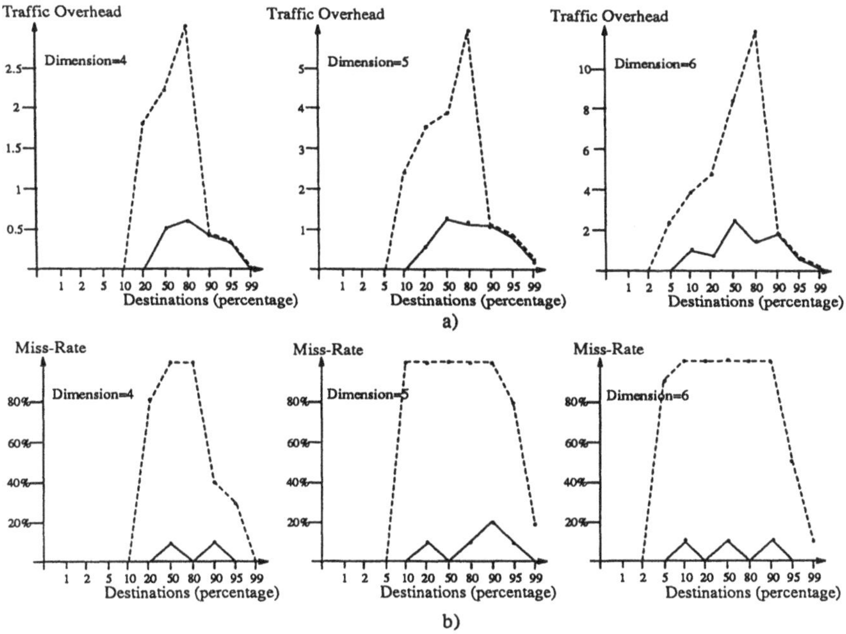

Figure 1: Greedy (solid line) and Increasing (dashed line) heuristic performance: a) Traffic overhead (≡traffic in heuristic - optimal traffic), b) Miss-rate.

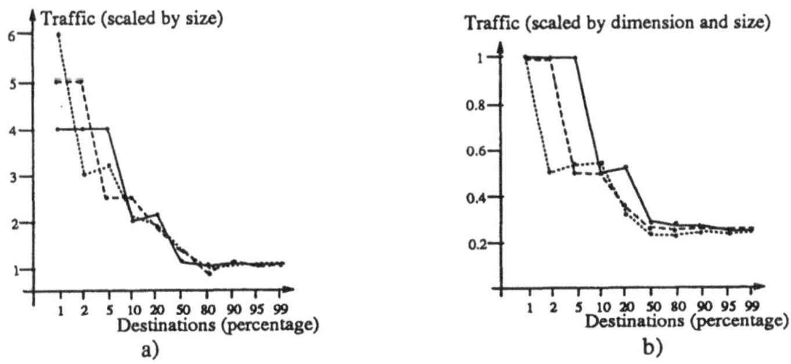

Figure 2: Optimum traffic (per destination node) variation with different multicast sizes (solid line for dimension=4, dashed line for dimension=5, dotted line for dimension=6).

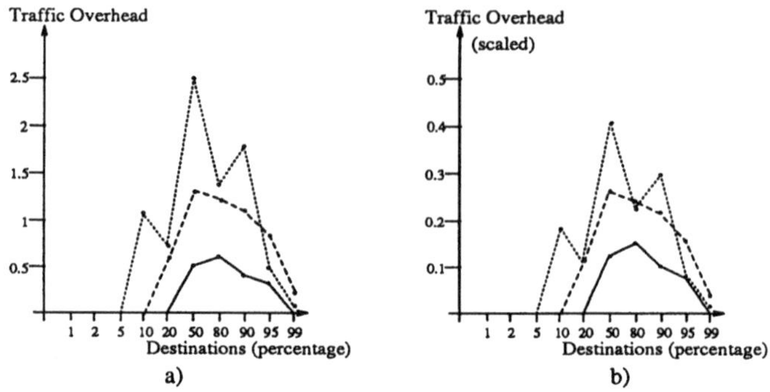

Figure 3: a) Traffic overhead of Greedy heuristic, b) Scaled traffic overhead (soild line for dimension=4, dashed for 5, dotted for 6).

1 out of 30 cases or at most 2 out of 30 cases). Thus it is an 'almost always optimum' algorithm. The number of mismatches increases as the dimension increases. However, the existing dimension ordering approaches (e.g., increasing or decreasing) have high miss-rate.

Observation 2: For small (or large) number of multicast destinations all three heuristics yield optimum (or near optimum) traffic solutions. This is because, for small number of destinations *expansion ratio* (refer Section 2.2) is almost always 1 and regardless of the actual heuristic used, a near-optimum strategy is effected. Similarly, with large fraction of nodes as destinations, the *expansion ratio* is almost always 2 and regardless of the actual heuristic used, a near-optimum strategy is effected.

Observation 3: Traffic overhead of greedy approach is superior to "increasing" approach.

Observation 4: Traffic overhead of greedy approach increases with dimension (Fig. 3a). This is expected since at higher dimensions, each source-destination multicast involves larger traffic amount. At the same time it can also be attributed to the inherent characteristic of the greedy approach, i.e., the greedy approach deviates from optimality more with increasing dimension.

These two factors are distinguished by scaling the traffic overhead using dimension value. The idea is to normalize the traffic overhead using the corresponding dimension value. The scaled traffic overhead (Fig. 3b) using the greedy heuristic for different dimensions also show that traffic overhead of greedy approach increases with dimension.

Fig.2 shows the optimum traffic load variation for different multicast sizes. The idea is to explore relationship (if any) between the total amount of traffic (T) required for a M destination multicast. We show the average traffic (i.e., traffic per destination node) reported from our simulations for cube sizes 4, 5 and 6. Then, we scale the traffic requirement using the corresponding dimension value. This plot also shows similar trend as the absolute traffic plot.

Observation 5: Optimum multicast traffic decreases with increasing multicast destination size. Initially, with small M, every destination requires nearly one unit of traffic per stage. This is because the destinations nodes are sparse and on an average no common message traffic can be shared by two destinations. However, with increasing M this ratio decreases until M reaches 50% of the cube size. Beyond this range of M the value of 'optimum traffic' per multicast destination becomes almost a constant indicating high availability of destination nodes and frequent ability to share message links most efficiently.

3 Multistage Communication Networks

We consider *type 2* MINs in this section. Among several topologies we choose the shuffle connection and the multistage binary cube connection. Without loss of generality we assume that $S_{0,0}$ is the source node, while the k destinations nodes are spread over the $(log_2N + 1)$ stages and N rows.

Type-1 MINs are unique path networks. This required us to re-order dimensions in order to have traffic reduction. The online dimension re-ordering led to practical feasibility questions (which is addressed in Section 2.1.1) and related issues. However, type-2 MINs are not unique path networks. They allow multiple paths between source and any destination. Hence, traffic reduction can be achieved even without any online topological reconfiguration.

Optimality Criterion

Given an $N \times log_2N$ type-2 multistage communication network (connected using a particular topology T) with a source node S and k destination nodes D_i $(1 \leq i \leq k)$, the objective is to find a path from the source node to each one of the destinations such that one of the following objective functions is minimized:

- Total *traffic*.
- *Time* (in hops) between source and each destination.

The first objective equates the problem to the *Steiner tree* problem for the topology T. The second objective has not been investigated so far.

3.1 Multistage Shuffle Network

We consider a shuffle network with $N \times log_2N$ PEs, arranged as log_2N stages of N PEs each. Let such a network be called **a** $\boldsymbol{log_2}N$-shuffle, Let $PE_{i,j}$ be the i-th row PE in the j-th stage, $0 \leq i \leq N-1$, $0 \leq j \leq log_2N - 1$. log_2N-Shuffle is a cyclical

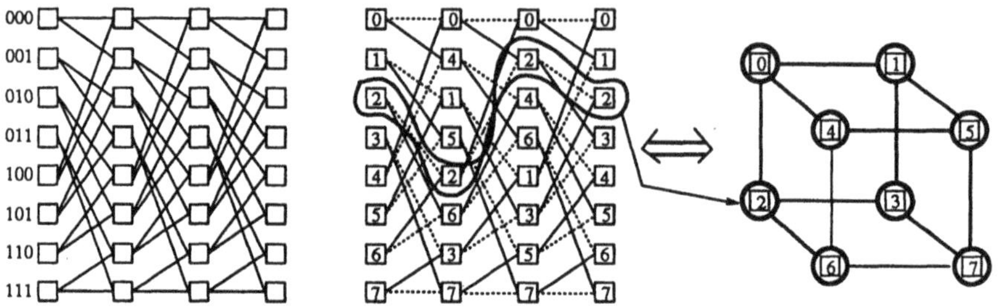

a) b) c)

Figure 4: NP-completeness: a) Example 3-stage shuffle network, b) Restricted instance of the shuffle graph - dotted lines indicate zero-cost edges - nodes which are grouped together are tagged with identical integer, c) Equivalence to a 3-cube multicast.

wrap around network, with log_2N-th stage $\equiv$ stage 0. Formally, the binary shuffle connectivity is defined as ($N = 2^n$) [10]

$$PE_{i,j} \text{ has } \textit{outgoing} \text{ links to } PE_{(2i+p) \bmod N, (j+1) \bmod n}, \text{ where } p \in \{0, 1\}$$

Theorem 3.1 *The problem of traffic optimal shuffle multicast tree generation is NP-complete.*

Proof sketch: It can be shown that a special case of this problem is the problem of *optimal traffic multicast for hypercube* (which is known as NP-complete[7]) by setting the costs of some edges to zero and thereby identifying some nodes together. Details can be found in [2]. Fig. 4 pictorially depicts the idea. □

3.2 Multistage Cube Network

We consider a multistage cube network with $N \times log_2N$ PEs, arranged as log_2N stages of N PEs each. Let such a network be called a log_2N-stage cube. Let $PE_{i,j}$ be the i-th row PE in the j-th stage, $0 \leq i \leq N-1$, $0 \leq j \leq log_2N - 1$. log_2N-stage cube is a cyclical wrap around network, with log_2N-th stage $\equiv$ stage 0. Formally, the binary multistage connectivity is defined as ($N = 2^n$)[8]

$$PE_{i,j} \text{ has } \textit{outgoing} \text{ links to } PE_{i,(j+1) \bmod n} \quad \textit{and}$$
$$PE_{i+(1-2r)\times 2^j,(j+1) \bmod n} \quad \text{where } r = i\text{-th bit in } j$$

Note that this multistage cube network is a particular instance of the generalized MIN_d considered in Section 2, where the dimensions are increasing from left to right. Also, intermediate stage nodes are active PEs here, unlike in the MIN_d of Section 2 (which are bare switches).

define $Dist(D,a) = \text{MIN} \{ H(D_j,a) : D_j \in D \}$
$\forall n, (n \notin D) \wedge (\exists D_j \in D : H(D_j,n) = 1)$ Do
$\quad count(n) = |\{d_k \in DL : Dist(D \cup \{n\}, d_k) < Dist(D, d_k)\}|$
select n such that $count(n)$ is maximum;

Figure 5: Greedy algorithm for type-2 MINs: selection of the next step message recipient node.

The proof of the following theorem is essentially similar to theorem 3.1. Details can be found in [2].

Theorem 3.2 *The optimal traffic multicast problem is NP-complete.*

3.3 Greedy Heuristic

We developed a greedy heuristic for *type-2* MINs. This heuristic is applied to both the type-2 shuffle MIN as well as type-2 multistage cube MIN. We describe this greedy heuristic first and then demonstrate its performance using simulation results.

The greedy heuristic is an iterative process selecting one node in each iteration. Every time a node is selected it is included in a set D (representing the set of nodes which received a copy of the message so far). Initially D set only includes S, i.e., the source node. The algorithm stops when $\forall i \; 1 \leq i \leq k \; D_i \in D$. Let DL denote the set of k destinations, and $H(a,b)$ equal the *shortest distance* between nodes a and b. Let $Dist(D,a)$ be a function indicating the shortest distance from any node in D to a. Each step of the greedy iteration chooses a node n as in Fig. 5.

3.4 Simulation Performance

We simulated the greedy algorithm in the multistage shuffle MIN as well as in the multistage cube MIN. Its performance is compared with the exhaustively generated optimum algorithm in the respective architectures. The number of destinations is varied as 1%, 2%, 5%, 10%, 20%, 50%, 80%, 90%, 95% and 99% of the total system size. In each case 50 random set of destinations were generated; both the greedy & optimum algorithms are run and their results compared.

Two metrics are used to characterize performance of the greedy heuristic: the *number of miss* and the *average overhead.* The former denote how often (out of the 50 runs) the greedy algorithm fails to produce optimal result, while the latter indicates average deviation of the greedy result when a miss occurs. A low miss rate indicates that the greedy algorithm is *almost always optimum*, while a low *average overhead* indicates that the greedy algorithm *almost always produces a near-optimum solution.* Let T_G (T_O) denote the traffic produced by the greedy (optimum) algorithm.

- *Number of miss* $= |T_G \neq T_O|$.
- *Average overhead* $= \sum \frac{T_G - T_O}{T_O}$, where $T_G \neq T_O$.

Table 2 shows the performance of the greedy algorithm in multistage shuffle MIN, while Table 3 shows the same for multistage cube MIN. As can be observed from these two tables, the greedy algorithm has low miss rate (particularly for the multistage cube MIN) and low traffic overhead.

Systems	Metrics	Destinations									
		0.01	0.02	0.05	0.1	0.2	0.5	0.8	0.9	0.95	0.99
3-stage Shuffle	Number of miss	1	3	6	8	11	24	16	14	12	9
	Average overhead	1.01	1.03	1.23	1.4	1.52	1.6	1.56	1.31	1.19	1.06
4-stage Shuffle	Number of miss	1	4	9	11	16	32	27	19	12	10
	Average overhead	1.12	1.15	1.33	1.47	1.63	1.82	1.58	1.39	1.28	1.17
5-stage Shuffle	Number of miss	2	5	10	14	21	37	29	23	17	11
	Average overhead	1.14	1.23	1.41	1.53	1.71	2.09	1.69	1.45	1.32	1.21

Table 2: Multistage Shuffle Multicast: Performance of the Greedy heuristic over the optimal solution.

Systems	Metrics	Destinations									
		0.01	0.02	0.05	0.1	0.2	0.5	0.8	0.9	0.95	0.99
3-stage Cube	Number of miss	1	2	3	3	4	7	5	3	1	1
	Average overhead	1.01	1.01	1.01	1.01	1.04	1.1	1.04	1.01	1.01	1.01
4-stage Cube	Number of miss	2	2	3	4	6	10	5	4	3	1
	Average overhead	1.01	1.01	1.01	1.03	1.11	1.18	1.09	1.03	1.01	1.01
5-stage Cube	Number of miss	2	3	6	7	9	12	8	6	3	2
	Average overhead	1.01	1.01	1.02	1.04	1.14	1.23	1.12	1.03	1.01	1.01

Table 3: Multistage Cube Multicast: Performance of the Greedy heuristic over the optimal solution.

4 Conclusion

Multistage networks are popular for parallel architecture and/or communication network applications. We formulate the traffic optimum multicasting problem for multistage networks. Optimum traffic multicasting problem is NP-complete. Several greedy heuristics are proposed and their performances are shown using analytical as well as simulation methods. This work considered just *traffic* optimality in MIN multicasting. Optimality issues of *time* metric in MIN multicasting is left as an interesting open problem.

Acknowledgments: We thank Gary Elsesser, Lionel M. Ni and Wei-Tek Tsai for helpful discussions.

References

[1] S. Bhattacharya, G. Elsesser, W.T. Tsai and D.Z. Du, Multicasting in Generalized Multistage Interconnection Networks, to appear in *Journal of Parallel and Distributed Computing*, 1993.

[2] S. Bhattacharya and L.M. Ni, Multicasting in Multistage Communication Networks, *preprint*, 1992.

[3] D. Comer and R. Sethi, Complexity of TRIE Index Construction, *17th Annual Symposium on FOCS*, 1976, pp 197-207.

[4] W. J. Dally and C. L. Seitz, Deadlock-Free Message Routing in Multiprocessor Interconnection Networks, *IEEE Transactions on Computers*, C-36 (1987), pp. 547-553.

[5] P. Dowd, Random Access Protocols for High-Speed Interprocessor Communication Based on an Optical Passive Star Topology, *Journal of Lightwave Technology*, 9 (1991), pp. 799-808.

[6] M.R. Garey M.R. and D.S. Johnson, *Computers and Intractability - A Guide to the Theory of NP-Completeness*, (Freeman, San Fransisco, CA, 1979).

[7] L.R. Foulds and R.L. Graham, The Steiner Problem in Phylogeny in NP-Complete, *Advances in Applied Mathematics*, 3 (1982), pp. 43-49.

[8] K. Hwang and F.A.Y. Briggs, *Computer architecture and parallel processing*, (New York : McGraw-Hill, c1984).

[9] K. Hwang and J. Ghosh, Hypernets: A Communication-Efficient Architecture for Constructing Massively Parallel Computers, *IEEE Transactions on Computers*, C-36 (1987), pp. 1450-1466.

[10] M.G. Hluchyj and M.J. Karol, ShuffleNet: An Application of Generalized Perfect Shuffles to Multihop Lightwave Networks, *Journal of Lightwave Technology*, 9 (1991).

[11] P. Kermani and L. Kleinrock, Virtual cut-through: A new computer communication switching technique, *Computer Networks*, 3 (1979), pp. 267-286.

[12] F.P. Preparata and J. Vuillemin, The Cube-Connected Cycles: A Versatile Network for Parallel Computation, CACM, 24 (1981), pp. 300-309.

www.ingramcontent.com/pod-product-compliance
Lightning Source LLC
LaVergne TN
LVHW051053210926
848486LV00002B/95
* 9 7 8 9 8 1 0 2 1 2 7 7 3 *